TURBULENCE IN THE ATMOSPHERE

This book provides a modern introduction to turbulence in the atmosphere and in engineering flows. Based on over 40 years of research and teaching, John Wyngaard's textbook is an excellent introduction to turbulence for advanced students and a reference work for researchers in the atmospheric sciences. Part I introduces the concepts and equations of turbulence. It includes a rigorous introduction to the principal types of numerical modeling of turbulent flows. Part II describes turbulence in the atmospheric boundary layer. Part III covers the foundations of the statistical representation of turbulence and includes illustrative examples of stochastic problems that can be solved analytically. Student exercises are included at the ends of chapters, and worked solutions are available online for use by course instructors. The book is an invaluable introduction to turbulence for advanced students and researchers in academia and industry in the atmospheric sciences and meteorology, as well as related fields in aeronautical, mechanical and environmental engineering, oceanography, applied mathematics, and physics.

JOHN WYNGAARD's experience in turbulence research and teaching spans the Air Force Cambridge Research Laboratories, the Wave Propagation Laboratory of the National Oceanographic and Atmospheric Administration (NOAA) in Boulder, the Atmospheric Analysis and Prediction Division of the National Center for Atmospheric Research (NCAR), and the Department of Meteorology at Pennsylvania State University, where he developed a sequence of courses on turbulence. This book is based on those courses. He has published over 100 refereed journal papers covering theoretical, observational, and numerical modeling aspects of engineering and geophysical turbulence.

TURBULENCE IN THE ATMOSPHERE

JOHN C. WYNGAARD

CAMBRIDGE
UNIVERSITY PRESS

CAMBRIDGE
UNIVERSITY PRESS

University Printing House, Cambridge CB2 8BS, United Kingdom

One Liberty Plaza, 20th Floor, New York, NY 10006, USA

477 Williamstown Road, Port Melbourne, VIC 3207, Australia

4843/24, 2nd Floor, Ansari Road, Daryaganj, Delhi - 110002, India

79 Anson Road, #06-04/06, Singapore 079906

Cambridge University Press is part of the University of Cambridge.

It furthers the University's mission by disseminating knowledge in the pursuit of education, learning and research at the highest international levels of excellence.

www.cambridge.org
Information on this title: www.cambridge.org/9781108445672

First published 2010
Reprinted 2012
First paperback edition 2017

A catalogue record for this publication is available from the British Library

Library of Congress Cataloging in Publication data
Wyngaard, John C.
Turbulence in the atmosphere / John C. Wyngaard.
p. cm.
ISBN 978-0-521-88769-4 (Hardback)
1. Atmospheric turbulence. I. Title.
QC880.4.T8W96 2009
551.55–dc22

2009035697

ISBN 978-0-521-88769-4 Hardback
ISBN 978-1-108-44567-2 Paperback

Additional resources for this publication at www.cambridge.org/9780521887694

Contents

Preface

I doubt if many students have started out to be a "turbulence person." I suspect it usually just happens, perhaps like meeting the person you marry. I was an engineering graduate student, nibbling at convective heat transfer, when a friend steered me to a turbulence course taught by John Lumley. It was not rollicking fun – we went through Townsend's *The Structure of Turbulent Shear Flow* page by page – but it was a completely new field.

I began to explore the turbulence literature, particularly that by the heavy hitters of theoretical physics and applied mathematics. To my engineering eyes it was impregnable; I would need much more coursework before I could even put it in a context. Today I can understand why those theoretical struggles continue. Phil Thompson, a senior scientist in the early NCAR† explained it this way:

> Lots of people have tried to develop a fundamental theory of turbulence. Some very well known people have given up on it. But I just can't give up on it – it's like a beautiful mistress. You know that she treats you badly, she's being ornery, but you just can't stay away from her. So periodically, this question comes up again in my mind, and I keep casting about for some different and simple and natural way of representing the motion of a fluid, and some way of treating the analytical difficulties. And I seem to get a little bit closer sometimes ...

When I was finishing my Ph.D. I began looking for a job. Hans Panofsky, whose 1964 book with John steered me toward atmospheric turbulence, kindly gave me a list of four prospects. The most intriguing was the Boundary-Layer Branch of Air Force Cambridge Research Laboratories, as it was then called. Led by the indefatigable Duane Haugen, they were in the final stages of planning a field measurement program on atmospheric surface-layer turbulence. It was as if the job were designed for me. I accepted their $12,822 per year offer.

One year later we carried out the 1968 Kansas experiment, perhaps the most ambitious such field program up to that time. The data analyses engaged us for

† National Center for Atmospheric Research, Boulder, Colorado.

several years. Owen Coté and I, with the plotting and programming assistance of Jack Izumi and Jean O'Donnell, mapped out perhaps the first observational analyses of the conservation equations for stress and scalar flux in a turbulent flow. Chandran Kaimal did his inimitable spectral analyses. Visitor Joost Businger oversaw the analysis of the flux-mean profile relations over the wide range of stability conditions. Henk Tennekes and I found that velocity derivative statistics in the huge Reynolds number Kansas turbulence were off the older charts, but in accord with the newer thinking of the Russian school. Those were heady times.

We returned to the field in 1973, in the very flat farming country of northwestern Minnesota. In collaboration with a British Met Office group we reached deep into the boundary layer with sensors on the tethering cable of a World War II surplus barrage balloon. It stayed up for several weeks, despite the best late-night efforts of rifle-toting cowboys, until it was brought down by a gust front. Part II discusses some of the insights gained from these field programs.

Today's main types of turbulent-flow models – large-eddy simulation and second-order closure (Chapters 5 and 6) – were in their infancy in 1970. I remember discovering with Owen Coté the myriad ways a second-order-closure model could misbehave – negative variances, violations of Schwartz's inequality, ... What we thought were obviously better closures gave poorer results. We saw the early hopes of universal second-order closures dashed in buoyancy-dominated flows. We developed a wariness about turbulence modeling. As Ronald Reagan later said, "Trust, but verify."

Anyone who has developed models of the second-moment equations (Chapter 5), discovered how poorly they can behave, and then in fatigue and discouragement wondered how Nature keeps variances positive, can appreciate this story:

> Some years ago, during the hall talk at a break in an NCAR meeting, a prominent senior scientist became impatient with a mathematician's fussing over obscure details of an equation. "Hell," he blurted, "in the atmospheric sciences we don't even know what the equations are."

The applied turbulence field seems different today. Numerical modeling of turbulent flows is a dominant technology used by a second- or even third-generation community. Programmers have ensured that the codes don't misbehave like they used to. Geophysical observations have not kept pace with the model predictions, nor could they have; modeling and observational work have cruelly different time scales. Now less likely to be rooted in personal experience, wariness of modeling seems to be diminishing.

Recently I previewed a video of the EPA Fluid-Modeling Facility before showing it in my class on atmospheric dispersion. The FMF, as it is called, is located in Raleigh, NC, and contains low-speed wind tunnels, a stratified towing tank, and a replica of the Deardorff–Willis convection tank (Chapter 11). The FMF is a world-class facility put together largely by Bill Snyder beginning about 1970.

The video shows visualizations of plume dispersion, wind-tunnel turbulence, and stably stratified flow around obstacles. Two students stopped briefly to watch, and as they left one said to the other, "That was before we had computer modeling."

Fortunately, EPA management recognizes the FMF's strengths and continues its funding for observations central to the testing and improving of dispersion models.

For a generation born into personal computing, numerical modeling is a natural research medium. Models are widely and instantly available, some through vendors, others being in the public domain. But in a time when observational work seems increasingly out of fashion, when a "sixth sense" about the behavior of turbulence is becoming rare, models can be easily misused and misinterpreted. We have no "Modeler General"; the models have no warning labels.

I suspect this lack of wariness about modeling is an experiential issue, not a generational one. I recall attending an AMS-EPA workshop on air-quality modeling in the 1980s. At one point the discussion focused on the performance of the standard Gaussian-plume air-quality model in fair weather, flat terrain, quasi-steady conditions. The question was asked: "How well do the model predictions of ground-level concentration downwind of a point source agree with one-hour measurements?" A crusty old air-quality "consultant," as they are called, who had enjoyed a long, successful practice, didn't hestitate in answering: "Within ten percent." No doubt that was his honest belief, but we now know it was wrong by more than an order of magnitude. The community hadn't yet focused on such considerations.

This book is based on the material in the graduate course in atmospheric turbulence I have taught for nearly 20 years at Penn State. Its four precepts are (1) engineering and geophysical turbulence have much in common; (2) our numerical models of turbulent flows, particularly those in the atmosphere, need effective representations of turbulence; (3) although the "turbulence problem" appears to be as unyielding as ever, we have learned much about *dealing with* turbulence; (4) users of turbulent-flow models should understand their foundations.

There are three self-contained parts. Part I, "A grammar of turbulence," covers the important attributes, concepts, rules, and tools of turbulence – those aspects that are common to all applications fields and are central to turbulence literacy. Done in a constant-density fluid, it begins with an overview of turbulence, including the contrast between its instantaneous and average properties, the averaging process and its convergence, the eddy velocity scale and turbulence spectrum, turbulent vorticity, and the eddy diffusivity. We then average the equations, over space or an ensemble of realizations, and discuss the turbulent fluxes this produces. There is a chapter each on the ensemble-average fluxes and their conservation equations, including their modeling by "second-order closure." A chapter on the space-averaged equations, the basis of large-eddy simulation, demonstrates the spectral energy cascade and explains the physical basis of the Kolmogorov hypotheses about the inertial subrange. The final chapter covers the dissipative range, both as hypothesized in 1941 by Kolmogorov and more recently through dissipation-intermittency models, and two-dimensional turbulence as described through the Kolmogorov-like notions of Kraichnan and Batchelor.

Part II covers turbulence in the atmospheric boundary layer (ABL). The first chapter generalizes the equations of Part I to a variable density environment in a standard way, using a background-plus-deviation representation for density, temperature, and pressure. The background state is hydrostatic, buoyancy is handled through the Boussinesq approximation, and a conserved temperature is used that in its ultimate form allows phase change. The four subsequent chapters survey the structure and dynamics of the ABL, emphasizing for a non-meteorological audience those features that make its turbulence different from that in engineering flows. They also cover turbulence in the surface layer, and discuss in depth the physics and the efficacy of the Monin–Obukhov similarity hypothesis for its turbulence structure. There is a chapter on the convective boundary layer, whose turbulence physics and structure have been extensively studied in the field and through large-eddy simulation. The final chapter covers the stable ABL, which has some regimes in which turbulence structure and dynamics have a reasonably simple interpretation.

Part III, "Statistical representation of turbulence," includes a number of important statistical tools and concepts – probability densities and distributions, covariances, autocorrelations, spectra, and local isotropy – that are used in turbulence and other stochastic problems. It has a number of illustrative examples of stochastic problems that can be solved analytically, including the wavenumber-space dynamics of turbulence spectra; relating spectra in the plane to traditional spectra; and the effects of spatial averaging, sensor separation, crosstalk, and probe-induced flow distortion on turbulence measurements.

In the course of writing this book I received valuable input on technical matters from Bob Antonia, Bob Beare, Craig Bohren, Frank Bradley, Peter Bradshaw, Jim Brasseur, Joost Businger, Steve Clifford, Steve Derbyshire, Diego Donzis, Carl Friehe, Steve Hatlee, Reg Hill, Bert Holtslag, Tom Horst, Mark Kelly, Don Lenschow, Charles Meneveau, Chin-Hoh Moeng, Parviz Moin, Ricardo Munoz, Laurent Mydlarski, Bill Neff, Ray Shaw, K. R. Sreenivasan, Peter Sullivan, Dennis Thomson, Chenning Tong, Zellman Warhaft, Jeff Weil, Keith Wilson, P. K. Yeung, and Sergej Zilitinkevich, for which I am most grateful. I'd like to thank Lori Mattina for expertly and patiently crafting the figures, Ned Patton for kindly setting up the LaTeX style files for me, and Peter Sullivan again for his generous and sustained assistance with LaTeX. I am grateful to the AFCRL group – Duane Haugen, Chandran Kaimal, Owen Coté, Jack Izumi, Jim Newman, Jean O'Donnell, and Don Stevens – for my once-in-a-lifetime experience in the 1968 Kansas experiment. Finally, I thank John Lumley for inspiring my career in turbulence.

Part I

A grammar of turbulence

1

Introduction

1.1 Turbulence, its community, and our approach

Even if you have not studied turbulence, you already know a lot about it. You have seen the chaotic, ever-changing, three-dimensional nature of chimney plumes and flowing streams. You know that turbulence is a good mixer. You might have come across an article that described the intrigue it holds for mathematicians and physicists.

Unless a fluid flow has a low Reynolds number or very stable stratification (less dense fluid over more dense fluid), it is turbulent. Most flows in engineering, in the lower atmosphere, and in the upper ocean are turbulent. Because of its "mathematical intractability" – turbulence does not yield exact mathematical solutions – its study has always involved observations. But over the past three decades numerical approaches have proliferated; today they are a dominant means of studying turbulent flows.

Turbulence has long been studied in both engineering and geophysics. G. I. Taylor's contributions spanned both (Batchelor, 1996). The Lumley and Panofsky (1964) work was my introduction to that breadth, but as Lumley later commented, their parts of that text "just ... touch." Today the turbulence field seems more coherent than it was in 1964, although it still has subcommunities and dialects (Lumley and Yaglom, 2001).

In Part I of this book we focus on the physical understanding of turbulence, surveying its key properties. We'll use its governing equations to guide our discussions and inferences. We shall also discuss the main types of numerical approaches to turbulence. You might be concerned by our use of little mathematical "tricks" – not because they're complicated or difficult, but because you've never seen them before and might not have thought of them yourself. Don't worry: we pass them on because they are some of the useful tools developed over the many years that scholars have pondered turbulence. You can pass them on too.

Figure 1.1 Instability of an axisymmetric jet. A laminar stream of air flows from a circular tube at the left at Reynolds number 10 000 and is made visible by a smoke wire. The edge of the jet develops axisymmetric oscillations, rolls up into vortex rings, and then abruptly becomes turbulent. Photograph courtesy Robert Drubka and Hassan Nagib. From Van Dyke (1982).

1.2 The origins and nature of turbulence

Turbulent rather than smooth, *laminar* flow of a fluid, liquid or gas, normally occurs if a dimensionless flow parameter called the *Reynolds number* $Re = UL/\nu$ exceeds a critical value. Here U and L are velocity and length scales of the flow[†] and ν is the kinematic viscosity (dynamic viscosity μ/density ρ) of the fluid. The atmospheric boundary layer is turbulent, but as we shall see in Part II stable density stratification can strongly modulate its depth and the intensity and scale of its turbulence. Winter sunrises here in central Pennsylvania often reveal laminar chimney plumes in the very stably stratified flow caused by the overnight cooling of the earth's surface. The turbulent eddies[‡] so prominent in cumulus clouds and flowing streams can be revealed in laboratory turbulence through flow-visualization techniques (Figure 1.1).

There are two types of turbulence with quite different physics. The most common type, *three-dimensional turbulence*, arises from the tendency of fluid motion of large Re to be turbulent and the tendency of turbulence to be three dimensional. But *two-dimensional turbulence* is also of interest; it causes the darting of the colors in soap films and is a model of the largest-scale motions of the atmosphere. We shall discuss it in Chapter 7.

† For example, in Figure 1.1 U is the velocity averaged over the tube cross section and L is the tube diameter.

‡ To paraphrase Batchelor (1950), "eddy" does not refer to any specific local distribution of velocity; it is simply a concise term for local turbulent motion with a certain length scale – an arbitrary local flow pattern characterized by size alone. A turbulent flow has a spectrum of eddies of different size, determined by an analysis of the velocity field into sinusoidal components of different wavelengths (Chapter 15).

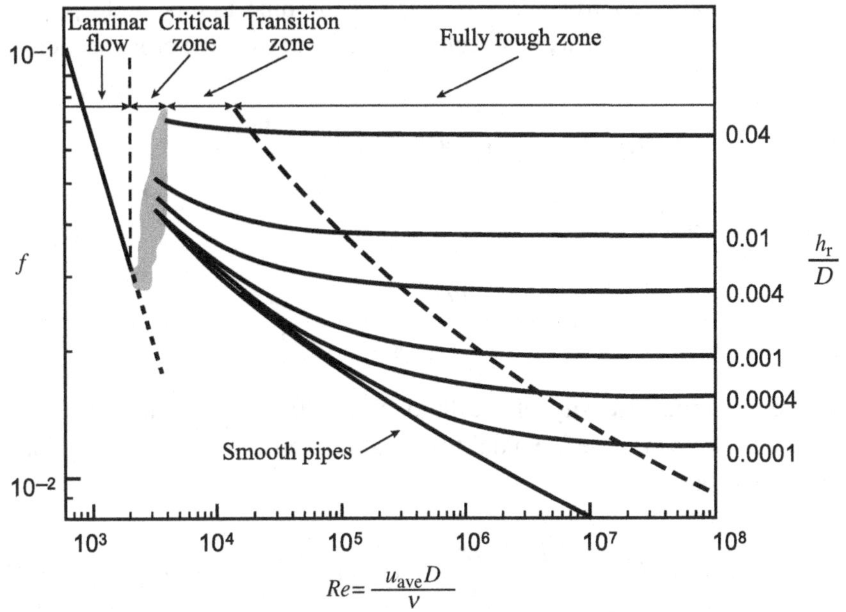

Figure 1.2 The *Moody chart*, which shows the behavior of the Darcy friction factor f, Eq. (1.5), in a circular pipe. In laminar flow $f \propto Re^{-1}$, Eq. (1.6); f jumps to larger values with the transition to turbulence at $Re \simeq 2000$, and in the region of equilibrium turbulence past the critical zone f depends also on the wall-roughness height h_r relative to D. Adapted from Moody (1944).

1.3 Turbulence and surface fluxes

An early motivation for the study of turbulence was to understand how it makes the fluxes of momentum, heat, and mass at a solid surface much larger than in the laminar case. This has important applications to both geophysical and engineering flows.

Fluid flowing through a long circular pipe becomes turbulent at some point downstream if the Reynolds number $Re = u_{ave}D/\nu$ (u_{ave} is the velocity averaged over the pipe cross section and D is the pipe diameter) exceeds about 2000. This *transition to turbulence*, as it is called, is marked by a jump in the shear stress (which is also interpretable as a momentum flux, Section 1.5) at the wall (Figure 1.2). There is a corresponding jump in the required pumping power (Problem 1.1).

To understand these abrupt changes at transition we need some background on pipe flow. In the steady, laminar case its velocity profile is parabolic (Problem 1.1),

$$u(r) = u_{max}\left(1 - \frac{r^2}{R^2}\right), \qquad (1.1)$$

where r is the radial coordinate, $R = D/2$ is the pipe radius, and $u_{\max}$ is the maximum (centerline) velocity. The velocity averaged over the cross section is

$$u_{\text{ave}} = \frac{1}{\pi R^2}\int_0^R u(r)2\pi r\, dr = \frac{u_{\max}}{2}. \tag{1.2}$$

The wall shear stress is

$$\tau_{\text{wall}} = -\mu \left.\frac{\partial u}{\partial r}\right|_{r=R} = 8\mu\frac{u_{\text{ave}}}{D}, \tag{1.3}$$

with μ the dynamic viscosity of the fluid. Since $\partial p/\partial x$ does not depend on x (Problem 1.1), we can write the axial force balance on a slug of fluid of length L and diameter D as

$$\tau_{\text{wall}}\,\pi DL = -\frac{\partial P}{\partial x}L\frac{\pi D^2}{4}, \quad \text{so that} \quad -\frac{\partial P}{\partial x}D = 4\tau_{\text{wall}}. \tag{1.4}$$

The mean pressure gradient nondimensionalized with $\rho(u_{\text{ave}})^2/2$ and D is called the *Darcy friction factor*,†

$$f \equiv \frac{-\frac{\partial P}{\partial x}D}{\rho(u_{\text{ave}})^2/2} = \frac{4\tau_{\text{wall}}}{\rho(u_{\text{ave}})^2/2}. \tag{1.5}$$

Thus f is, from Eq. (1.3),

$$f_{\text{lam}} = \frac{64\mu u_{\text{ave}}}{D\rho(u_{\text{ave}})^2} = \frac{64}{Re}. \tag{1.6}$$

Figure 1.2 shows this inverse-Re dependence of f in the laminar-flow regime.

Past the critical zone, Figure 1.2, u_{ave} and τ_{wall} are turbulent quantities, so (as we'll discuss in detail in Chapter 2) we work with their mean values $\overline{u}_{\text{ave}}$ and $\overline{\tau}_{\text{wall}}$. In the turbulent regime Eq. (1.5) implies $\overline{\tau}_{\text{wall}} = f_{\text{turb}}\,\rho(\overline{u}_{\text{ave}})^2/8$. Therefore the ratio of the mean wall stress in turbulent pipe flow and the wall stress in laminar flow at the same average velocity is

$$\frac{\overline{\tau}_{\text{wall}}}{\tau_{\text{wall}}(\text{laminar flow})} = \frac{f_{\text{turb}}}{f_{\text{lam}}} = \frac{f_{\text{turb}}\,Re}{64}. \tag{1.7}$$

† The *Fanning friction factor* is the wall stress nondimensionalized with $\rho(u_{\text{ave}})^2/2$. The Darcy friction factor, Eq. (1.5), is larger by a factor of four.

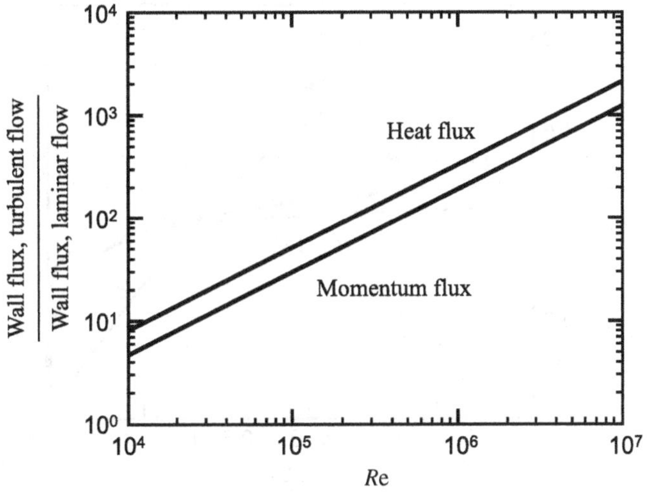

Figure 1.3 The ratios of mean fluxes at the wall in turbulent and laminar flow through smooth pipes. The momentum-flux ratio is Eq. (1.7) evaluated with f data from Figure 1.2; the heat-flux ratio is Eq. (1.16) evaluated with Nu data from Dittus and Boelter (1930), as summarized by Turns (2006).

This ratio is plotted for smooth pipes in Figure 1.3. It has very large values at large Re, indicating the strong influence of turbulence on the wall stress.

Turns (2006) shows that a good fit to the classical mean-velocity measurements of Nikuradse (1933) in turbulent pipe flow is

$$\frac{\overline{u}(r)}{\overline{u}_{\text{ave}}} = \frac{f^{1/2}}{\sqrt{2}}\left(2.5\ln\left[\frac{Re\, f^{1/2}}{2\sqrt{2}}\left(1-\frac{r}{R}\right)\right]+5.5\right). \tag{1.8}$$

Figure 1.4 shows that this profile is much "flatter" in the core region than the laminar profile (1.1). At large Re the mean-velocity gradient is significant only adjacent to the wall, where it is much larger than in laminar flow of the same bulk fluid velocity. The wall stress in turbulent flow is still defined by the velocity gradient at the wall, Eq. (1.3), but that gradient, and therefore the wall shear stress, fluctuates chaotically with time and with position. The mean value (which we designate by an overbar) of the wall stress is

$$\overline{\tau}_{\text{wall}} = -\mu\overline{\left.\frac{\partial u}{\partial r}\right|_{r=R}} = -\mu\left.\frac{\partial \overline{u}}{\partial r}\right|_{r=R}. \tag{1.9}$$

We have used the property that the differentiation and averaging can be done in either order (Problem 1.3). The sharp increase in the mean-velocity gradient at the wall at transition (Figure 1.4) causes a sharp increase in wall stress (Figure 1.2).

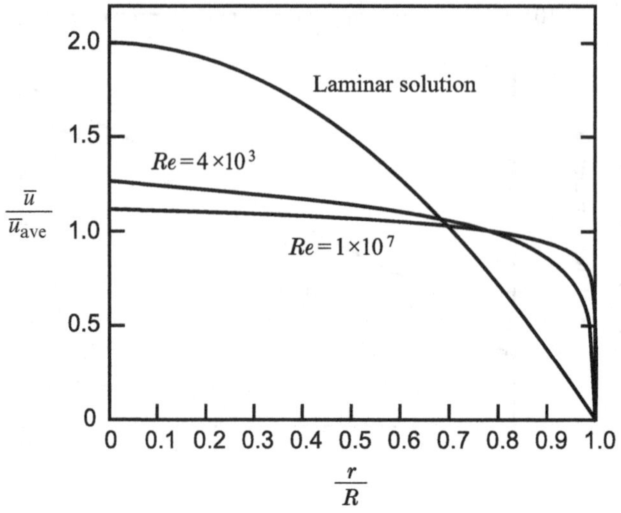

Figure 1.4 Profiles of $\overline{u}/\overline{u}_{\text{ave}}$ in turbulent pipe flow, Eq. (1.8), "flatten" as the Reynolds number increases, making the mean shear and mean stress at the wall much larger than in laminar flow with the same average velocity.

Pipes typically have some wall roughness, and Figure 1.2 indicates that the mean wall stress increases with that roughness. The explanation (Kundu, 1990) is that immediately adjacent to the wall in turbulent pipe flow is a *laminar sublayer* of thickness $\delta \sim 5\nu/u_*$, with $u_* = (\overline{\tau}_{\text{wall}}/\rho)^{1/2}$ the *friction velocity*. If the typical height h_r of the individual "bumps" or *roughness elements* on the wall is much less than δ, wall roughness has minimal effect and the mean wall stress is the viscous one given by Eq. (1.9). But as h_r approaches δ the roughness elements cause *form drag* through the pressure distribution on their surface, which adds to the viscous drag and increases the friction factor f. When h_r is large enough this form drag dominates and f ceases to change with Re, as indicated in Figure 1.2.

An analogous situation exists for the wall heat flux H_{wall} (watts m^{-2}). It is carried entirely by the molecular diffusion process called *conduction heat transfer*:

$$H_{\text{wall}} = -k \left. \frac{\partial T}{\partial r} \right|_{r=R}, \tag{1.10}$$

with k the thermal conductivity (watts m^{-1} K^{-1}). The heat flux is continuous at the fluid–wall interface, but the temperature gradient there is discontinuous if k of the wall material and the fluid differ. We shall consider the fluid side.

The temperature profile in fully developed laminar pipe flow depends on the temperature boundary conditions. We'll consider the analytically simple case where the fluid and wall temperatures vary linearly with x but their difference, and the wall heat flux, are independent of x. Its temperature profile is (Problem 1.2)

$$T(0,x) - T(r,x) = -\frac{R^2 u_{\text{ave}}}{\alpha}\frac{\partial T}{\partial x}\left[\frac{r^2}{2R^2}\left(1 - \frac{r^2}{4R^2}\right)\right], \tag{1.11}$$

with $\alpha = k/(\rho c_p)$ the thermal diffusivity of the fluid. The relation between the wall heat flux and $\partial T/\partial x$ is (Problem 1.2)

$$H_{\text{wall}} = -\frac{D u_{\text{ave}} \rho c_p}{4}\frac{\partial T}{\partial x}, \tag{1.12}$$

so the temperature profile (1.11) can be rewritten as

$$T(0,x) - T(r,x) = \frac{H_{\text{wall}} D}{k}\left[\frac{r^2}{2R^2}\left(1 - \frac{r^2}{4R^2}\right)\right]. \tag{1.13}$$

The wall heat flux made dimensionless with pipe diameter D, fluid thermal conductivity k, and a temperature difference ΔT is called a Nusselt number Nu:

$$Nu = \frac{H_{\text{wall}} D}{k \Delta T}. \tag{1.14}$$

ΔT is defined through the wall temperature $T_{\text{w}}(x)$ and the "bulk fluid temperature" at that position, $T_{\text{b}}(x)$:

$$T_{\text{b}}(x) = \frac{\int_0^R u(r)\, T(r,x)\, 2\pi r\, dr}{\pi R^2 u_{\text{ave}}}, \qquad Nu = \frac{H_{\text{wall}} D}{k(T_{\text{b}} - T_{\text{w}})}. \tag{1.15}$$

Turns (2006) shows that in the laminar case in this problem $Nu \simeq 4.4$.

In the turbulent case the heat flux at a point on the wall, like the stress there, fluctuates chaotically in time. The turbulent mixing makes the mean temperature gradient relatively small over most of the cross section; it is large only near the wall, as for velocity (Figure 1.4). The mean wall heat flux, the product of the fluid thermal conductivity k and the mean temperature gradient at the wall, is much larger than in the laminar case.

From Eq. (1.14) in this problem we can write the ratio of wall heat fluxes as, for given values of D, k, and ΔT,

$$\frac{\overline{H}_{\text{wall}}}{H_{\text{wall}}(\text{laminar flow})} = \frac{Nu}{Nu(\text{laminar flow})} = \frac{Nu}{4.4}. \tag{1.16}$$

Figure 1.3 shows the wall-flux ratios in the turbulent regime. The ratios for heat and momentum differ by a constant factor of about 1.5–2 (a manifestation of the *Reynolds analogy* between heat and momentum transfer) as they increase sharply with Reynolds number.

If the flow over the earth's surface were laminar, not turbulent, the environmental effects would be profound. In clear summer weather, for example, the earth's surface temperature could routinely approach 100 °C during the day and 0 °C at night (Chapter 9).

1.4 How do we study turbulence?

Turbulence has long had a special attraction for physicists and mathematicians; it has been called "the last great unsolved problem of classical physics."† In practical terms this means that we cannot analytically solve the equations of turbulent fluid motion. The difficulty stems from their nonlinearity.

Leonardo da Vinci sketched turbulent water flows, and reportedly gave the sage advice: "Remember when discoursing on the flow of water to adduce first experience and then reason."‡ Even today, some 500 years after da Vinci, much of our understanding of turbulence is rooted in observations.

Since the 1960s turbulence has been studied numerically as well. One early study had a revolutionary impact. Lorenz (1963) discovered the profound effects of very small changes in initial conditions on the behavior of a very simplified, three-equation, nonlinear model of turbulent convection. He found that two solutions with slightly different initial conditions diverged with time. This *sensitive dependence on initial conditions* is now recognized as a fundamental property of turbulence. Gleick (1987) describes Lorenz' findings as the beginning of the field now called *chaos*.

The advances in digital computers and numerical techniques for solving differential equations after Lorenz' early work soon allowed the *numerical simulation* of turbulence. There are two varieties. *Direct numerical simulation* (DNS) is the numerical solution of the governing fluid equations. It is (within the numerical approximations used) exact, but it is possible only in low Reynolds number flows (Problem 1.9). The Orszag–Patterson (1972) $32 \times 32 \times 32$ (32^3) calculation of isotropic turbulence is considered the first DNS. *Large-eddy simulation* (LES) is an approximate technique that solves for the largest-scale structure of turbulence fields; its underlying concepts were laid out by Lilly (1967). Deardorff's (1970a) study of turbulent channel flow on a $24 \times 14 \times 20$ grid mesh (6720 grid points) is widely

† According to Holmes *et al.* (1996), precise references to such remarks are elusive. They have been attributed to Sommerfeld, Einstein, and Feynman, and beginning in 1895 Horace Lamb expressed similar sentiments in his *Hydrodynamics*.

‡ Rouse and Ince (1957) state that this quote appears in the Carusi and Favaro (1924) republication of da Vinci's writings.

cited as the first LES. DNS and LES, which now use as many as $4096^3 \simeq 6 \times 10^{10}$ grid points, are leading research tools in turbulence today.

1.5 The equations of turbulence

In the *Eulerian* description one expresses the fluid velocity $\mathbf{u}$, for example, as a function of position $\mathbf{x}$ in a fixed (relative to the earth) coordinate system, and time t. One then seeks $\mathbf{u}(\mathbf{x}, t)$ in the flow domain of interest. In the *Lagrangian* description, one labels each fluid parcel (with its initial position $\mathbf{a}$, for example) and seeks its velocity history $\mathbf{v}(\mathbf{a}, t)$. We will use the Eulerian description almost exclusively; the one exception is Taylor's solution for dispersion of effluent from a continuous source, Chapter 4.

In Part I we shall use the equations for fluids of time-independent, uniform density (which we shall call simply *constant density*), because they contain the essence of the physics of turbulence. These equations are derived and discussed in graduate-level fluid mechanics texts (e.g., Kundu, 1990). Buoyancy effects stemming from density variations due to heat transfer or phase change can strongly influence turbulence, particularly in the atmosphere; this is the focus of Part II.

In cartesian tensor notation the fluid continuity or mass-conservation equation is

$$\frac{\partial \rho}{\partial t} + \frac{\partial \rho u_i}{\partial x_i} = 0, \tag{1.17}$$

where ρ is fluid density, $x_i = (x_1, x_2, x_3)$ is spatial position, and $u_i = (u_1, u_2, u_3)$ is velocity. We use the convention that repeated Roman indices are to be summed over 1, 2, and 3; if we do not wish to sum we use Greek indices. Equation (1.17) says that at any point in space the time rate of change of fluid density plus the divergence of the fluid mass flux ρu_i is zero. We call ρu_i an advective flux, a vector that represents the amount of fluid mass flowing through unit area per unit time. In general there is also a *molecular* flux that represents the diffusive effect of the random molecular motion, but there is no molecular diffusion of fluid density.

When the fluid density is constant Eq. (1.17) reduces to

$$\frac{\partial u_i}{\partial x_i} = \frac{\partial u_1}{\partial x_1} + \frac{\partial u_2}{\partial x_2} + \frac{\partial u_3}{\partial x_3} = 0, \tag{1.18}$$

meaning that the velocity divergence is zero. A fluid that satisfies Eq. (1.18) is called *incompressible* – its density does not change with pressure. Gases at low speeds and liquids are usually treated as incompressible.

Newton's Second Law for a fluid is

$$\rho \frac{Du_i}{Dt} = \rho \left(\frac{\partial u_i}{\partial t} + u_j \frac{\partial u_i}{\partial x_j} \right) = -\frac{\partial p}{\partial x_i} + \rho g_i + \frac{\partial \sigma_{ij}}{\partial x_j}, \tag{1.19}$$

with p the pressure, g_i the gravity vector and σ_{ij} the viscous stress tensor. The left side of Eq. (1.19) is density times the total acceleration following the motion, the sum of local and advective accelerations; the right side is the sum of the pressure-gradient, gravity, and viscous forces. Equation (1.19) requires a nonaccelerating coordinate system, but our earth-based coordinate system is accelerating because the earth rotates; as a result (1.19) also needs a *Coriolis* term. This can be important in atmospheric turbulence (Part II) but we shall ignore it in Part I.

Batchelor (1967) points out that when ρ is uniform one can define a pressure ($p^{\rm s}$, say) whose gradient exactly balances the gravity force:

$$0 = -\frac{\partial p^{\rm s}}{\partial x_i} + \rho g_i. \tag{1.20}$$

It follows that $p^{\rm s} = \rho g_j x_j + p_0$, with p_0 a constant. If we write pressure as

$$p = p^{\rm s} + p^{\rm m}, \tag{1.21}$$

with $p^{\rm m}$ a *modified pressure* that is due to the fluid motion, then we can write Eq. (1.19) as

$$\rho\left(\frac{\partial u_i}{\partial t} + u_j\frac{\partial u_i}{\partial x_j}\right) = -\frac{\partial p^{\rm m}}{\partial x_i} + \frac{\partial \sigma_{ij}}{\partial x_j}, \tag{1.22}$$

so the gravity term does not appear explicitly. In Part I we shall use the form (1.22) and drop the superscript m on pressure.

Using Eq. (1.18) we can write the momentum equation (1.22) as

$$\rho\frac{\partial u_i}{\partial t} = -\frac{\partial p}{\partial x_i} + \frac{\partial}{\partial x_j}\left(-\rho u_i u_j + \sigma_{ij}\right). \tag{1.23}$$

The final term in Eq. (1.23) is in *flux form*. It can be interpreted as the divergence of the total flux of momentum, the sum of advective and viscous parts. Momentum is mass times velocity; it is a vector. The momentum flux is the amount of momentum passing through a unit area per unit time. It is a second-order tensor quantity; it involves two directions, that of the unit normal to the area and that of the momentum. Its units, density times velocity squared, are equivalent to (newtons/m^2), or stress. Thus, we can also interpret the final term in Eq. (1.23) as the divergence of a generalized stress.

In a incompressible *Newtonian* fluid the viscous stress tensor σ_{ij} is a linear function of the *strain-rate tensor* s_{ij}. We write this as

$$\sigma_{ij} = \mu\left(\frac{\partial u_i}{\partial x_j} + \frac{\partial u_j}{\partial x_i}\right) = 2\mu s_{ij}, \tag{1.24}$$

where μ is the dynamic viscosity and s_{ij} is

$$s_{ij} = \frac{1}{2}\left(\frac{\partial u_i}{\partial x_j} + \frac{\partial u_j}{\partial x_i}\right). \tag{1.25}$$

It is usual to divide by density and use Eqs. (1.24) and (1.25) to write Eq. (1.19) for a Newtonian fluid with constant viscosity as

$$\frac{Du_i}{Dt} = \frac{\partial u_i}{\partial t} + u_j \frac{\partial u_i}{\partial x_j} = -\frac{1}{\rho}\frac{\partial p}{\partial x_i} + \nu \frac{\partial^2 u_i}{\partial x_j \partial x_j}. \tag{1.26}$$

Quantities divided by density are called *kinematic*, so $\mu/\rho = \nu$ is called the kinematic viscosity. Equation (1.26) is called the *Navier–Stokes equation.*

The vorticity ω_i is the curl of velocity; in tensor notation it is

$$\omega_i = \epsilon_{ijk} \frac{\partial u_k}{\partial x_j}. \tag{1.27}$$

Its conservation equation is

$$\frac{D\omega_i}{Dt} = \frac{\partial \omega_i}{\partial t} + u_j \frac{\partial \omega_i}{\partial x_j} = \omega_j \frac{\partial u_i}{\partial x_j} + \nu \frac{\partial^2 \omega_i}{\partial x_j \partial x_j}. \tag{1.28}$$

This says that the total time derivative of vorticity is the sum of a term representing the interaction of vorticity and the velocity gradient, which we will interpret shortly, and a molecular-diffusion term.

The statement of mass conservation for a scalar c that has no sources or sinks (such as the mass density of a nonreacting trace constituent in the fluid) is

$$\frac{\partial c}{\partial t} + \frac{\partial c u_i}{\partial x_i} = \gamma \frac{\partial^2 c}{\partial x_i \partial x_i}, \tag{1.29}$$

where γ is the molecular diffusivity of c in the fluid. We can also write Eq. (1.29) in flux form,

$$\frac{\partial c}{\partial t} = -\frac{\partial}{\partial x_i}\left(c u_i - \gamma \frac{\partial c}{\partial x_i}\right), \tag{1.30}$$

which says that local time changes in c are due to the divergence of the total flux of c, the sum of advective and molecular components.

In Part I we are considering constant-density fluids, in which the velocity divergence vanishes, so Eq. (1.29) can also be written

$$\frac{Dc}{Dt} = \frac{\partial c}{\partial t} + u_i \frac{\partial c}{\partial x_i} = \gamma \frac{\partial^2 c}{\partial x_i \partial x_i}. \tag{1.31}$$

This says that following the fluid motion and neglecting molecular diffusion, c does not change. We call such a scalar a *conserved* scalar.[†]

In flows where heating due to radiation, phase change, chemical reactions, and viscous effects is negligible the thermal energy equation reduces to the same form as Eq. (1.31),

$$\frac{DT}{Dt} = \frac{\partial T}{\partial t} + u_i \frac{\partial T}{\partial x_i} = \alpha \frac{\partial^2 T}{\partial x_i \partial x_i}, \tag{1.32}$$

where T is temperature and α is the thermal diffusivity of the fluid. Equation (1.32) says that under these conditions temperature is a conserved variable, changing only through conduction heat transfer.

1.6 Key properties of turbulence

Equations (1.17), (1.26), (1.28), (1.31), and (1.32) govern the evolution of the fluid mass, velocity, vorticity, conserved scalar constituent, and temperature fields in a constant-density, Newtonian fluid. Their turbulent solutions have properties that distinguish them from other three-dimensional, time-dependent flow fields.[‡]

1.6.1 Vortex stretching and tilting: viscous dissipation

Vortex stretching is one of the mechanisms contained in the first term on the far right of the vorticity equation (1.28). To illustrate, let's consider a vortex with its axis in the x_1 direction, say, so the initial vorticity is $\omega_i = (\omega_1, 0, 0)$. Equation (1.28) says that ignoring viscous effects the vorticity initially evolves as

$$\frac{D\omega_1}{Dt} = \omega_1 \frac{\partial u_1}{\partial x_1}, \quad \frac{D\omega_2}{Dt} = \omega_1 \frac{\partial u_2}{\partial x_1}, \quad \frac{D\omega_3}{Dt} = \omega_1 \frac{\partial u_3}{\partial x_1}. \tag{1.33}$$

If $\partial u_1/\partial x_1$ is positive (1.33) says the vortex is stretched in the x_1 direction, increasing the magnitude of ω_1. $\partial u_2/\partial x_1$ and $\partial u_3/\partial x_1$ can generate ω_2 and ω_3 from ω_1; this is sometimes called *vortex tilting*.

In two-dimensional turbulence the velocity field is $u_i = [u_1(x, y), u_2(x, y), 0]$, say. Then $\omega_i = (0, 0, \omega_3)$ and the vortex-stretching term in Eq. (1.28) for ω_3 is $\omega_3 \partial u_3/\partial x_3 = 0$. This demonstrates that three dimensionality is necessary for vortex stretching.

A cascade of kinetic energy through eddies of diminishing size (Chapters 6 and 7) that terminates in *viscous dissipation* – the conversion of kinetic energy into internal energy by viscous forces in the smallest eddies – is a defining feature

[†] Unfortunately, if the velocity divergence is nonzero the density of a mass-conserving species is not a conserved scalar. The ratio of its density and the fluid density is a conserved scalar, however (Part II).

[‡] Parts of this discussion are adapted from Lumley and Panofsky (1964).

of three-dimensional turbulence. Without viscous dissipation the kinetic energy of turbulence could grow without bound (Problem 1.5). This kinetic-energy cascade is a statistical concept, but it has direct implications for instantaneous turbulence fields: it says they not only have large, prominent, energetic eddies that we can see in clouds and smoke plumes, but they also have very much smaller eddies whose viscous forces dissipate kinetic energy at the required rate. It is generally accepted that vortex stretching is one physical process responsible for the generation of this wide range of smaller eddies in three-dimensional turbulence.

1.6.2 Random, stochastic

Imagine generating turbulent flow of a certain geometry (in the laboratory, say) any number of times. Because of the sensitivity of turbulence to inevitable small differences in its initial state, each resulting flow, called a *realization*, is unique. We call such a flow *random*, by which we mean *different in every realization*.

A property at any point in a turbulent flow has a mean value and fluctuations about that mean. As we shall discuss in Chapter 2, the mean can in principle be an ensemble mean, the average of values at the point over many realizations of the flow; a time mean, which we can use in statistically steady cases to approximate the ensemble mean; or a spatial mean, which we can use in spatially homogeneous cases. The fluctuations about this mean are *stochastic*, which we define as varying irregularly in space or time in a given realization.†

Numerical models of turbulent flows generally provide only such mean values. The most common atmospheric-diffusion models estimate the ensemble-mean concentration downwind of an effluent release (although some users' manuals call it a time mean). But an effluent plume in the daytime atmospheric boundary layer will fluctuate in time and space about that ensemble mean, and will behave differently in each realization. Such concentration excursions can induce dangerously high concentrations of toxic effluents over short periods.

1.6.3 The effective diffusivity

In turbulent flows the advective fluxes in the Navier–Stokes (1.26) and conserved scalar constituent (1.31) equations are spatially and temporally chaotic, three dimensional, and generally (except at solid surfaces) much larger than the molecular fluxes. This gives turbulent flow a much greater "mixing power" or "effective diffusivity" than laminar flow. But these are, to use G. I. Taylor's term, *virtual properties* of turbulence – properties of the mean flow, not of a realization. Effluents do disperse

† This distinction between *random* and *stochastic* is not always made in turbulence, but it is useful.

far more rapidly in the *mean* chimney plume than in a laminar plume, but within any realization local effluent concentrations can be far higher than in the mean plume.

1.6.4 Range of scales: the turbulence Reynolds number

In turbulent flows made visible by a tracer – such as finely dispersed water droplets in clouds and aircraft contrails, and smoke in chimney plumes – we see *eddies*, locally coherent structures in the velocity field. We can also see eddies through remote sensors such as radar, sodar, and lidar, and through computer graphics in numerically calculated turbulent flows.

The larger, *energy-containing* eddies of typical size ℓ and velocity u relative to the mean contain most of the kinetic energy of the turbulent motion. ℓ scales with, and is roughly the magnitude of, a characteristic flow dimension L. In the atmospheric boundary layer, for example, ℓ scales with the boundary-layer depth; in a pipe flow it scales with the pipe diameter. Likewise u scales with a flow speed U. The turbulence time scale ℓ/u, also called the *eddy turnover time*, is an order-of-magnitude estimate of the typical lifetime of an energy-containing eddy (Problem 1.10).

Since turbulence occurs in flows of large Reynolds number $Re = UL/\nu$, the *turbulence Reynolds number* $R_t = u\ell/\nu$, while smaller than Re, is $\gg 1$. Thus to an excellent approximation the energy-containing eddies are not directly influenced by viscosity. Whenever we refer to "turbulence Reynolds number" we shall mean R_t.

Two important findings enable us to estimate the range of eddy sizes in any turbulent flow. The first is a paradox: the expression for ϵ, the rate of viscous dissipation of kinetic energy per unit mass, contains the kinematic viscosity ν (Problem 1.4), but ϵ does not depend on ν. It is determined by the inviscid, energy-containing eddies: $\epsilon = \epsilon(u, \ell) \sim u^3/\ell$.† Second, as hypothesized by Kolmogorov (1941)‡ the velocity and length scales υ and η of the dissipative eddies depend only on ϵ and the kinematic viscosity ν of the fluid.§ If so, then it follows that the scales of the dissipative eddies are

$$\text{Kolmogorov velocity scale } \upsilon = (\nu\epsilon)^{1/4}, \qquad \text{Kolmogorov length scale } \eta = \left(\frac{\nu^3}{\epsilon}\right)^{1/4}, \tag{1.34}$$

† We use $\sim$ in the sense of Tennekes and Lumley (1972), the implied proportionality coefficient being between 1/5 and 5, say.

‡ This and other classic papers on turbulence are included in a collection by Friedlander and Topper (1961).

§ As we discuss in Chapter 7, Kolmogorov and others subsequently modified this hypothesis to allow for the effects of dissipation intermittency.

for these are the only combinations of ϵ and ν that produce a length scale and a velocity scale. The Reynolds number of the dissipative eddies is therefore $\sim \upsilon\eta/\nu = 1$, which confirms that they are strongly influenced by viscosity.

Using $\epsilon \sim u^3/\ell$ we can write, using Eq. (1.34)

$$\begin{aligned} \frac{\ell}{\eta} &= \frac{\ell\epsilon^{1/4}}{\nu^{3/4}} \sim \frac{\ell u^{3/4}}{\ell^{1/4}\nu^{3/4}} = \left(\frac{u\ell}{\nu}\right)^{3/4} = R_t^{3/4}, \\ \frac{u}{\upsilon} &= \frac{u}{(\epsilon\nu)^{1/4}} \sim \frac{u}{(u^3\nu/\ell)^{1/4}} = R_t^{1/4}. \end{aligned} \tag{1.35}$$

Equation (1.35) implies that in large-R_t turbulent flows the dissipative eddies are quite weak and quite small compared to the energy-containing eddies. If, for example, $u \sim 1$ m s^{-1} and $\ell \sim 10^3$ m, as is typical in the atmospheric boundary layer, then $R_t \sim 10^8$ so that $\upsilon \sim 10^{-2}$ m s^{-1} and $\eta \sim 10^{-3}$ m.

The ratio of vorticities typical of the dissipative and energy-containing eddies is

$$\frac{\text{vorticity of dissipative eddies}}{\text{vorticity of energy-containing eddies}} \sim \frac{\upsilon/\eta}{u/\ell} = \frac{\upsilon}{u}\frac{\ell}{\eta} \sim R_t^{1/2}; \tag{1.36}$$

at large R_t the dissipative eddies contain essentially all the turbulent vorticity.

The smallness of η/ℓ in large-R_t turbulence puts severe limits on the R_t that can be reached in direct numerical simulations of turbulence (Problem 1.9). Although few turbulent flows of practical importance have R_t values small enough to be calculated in this way, the concept of "Reynolds number similarity" (Chapter 2) does make them useful.

The Kolmogorov microscale η, which from Eq. (1.35) can be written as

$$\eta \sim \frac{\nu^{3/4}\ell^{1/4}}{u^{3/4}}, \tag{1.37}$$

is seldom smaller than 10^{-4} m in engineering flows and about 10^{-3} m in the atmosphere. This is almost always large enough to ensure the applicability of continuum fluid mechanics.

1.6.5 Mathematical intractability

Generally speaking only linear differential equations can be directly and straightforwardly solved by analytical means. The advective acceleration term in the Navier–Stokes equation (1.26) involves a product of velocity and velocity gradient, making the equation nonlinear and mathematically *intractable.* This has a simple physical intepretation. This nonlinear term produces the vortex stretching term in Eq. (1.28); vortex stretching is believed to be a principal mechanism in the cascade of kinetic energy from large scales to small (Chapters 6, 7). This cascade dynamically "couples" all the eddies in a turbulent flow. Equation (1.35) shows that the ratio of energy-containing and dissipative eddy sizes increases as the 3/4 power of the large-eddy Reynolds number R_t, so a turbulent flow of large R_t has a huge number of interacting eddies. This has thwarted all attempts to solve the turbulence equations analytically. Even at relatively small R_t the "bookkeeping" for these interactions overwhelms even our largest computers.

1.7 Numerical modeling of turbulent flows

Today's numerical calculations of large-R_t turbulent flows (our daily weather forecasts, for example) do not use the basic fluid equations. Instead they use approximate forms of the averaged equations first derived by Osborne Reynolds (1895). Reynolds averaged over a region of space surrounding a point; the average over an ensemble of realizations of the flow was introduced later. For now we shall not be specific about the type of averaging; it can be time, space, or ensemble averaging. If the average commutes with differentiation (as we shall see, most averages do) we can write the averaged form of the Navier–Stokes equation (1.26) as

$$\frac{\partial \overline{u}_i}{\partial t} + \frac{\partial \overline{u_i u_j}}{\partial x_j} = -\frac{1}{\rho}\frac{\partial \overline{p}}{\partial x_i}, \tag{1.38}$$

with the overbar denoting the average. We have assumed the averaged viscous term is negligible (Problem 1.8). In writing (1.38) we have used the zero-divergence property of u_j to bring it into the derivative. But the averaging has created a $\overline{u_i u_j}$ term in (1.38); in turbulent flow it differs from $\overline{u_i}\;\overline{u_j}$ (Problem 1.17) and, hence, is an unknown. If we write $\overline{u_i u_j}$ as

$$\overline{u_i u_j} = \overline{u}_i \overline{u}_j + \left(\overline{u_i u_j} - \overline{u}_i \overline{u}_j\right) = \overline{u}_i \overline{u}_j - \frac{\tau_{ij}}{\rho}, \tag{1.39}$$

the averaged equation (1.38) becomes

$$\frac{\partial \overline{u}_i}{\partial t} + \frac{\partial \overline{u}_i \overline{u}_j}{\partial x_j} = -\frac{1}{\rho}\frac{\partial \overline{p}}{\partial x_i} + \frac{1}{\rho}\frac{\partial \tau_{ij}}{\partial x_j}. \tag{1.40}$$

Averaging the Navier–Stokes equation has produced Eq. (1.40) for the average velocity field, but the equation contains a new term involving a turbulent stress τ_{ij}. We'll see that this is a subtly different quantity for ensemble and space averages, but in each case it is typically called the *Reynolds stress*. Experience has shown that in turbulent flow it is always important.

In the same way, averaging Eq. (1.29) for a conserved scalar and neglecting its averaged molecular term produces

$$\frac{\partial \overline{c}}{\partial t} + \frac{\partial \overline{c}\,\overline{u}_i}{\partial x_i} = -\frac{\partial f_i}{\partial x_i}, \tag{1.41}$$

where $f_i = \overline{cu_i} - \overline{c}\,\overline{u}_i$ is the *turbulent flux* of the scalar.

In Chapter 4 we'll derive conservation equations for these turbulent fluxes. They involve further unknowns, and so in practice the turbulent fluxes (or their conservation equations) are *modeled* – approximated in some way. The two broad categories of turbulent flow models correspond to the type of averaging that produced them. Although the type of averaging used is often not indicated explicitly, you can often infer it from careful reading of the model description.

In ensemble averaging the velocity and scalar fields are broken into ensemble-mean and fluctuating (turbulent) parts, and the flux produced is due to all the turbulence. In space averaging the filter separates the fields into resolvable and unresolvable (also called subgrid-scale or subfilter-scale) parts. Here the flux is due to the unresolvable turbulence.

1.7.1 Ensemble-averaged turbulence models

The earliest approximation for turbulent fluxes was a gradient-diffusion model like that used for the molecular fluxes, but with a much larger, "eddy" diffusivity. The "mixing-length" model used by Prandtl and Taylor is the simplest example, but it is now seldom used in computations. We discuss it briefly in Chapter 4 to motivate your physical understanding of turbulent fluxes.

A more recent approach, which we discuss in Chapter 5, is to compute the turbulent fluxes through models of their evolution equations. This *second-order-closure* modeling, as it is often called, has been computationally feasible since the late 1960s.

"Pdf modeling" is a type of ensemble-average modeling. While the technique for deriving evolution equations for turbulent fluxes has been known for many decades, that for deriving the evolution equation for the probability density function, or pdf, of a turbulent flow variable is much newer, being presented first by Lundgren (1967). We shall discuss the pdf and its evolution equation in Part III.

1.7.2 Space-averaged turbulence models

The space averaging introduced by Reynolds in 1895 has been generalized recently to *spatial filtering* of the governing fluid equations. *Averaging* implies a uniform weighting of all contributions; in *filtering* the weighting function can vary with position. The concept is to apply to the governing equations a spatial filter that removes the smallest eddies and leaves the energy-containing ones unaffected, producing equations for the larger-scale part of each variable. If L is the spatial scale of the computational domain and Δ_f is the *cutoff scale* of the filter – that is, the filter removes spatial variations of scale less than Δ_f – then of order $N = L/\Delta_f$ computational grid points are required in each direction to resolve the filtered fields, of order N^3 grid points in all. For this reason the filtered variables are often called *resolvable-scale* variables. Today N is typically in the range 30–300. This modeling approach now goes by the apt name *large-eddy simulation*, or LES. It was proposed by Lilly (1967) and first used successfully by Deardorff for turbulent channel flow (Deardorff, 1970a) and then for the atmospheric boundary layer (Deardorff, 1970b).

1.8 Physical modeling of turbulent flows

In a review paper on turbulence written at the dawning of the computer age, Corrsin (1961) estimated the number of grid points required in a numerical calculation of a modest Reynolds number ($R_t \simeq 10^4$) turbulent flow.† Upon presenting his result, 4×10^{14} grid points (which is still well out of reach today) he wrote:

> The foregoing estimate is enough to suggest the use of analog rather than digital computation; in particular, how about an analog consisting of a tank of water?

Corrsin's suggestion of "an analog consisting of a tank of water" is now called *physical modeling* or *fluid modeling*. It allows the structure of both convective and "mechanical" turbulence to be observed in scaled-down, laboratory flows. Some of its most successful applications have been to turbulent dispersion of effluents. Early studies in a 1-m scale convection tank revealed for the first time some of the unusual dispersion properties of convective turbulence (Deardorff and Willis, 1975). Another successful application is the turbulent dispersion of effluents from sources in complex terrain (Snyder, 1985).

1.9 The impact of Kolmogorov

Of the many scientists who have worked in turbulence, none stands taller than Andrei Nikolaevich Kolmogorov. In a brief paper published in 1941 he laid out the basis for our present-day understanding of turbulence as a dynamical system.‡

† To put this example in more physical terms, the Reynolds number ud/ν of turbulent flow in a stirred (at $u = 10^{-1}$ m s^{-1}) cup (of diameter $d = 10^{-1}$m) of tea ($\nu = 10^{-6}$ m^2 s^{-1}) is 10^4. That flow, which has an R_t value somewhat smaller than 10^4 because $\ell < d$, probably can be computed through DNS today.

‡ The Frisch (1995) monograph on turbulence is subtitled "The Legacy of A. N. Kolmogorov."

Earlier work (as summarized by Taylor (1935), for example) had shown that a turbulent flow has large eddies that interact with the mean flow, contain most of the kinetic energy of the turbulence, and lie at one end of a wide range of eddies that interact nonlinearly through mechanisms such as vortex stretching. It was understood that in equilibrium the rate of working of the fluid against the viscous stresses in the smallest of these eddies dissipates kinetic energy at the same mean rate it is extracted from the mean flow by the large eddies. But Taylor (1935) misidentified the spatial scale of this dissipative *microturbulence*, as he called it. He chose the scale (Chapter 7)

$$\lambda = \left(\frac{\nu u^2}{\epsilon}\right)^{1/2}, \tag{1.42}$$

but as we discussed in Section 1.6.4, Kolmogorov argued (and we now accept) that the dissipative-eddy scale is $\eta = (\nu^3/\epsilon)^{1/4}$. They are related by

$$\frac{\lambda}{\eta} \sim R_t^{1/4} \tag{1.43}$$

(Problem 1.19), so that λ can be considerably larger. We now call λ the Taylor microscale.

According to Batchelor (1996), Taylor's "The spectrum of turbulence" (1938) was his "last paper on turbulence before the needs of World War II took him away from academic research." In this paper Taylor connected the power spectral density, or spectrum, to the two-point correlation function in physical space through the Fourier transform (Part III) and, in a casual aside, introduced what is now known as "Taylor's hypothesis" for interpreting a time series at a spatial point as a spatial record in the upstream direction.

But until Kolmogorov (1941) there was no unifying view of turbulence dynamics across the entire scale range. He postulated that for scales beyond the energy-containing range there are but two governing parameters: the mean rate of transfer of kinetic energy per unit mass from larger eddies to smaller (the energy *cascade rate*, Part III) and the fluid kinematic viscosity. Therefore (as we showed in Section 1.6.4) this extensive scale range yields readily to dimensional analysis. The dimensional analysis is even simpler in what we now call the inertial subrange, the larger-scale end of this range. Here the local Reynolds number is large so the viscosity is not important, and so the turbulence spectrum – the mean-squared amplitude of velocity fluctuation as a function of spatial wavenumber (inverse scale) – is then determined solely by the energy cascade rate. Its analytical form, the famous Kolmogorov spectrum (Chapter 7), emerges directly from dimensional analysis. In large-R_t turbulence this inertial subrange can be extensive; in the atmospheric boundary layer it can span four decades.

1.9.1 The conceptual model

Kolmogorov's seminal work enables us to view turbulence as a nonlinear system of interacting eddies that dynamically determines some of its properties and takes others from its fluid-mechanical environment.

The velocity scale u and the spatial scale ℓ of the energy-containing turbulent eddies are of the order of, but typically smaller than, the velocity and length scales of the mean flow. The turbulence Reynolds number $R_t = u\ell/\nu$ being large, these energy-containing eddies are essentially inviscid. They determine the viscous dissipation rate, the rate of conversion of turbulence kinetic energy into internal energy; it is independent of the value of the viscosity (and, as Holmes *et al.* (1996) suggest, even independent of the mechanism of the dissipation).

Since the rates of production and dissipation of turbulence kinetic energy per unit mass (TKE) depend only on u and ℓ, on dimensional grounds they are of order u^3/ℓ. But the dissipation process itself is a viscous one, so the velocity and length scales of the eddies in which it occurs do depend on ν. By Kolmogorov's 1941 hypotheses the velocity scale υ and the length scale η of the dissipative motion are $\eta = \eta(\epsilon, \nu) \sim (\nu^3/\epsilon)^{1/4}$, $\upsilon = \upsilon(\epsilon, \nu) \sim (\nu\epsilon)^{1/4}$.

Turbulence obtains its kinetic energy by direct transfer from the mean flow; in equilibrium it loses kinetic energy at that rate through the viscous dissipation into internal energy occurring in its smallest eddies. The turbulence field adjusts the size and intensity of those dissipative eddies in order to achieve the required energy dissipation rate. This viscous dissipation rate is proportional to the third power of flow speed, and the fluid heating it causes can be important in hurricanes (Bister and Emanuel, 1998) and more generally in storms (Businger and Businger, 2001).

Questions on key concepts

1.1 Explain the physical mechanism by which turbulence increases surface fluxes. Can you give some insight into the environmental implications of this?

1.2 Write out the components of the Navier–Stokes equation in turbulent flow.

1.3 What do we mean by a *conserved* scalar? Write out its equation in turbulent flow. Can a conserved scalar mix? Explain.

1.4 What are vortex stretching and vortex tilting? Why are they so important in turbulence?

1.5 What do we mean by the "velocity and length scales" u and ℓ of a turbulent flow? To which set of eddies do they refer? Why do the dissipative eddies have their own scales? How are the two sets of scales related?

1.6 What old saying about snowflakes applies also to turbulent flows? Explain.

1.7 Explain why most practically important turbulent flows cannot be calculated numerically through their governing equations.

1.8 Explain how the Reynolds stress emerges when the equation of motion for turbulent flow is averaged. Interpret the Reynolds stress physically.

1.9 What is a turbulence model? Why are they necessary? What are the two broad types, and how do they differ?

1.10 Why do you think only the lowest part of the atmosphere is continuously turbulent?

1.11 Explain the statement "turbulence is an unsolved problem."

1.12 We say turbulence is *random* and *stochastic*. What do those terms mean?

1.13 Explain how the rate of viscous dissipation can be independent of the fluid viscosity when that viscosity appears in its definition.

1.14 Explain why the density of a mass-conserving constituent in a turbulent flow need not be a conserved variable.

1.15 Discuss what properties of the ordinary derivative are shared by the substantial derivative. What property is not shared?

Problems

1.1 Consider steady, fully developed laminar flow in a circular pipe of diameter D. The axial equation of motion is

$$0 = -\frac{1}{\rho}\frac{\partial p(x)}{\partial x} + \frac{\nu}{r}\frac{\partial}{\partial r} r \frac{\partial u(r)}{\partial r}.$$

(a) Why does p not depend on r? u not depend on x? $\partial p/\partial x$ not depend on x?

(b) Solve this differential equation for $u(r)$.

(c) Find expressions for u_{ave}, Eq. (1.2), and the Darcy friction factor f, Eq. (1.5).

(d) Express the required pumping power per unit mass of fluid in terms of f.

1.2 Generalize Problem 1.1 to include conduction heat transfer in the radial direction. The temperature equation is

$$u(r)\frac{\partial T(x,r)}{\partial x} = \frac{\alpha}{r}\frac{\partial}{\partial r} r \frac{\partial T(x,r)}{\partial r}.$$

Here $\alpha = k/(\rho c_p)$ is the thermal diffusivity. Consider the case when $\partial T/\partial x$ does not depend on x.

(a) By differentiating this temperature equation with respect to x show that $\partial T/\partial x$ does not depend on r. Then derive the solution (1.11) that we wrote for this equation.

(b) Show that the wall heat flux H satisfies

$$H = -\frac{D u_{\text{ave}} \rho c_p}{4} \frac{\partial T}{\partial x}.$$

1.3 Define a time average for use in Eq. (1.9). Show that it commutes with differentiation.

1.4 Dot the Navier–Stokes equation (1.26) with velocity to form a kinetic energy equation. Where possible write terms as divergences. Integrate the equation over the entire volume of a turbulent flow, assuming the velocity vanishes on the bounding surface, and thereby eliminate them. Show that the volume-integrated kinetic energy must decay with time. Identify the decay mechanism.

1.5 Suppose in Problem 1.4 we applied a body-force field to the fluid. When can the volume-integrated balance of kinetic energy now be steady in time? Interpret the steady energy balance in terms of the first law of thermodynamics. What can you conclude about the role of viscous dissipation in turbulence?

1.6 It has been found that the viscous dissipation rate of kinetic energy per unit mass is of order u^3/ℓ. Using the expression for viscous dissipation found in Problem 1.4, show that the velocity and length scales of the dissipative eddies cannot be u and ℓ.

1.7 Write the rate of viscous dissipation per unit mass as the scalar product of the viscous stress tensor and the strain-rate tensor.

1.8 Explain why the viscous term in the averaged Navier–Stokes equation (1.38) is negligible. What restrictions must you place on the averaging scale for this to be true in the case of spatial averaging?

1.9 How does the number of grid points needed to calculate a turbulent flow directly from the governing equations depend on R_t?

1.10 If a cumulus cloud is turbulent, why does it appear "frozen"?

1.11 Derive the equation for the evolution of the gradient of a conserved scalar. Can its production term operate in two-dimensional turbulence?

1.12 Show that

$$\frac{Dab}{Dt} = a\frac{Db}{Dt} + b\frac{Da}{Dt}.$$

Use this property to show that the dot product of the gradient of a conserved scalar and vorticity is a conserved scalar.

1.13 Show that if $c_1(\mathbf{x}, t)$ and $c_2(\mathbf{x}, t)$ are solutions of Eq. (1.31), then $c_1 + c_2$ is also a solution. How is this used in determining the dispersion of effluents in the lower atmosphere? Why is it not valid for the Navier–Stokes equation?

1.14 Two geometrically identical turbulent flows, one using water and the other air, have the same u and ℓ. By what factor do their total rates of dissipation of kinetic energy differ?

1.15 If S_{ij} and A_{ij} are symmetric and antisymmetric tensors, respectively, show that their contraction $S_{ij}A_{ij}$ vanishes.

1.16 Resolve the paradox that ϵ is independent of ν but the expression for it (Problem 1.4) involves ν.

1.17 Explain why, in connection with Eq. (1.40), $\overline{u_i u_j}$ is different from $\overline{u_i}\,\overline{u_j}$ in a turbulent flow.

1.18 Derive the vorticity equation (1.28) from the Navier–Stokes equation (1.26). Hint: use the vector identities $\vec{\omega} \times \vec{u} = (\vec{u} \cdot \vec{\nabla})\vec{u} - \vec{\nabla}(\vec{u} \cdot \vec{u})/2$ and $(\vec{u} \cdot \vec{\nabla})\vec{u} = \vec{\omega} \times \vec{u} + \vec{\nabla}(\vec{u} \cdot \vec{u})/2$.

1.19 Derive Eq. (1.43) for the ratio of the Taylor and Kolmogorov microscales.

1.20 Do Problem 1.4 for steady flow in a pipe, integrating between two cross sections. Interpret your result.

1.21 Show that the total time derivative need not commute with other derivatives.

References

Batchelor, G. K., 1950: The application of the similarity theory of turbulence to atmospheric diffusion. *Quart. J. Roy. Meteor. Soc.*, **76**, 133–146.

Batchelor, G. K., 1967: *An Introduction to Fluid Dynamics.* Cambridge University Press.

Batchelor, G. K., 1996: *The Life and Legacy of G. I. Taylor.* Cambridge University Press.

Bister, M., and K. A. Emanuel, 1998: Dissipative heating and hurricane intensity. *Meteor. Atmos. Phys.*, **65**, 233–240.

Businger, S., and J. A. Businger, 2001: Viscous dissipation of turbulence kinetic energy in storms. *J. Atmos. Sci.*, **58**, 3793–3796.

Carusi, E., and A. Favaro, Eds., 1924: Leonardo da Vinci's *Del Moto e Misura dell' Acqua.* Bologna.

Corrsin, S., 1961: Turbulent flow. *Am. Scientist*, **49**, 300–325.

Deardorff, J. W., 1970a: A numerical study of three-dimensional turbulent channel flow at large Reynolds numbers. *J. Fluid Mech.*, **41**, 453–480.

Deardorff, J. W., 1970b: A three-dimensional numerical investigation of the idealized planetary boundary layer. *Geophys. Fluid Dyn.*, **1**, 377–410.

Deardorff, J. W., and G. E. Willis, 1975: A parameterization of diffusion into the mixed layer. *J. Appl. Meteorol.*, **14**, 1451–1458.

Dittus, F. W., and L. M. K. Boelter, 1930: Heat transfer in automobile radiators of the tubular type. *Univ. Calif., Berkeley, Publ. Eng.*, **2**, 443–461.

Friedlander, S. K., and L. Topper, Eds., 1961: *Turbulence: Classical Papers on Statistical Theory*. New York: Interscience.

Frisch, U., 1995: *Turbulence: The Legacy of A. N. Kolmogorov.* Cambridge University Press.

Gleick, J., 1987: *Chaos.* New York: Viking Penguin.

Holmes, P., J. L. Lumley, and G. Berkooz, 1996: *Turbulence, Coherent Structures, Dynamical Systems and Symmetry.* Cambridge University Press.

Kolmogorov, A. N., 1941: The local structure of turbulence in incompressible viscous fluid for very large Reynolds numbers. *Doklady ANSSSR*, **30**, 301–305.
Kundu, P. K., 1990: *Fluid Mechanics*. San Diego: Academic Press.
Lilly, D. K., 1967: The representation of small-scale turbulence in numerical simulation experiments. *Proceedings of the IBM Scientific Computing Symposium on Environmental Sciences*, IBM Form no. 320-1951, pp. 195–210.
Lorenz, E., 1963: Deterministic nonperiodic flow. *J. Atmos. Sci.*, **20**, 130–141.
Lumley, J. L., and H. A. Panofsky, 1964: *The Structure of Atmospheric Turbulence*. New York: Interscience.
Lumley, J. L., and A. M. Yaglom, 2001: A century of turbulence. *Flow, Turbulence, and Combustion*, **66**, 241–286.
Lundgren, T. S., 1967: Distribution functions in the statistical theory of turbulence. *Phys. Fluids*, **10**, 969–975.
Moody, L. F., 1944: Friction factors for pipe flow. *Trans. ASME*, **66**, 671–684.
Nikuradse, J., 1933: Strömungsgesetze in rauhen Röhren. *VDI-Forschungsheft*, **361**, 1933. English translation: Laws of Flow in Rough Pipes. NACA Technical Memorandum 1292, 1950.
Orszag, S. A., and G. S. Patterson Jr., 1972: Numerical simulation of three-dimensional homogeneous isotropic turbulence. *Phys. Rev. Lett.*, **28**, 76–79.
Reynolds, O., 1895: On the dynamical theory of incompressible viscous fluids and the determination of the criterion. *Philos. Trans. R. Soc. London, Ser. A*, **186**, 123–164.
Rouse, H., and S. Ince, 1957: *History of Hydraulics*. Ann Arbor: Edwards Brothers, Inc., p. 47.
Snyder, W. H., 1985: Fluid modeling of pollutant transport and diffusion in stably stratified flows over complex terrain. *Annu. Rev. Fluid Mech.*, **17**, 239–266.
Taylor, G. I., 1935: Statistical theory of turbulence. *Proc. R. Soc.*, **A151**, 421–478.
Taylor, G. I., 1938: The spectrum of turbulence. *Proc. R. Soc.*, **A164**, 476–490.
Tennekes, H., and J. L. Lumley, 1972: *A First Course in Turbulence*. Cambridge, MA: MIT Press.
Turns, S. R., 2006: *Thermal-Fluid Sciences: An Integrated Approach*. Cambridge University Press.
Van Dyke, M., 1982: *An Album of Fluid Motion*. Stanford: Parabolic Press.

2

Getting to know turbulence

2.1 Average and instantaneous properties contrasted

Figure 2.1 is a famous snapshot of a turbulent wake, the region downstream of a body in a moving fluid. The instantaneously thin, irregular boundary between the turbulent and nonturbulent flow is continuously deformed by the turbulent eddies, so that under averaging it becomes a broad, smooth transition region. Figure 2.2 illustrates this same feature at the top of the atmospheric boundary layer.

We have had access to instantaneous turbulence fields through remote sensing and numerical simulation only since the 1970s. Perhaps that is why our descriptive terms for turbulence tend to refer to its statistical properties, not its instantaneous ones. For example,

- *Homogeneous* turbulence has spatially uniform statistical properties (with the exception of mean pressure). A turbulent flow can be homogeneous in zero, one, two, or three directions. A sphere wake is an example of the first. The turbulent boundary layer near the leading edge of a flat plate, or downstream of a change in surface conditions, can be homogeneous in one direction (the lateral) but is inhomogeneous in the wall-normal and streamwise directions. The turbulent boundary layer over a uniform surface can be homogeneous in two directions, those in the plane parallel to the surface, but is necessarily inhomogeneous in the normal direction. The *grid turbulence* produced by a grating of bars spanning the cross section of a wind tunnel is homogeneous in the cross-stream plane but inhomogeneous in the streamwise direction because it decays as it goes downstream (Chapter 5). A carefully tailored *homogeneous shear flow* in a wind tunnel (Tavoularis and Corrsin, 1981) is, to a good approximation, homogeneous in all three directions.
- *Steady* (also called *stationary*) turbulence – turbulence in a laboratory flow driven by a constant-speed blower, for example – has statistics that are independent of time. Turbulence in the atmospheric boundary layer can be steady for up to a few hours, but near sunrise and sunset in clear weather it is made unsteady (nonstationary) by the changing surface energy budget (Chapter 9).

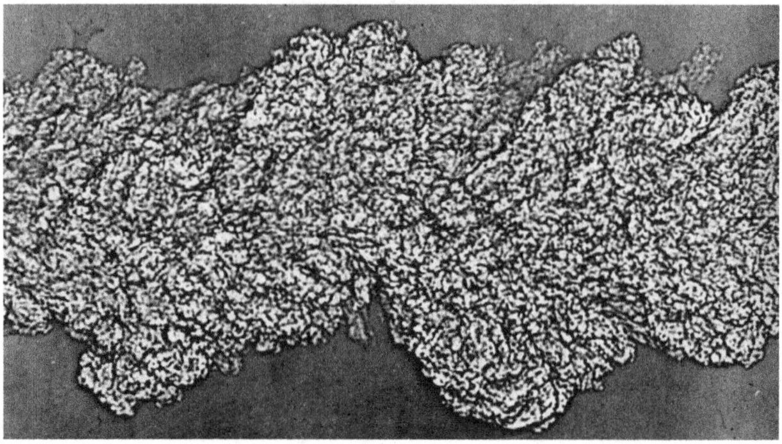

Figure 2.1 The turbulent wake of a bullet, which is several hundred wake diameters to the left. This shadowgraph shows the strikingly sharp, irregular boundary between the turbulent fluid and the almost motionless fluid outside. Photograph courtesy Army Research Laboratory. From Van Dyke (1982).

- *Isotropic turbulence* has statistical properties that are independent of translation, rotation, and reflection of the coordinate axes. It must decay in time, for the production mechanisms that maintain stationary turbulence are anisotropic (Chapter 6). Isotropy of the smallest spatial scales is called *local isotropy* (Chapter 14).
- The *logarithmic profile* (or *law of the wall*) and the *constant-stress layer* refer to the height variation of mean velocity and the Reynolds shear stress, respectively, in the turbulent boundary layer over a flat surface. Like the Reynolds stresses themselves (which G. I. Taylor (1935) called "virtual mean stresses") they exist only as averages.
- The *Gaussian plume* refers to a mean effluent plume in homogeneous turbulent flow, not an instantaneous plume.
- A *well-mixed* state of a convective atmospheric boundary layer (Chapter 11) has *mean*, not instantaneous, profiles of potential temperature and water-vapor mixing ratio that are essentially uniform with height.
- A *turbulent flux* of a property is the *mean*, not instantaneous, amount of that property flowing through unit area per unit time due to the turbulent velocity.

2.2 Averaging

All the dependent variables in a turbulent flow – velocity, vorticity, temperature (if there is heat transfer), density of an advected constituent (if there is mass transfer), pressure – are turbulent. At any instant they are distributed irregularly in space, at any point in space they fluctuate chaotically in time, and at given point and a given time they vary randomly from realization to realization. Since Osborne Reynolds'

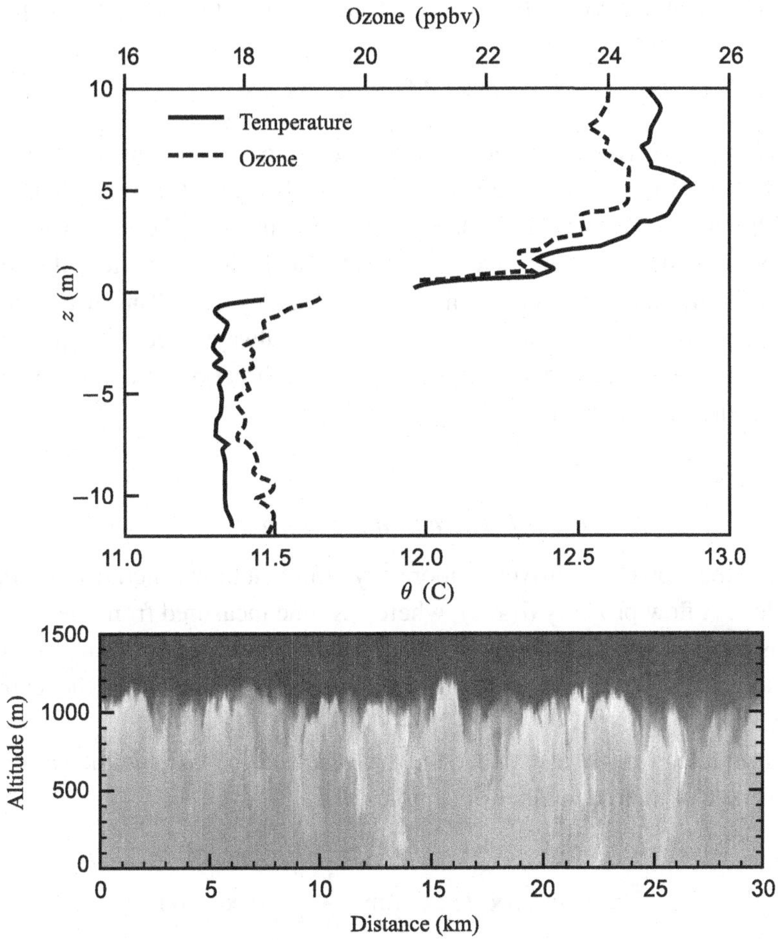

Figure 2.2 Upper: Profiles of temperature and ozone mixing ratio at the top of a cloud-capped convective boundary layer. z is measured from the mean top. From Lenschow *et al.* (1988). Lower: Airborne-lidar measurements of aerosol concentration in a vertical plane of a clear convective boundary layer. The instantaneous top is quite thin and variable in space. Courtesy C. Kiemle, DLR Germany, and J. Grabon, Penn State.

time it has been traditional to separate a turbulent flow variable $a(\mathbf{x}, t)$ into mean and fluctuating parts distinguished by the overbar and prime:

$$a(\mathbf{x}, t) = \overline{a}(\mathbf{x}, t) + a'(\mathbf{x}, t). \tag{2.1}$$

The prime can make this notation cumbersome, so we shall use instead the notation of Tennekes and Lumley (1972), denoting the "full" turbulent flow variable with a tilde, $\tilde{a}(\mathbf{x}, t)$, and representing its mean and fluctuating parts with upper- and lower-case symbols:

$$\tilde{a}(\mathbf{x}, t) = A(\mathbf{x}, t) + a(\mathbf{x}, t). \tag{2.2}$$

Several types of averages have been used to define mean values in turbulence. Reynolds (1895) used a volume average. Somewhat later (in the 1930s, according to Monin and Yaglom (1971)), Kolmogorov and his school, and Kampé de Fériet, brought the ensemble average of statistical physics to turbulence; it is conceptually the most elegant. Tennekes and Lumley (1972) used a time average in steady conditions. A time average is almost always used with quasi-steady observations, and space averages in homogeneous directions are convenient with numerical-simulation results.

2.2.1 The ensemble average

Turning on the blower that drives a laboratory turbulent flow generates a *realization* of that flow. A flow property $\tilde{a}(\mathbf{x}, t)$, where t is time measured from the instant the blower is turned on, say, is random – i.e., different in every realization. We indicate this randomness by writing the flow property as $\tilde{a}(\mathbf{x}, t; \alpha)$, α denoting the realization number.

The ensemble average (also called the *expected value*) of $\tilde{a}$ is defined as the limit of the average of a large number of samples of $\tilde{a}$:

$$\overline{\tilde{a}}(\mathbf{x}, t) \equiv A(\mathbf{x}, t) \equiv \lim_{N\to\infty} \frac{1}{N} \sum_{\alpha=1}^{N} \tilde{a}(\mathbf{x}, t; \alpha). \tag{2.3}$$

As indicated in Eq. (2.3), the ensemble average can depend on both position and time. Being linear, it commutes with other linear operations such as differentiation and integration,

$$\overline{\frac{\partial \tilde{a}(\mathbf{x}, t; \alpha)}{\partial t}} = \frac{\partial}{\partial t}\overline{\tilde{a}}(\mathbf{x}, t), \qquad \overline{\int_a^b \tilde{a}(\mathbf{x}, t; \alpha)\, dt} = \int_a^b \overline{\tilde{a}}(\mathbf{x}, t)\, dt, \tag{2.4}$$

and so forth (Problem 2.6).

In the literature the ensemble parameter is often not explicitly indicated, any unaveraged quantity being taken as an arbitrary member of the ensemble. In later chapters we shall follow this convention, suppressing the ensemble index. Unless stated otherwise, by *average* or *mean* we generally intend the ensemble average.

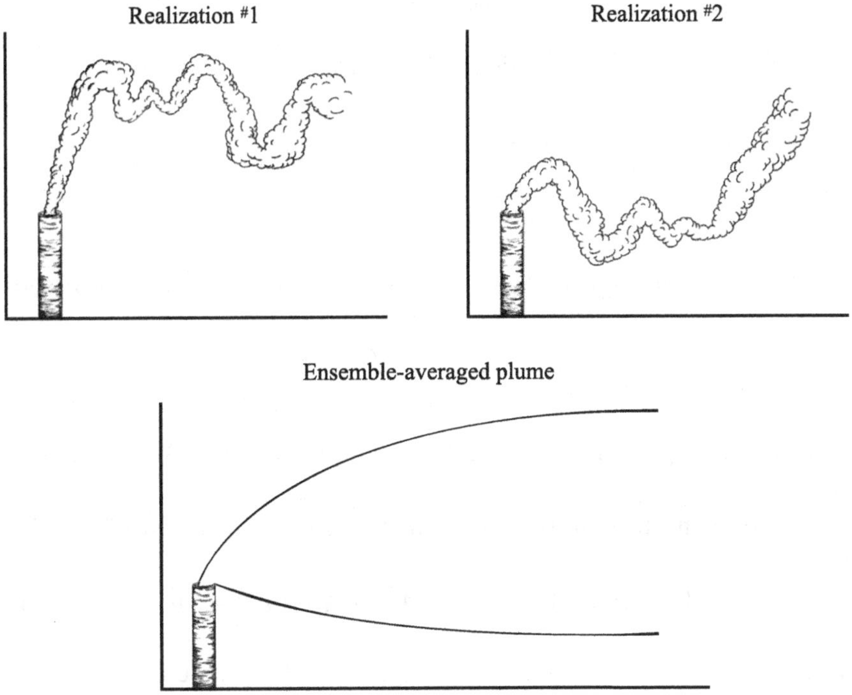

Figure 2.3 Realizations of effluent plumes contrasted with the ensemble-averaged plume.

2.2.2 Does the ensemble-averaged field exist?

An instantaneous effluent plume from a point source (Figure 2.3) is sinuous and irregular. Measurements show that near the source the effluent concentration is roughly uniform within it and zero outside it. But the ensemble-averaged plume is diffuse and smooth; in homogeneous, wind-tunnel turbulence its mean concentration profile is found to be Gaussian. This suggests that the ensemble-averaged field is unlikely to exist in any realization of a turbulent flow, even for an instant.

2.2.3 The ensemble-averaging rules

The ensemble average (we shall use *average* and *mean* interchangeably) has convenient properties that are commonly called the *Reynolds averaging* rules:†

- The average of a sum is the sum of the averages (the *distributive* property):

$$\overline{\tilde{a}+\tilde{b}} = \overline{\tilde{a}} + \overline{\tilde{b}}. \tag{2.5}$$

† Reynolds (1895) used a volume average but assumed it follows the rules for the ensemble average, so in the turbulence community they now bear his name. We'll discuss the volume averaging rules in Chapter 3.

- The average of an average is the average:

$$\overline{\overline{\tilde{a}}} = \overline{\tilde{a}} \quad (\overline{A} = A). \tag{2.6}$$

- The average of a fluctuation is zero:

$$\overline{(\tilde{a} - \overline{\tilde{a}})} = 0 \quad (\overline{a} = 0). \tag{2.7}$$

- The average of a derivative is the derivative of the average (the *commutative* property):

$$\overline{\frac{\partial \tilde{a}}{\partial x_i}} = \frac{\partial \overline{\tilde{a}}}{\partial x_i}; \qquad \overline{\frac{\partial \tilde{a}}{\partial t}} = \frac{\partial \overline{\tilde{a}}}{\partial t}. \tag{2.8}$$

These can be easily proved (Problem 2.1) from the definition (2.3) of the ensemble average.

It follows from the distributive property that the ensemble average of a product is

$$\overline{\tilde{a}\tilde{b}} = \overline{(A + a)(B + b)} = \overline{AB} + \overline{Ab} + \overline{aB} + \overline{ab}. \tag{2.9}$$

From the definition of the ensemble average, Eq. (2.3), the cross terms vanish because the mean is a constant in the averaging process (Problem 2.1):

$$\overline{Ab} = \overline{aB} = 0. \tag{2.10}$$

Thus, using the decomposition $\tilde{u}_i = U_i + u_i$, $\tilde{c} = C + c$ the Reynolds stress τ_{ij} in the ensemble-averaged Navier–Stokes equation (1.40) and the corresponding turbulent scalar flux f_i in (1.41) can be written as

$$-\frac{\tau_{ij}}{\rho} = \overline{\tilde{u}_i\tilde{u}_j} - \overline{\tilde{u}_i}\,\overline{\tilde{u}_j} = \overline{u_iu_j}, \qquad f_i = \overline{\tilde{c}\tilde{u}_i} - \overline{\tilde{c}}\,\overline{\tilde{u}_i} = \overline{cu_i}. \tag{2.11}$$

These fluxes produced by ensemble averaging are *covariances* of the turbulence field. If a covariance $\overline{ab}$ is nonzero, a and b are said to be *correlated*. We'll see in Chapter 4 that any two turbulent variables tend to be correlated unless mean-flow symmetry requires otherwise.

2.2.4 A simple example of ensemble averaging

We can demonstrate ensemble averaging with realizations of a random sine wave in one dimension,

$$u(x, \alpha) = \sin(\kappa x + \phi_\alpha), \qquad \alpha = 1, 2, \ldots, N, \tag{2.12}$$

where α is the realization index, κ is spatial wavenumber (2π/wavelength) and ϕ_α is the phase angle in realization α. We choose the ϕ_α from random numbers that vary from 0 to 2π with uniform probability, giving u a different phase in each realization. Thus $u(x)$ is a random variable that is statistically homogeneous in x.

The ensemble average of u is

$$\overline{u} = \overline{\sin(\kappa x + \phi)} = \lim_{N\to\infty} \frac{1}{N} \sum_{\alpha=1}^{N} \sin(\kappa x + \phi_\alpha). \tag{2.13}$$

Expanding $\sin(\kappa x + \phi)$ and using the distributive property (2.5) of the ensemble average yields

$$\overline{u} = \overline{\sin \kappa x \cos \phi + \cos \kappa x \sin \phi} = \overline{\sin \kappa x \cos \phi} + \overline{\cos \kappa x \sin \phi}. \tag{2.14}$$

Position x is fixed in the ensemble average, so $\sin \kappa x$ and $\cos \kappa x$, being constants in the averaging process, can be taken out from under the overbar:

$$\overline{u} = \sin \kappa x \, \overline{\cos \phi} + \cos \kappa x \, \overline{\sin \phi}. \tag{2.15}$$

Since the random phase ϕ varies from 0 to 2π with uniform probability, we expect that

$$\overline{\cos \phi} = \overline{\sin \phi} = 0. \tag{2.16}$$

We shall prove this in Part III. We conclude from Eqs. (2.15) and (2.16) that our test field has zero ensemble mean, $\overline{u} = 0$.

The same process yields the variance $\overline{u^2}$:

$$\begin{aligned}
\overline{u^2} = \overline{\sin^2(\kappa x + \phi)} &= \overline{\frac{1}{2}[1 - \cos 2(\kappa x + \phi)]} = \frac{1}{2}\left[1 - \overline{\cos 2(\kappa x + \phi)}\right] \\
&= \frac{1}{2}\left[1 - \overline{\cos 2\kappa x \cos 2\phi} + \overline{\sin 2\kappa x \sin 2\phi}\right] \\
&= \frac{1}{2}\left[1 - \cos 2\kappa x \, \overline{\cos 2\phi} + \sin 2\kappa x \, \overline{\sin 2\phi}\right] = \frac{1}{2}.
\end{aligned} \tag{2.17}$$

The derivative is $\partial u/\partial x = \kappa \cos(\kappa x + \phi)$. The same process shows that $\overline{\partial u/\partial x} = 0$. We can prove this more directly by using property (2.8), the interchangeability of differentiation and averaging, which implies

$$\overline{\frac{\partial u}{\partial x}} = \frac{\partial \overline{u}}{\partial x} = 0, \tag{2.18}$$

since in this example $\overline{u} = 0$.

The mean-square derivative is

$$\overline{\left(\frac{\partial u}{\partial x}\right)^2} = \kappa^2 \overline{\cos^2(\kappa x + \phi)} = \frac{\kappa^2}{2}, \tag{2.19}$$

so it increases as the square of wavenumber. We shall see in Chapter 5 that this property of a derivative allows the small (large wavenumber), weak, Kolmogorov-microscale eddies to keep turbulence in equilibrium by dissipating kinetic energy and diffusing away scalar fluctuations at the same mean rate they are produced at the large scales.

The mean of the product of u and $\partial u/\partial x$ is

$$\overline{u\frac{\partial u}{\partial x}} = \kappa \overline{\sin(\kappa x + \phi)\cos(\kappa x + \phi)} = \frac{\kappa}{2}\overline{\sin 2(\kappa x + \phi)} = 0. \tag{2.20}$$

Thus, here u and $\partial u/\partial x$ are *uncorrelated*. This also follows directly from the homogeneity of u:

$$\overline{u\frac{\partial u}{\partial x}} = \frac{1}{2}\frac{\partial \overline{u^2}}{\partial x} = 0. \tag{2.21}$$

2.3 Ergodicity

Let's assume we want the mean value of a variable $\tilde{u}(x, t; \alpha)$ that has stochastic variations in both x and t and is also random – different in each realization α. We assume further that its ensemble mean U, which in the most general case depends on both x and t,

$$U(x, t) = \lim_{N\to\infty} \frac{1}{N} \sum_{\alpha=1}^{\alpha=N} \tilde{u}(x, t; \alpha), \tag{2.22}$$

is impractical to determine. But we can determine its time average at position x in a single realization n, say,

$$U^T(x, t, T; n) = \frac{1}{T}\int_0^T \tilde{u}(x, t + t'; n)\, dt', \tag{2.23}$$

and its spatial average at time t in a single realization m:

$$U^L(x, t, L; m) = \frac{1}{L}\int_0^L \tilde{u}(x + x', t; m)\, dx'. \tag{2.24}$$

If $\tilde{u}$ is stationary in time, so that $U = U(x)$, we intuitively expect this time average in a single realization to converge to the ensemble average as the averaging time T increases:

$$\lim_{T\to\infty} U^T(x, t, T; n) = U(x). \tag{2.25}$$

Similarly, if $\tilde{u}$ is homogeneous in x, so that $U = U(t)$, we intuitively expect this spatial average in a single realization to converge to the ensemble average as the averaging distance L increases:

$$\lim_{L\to\infty} U^L(x, t, T; m) = U(t). \tag{2.26}$$

If $\tilde{u}$ is both homogeneous and stationary, so that U does not depend on x or t, we expect both the time and spatial averages to converge to the ensemble average U.

The property that the time average of a stationary random variable and the space average of a homogeneous random variable converge to the ensemble average is called *ergodicity.* Physically, ergodicity means that any unbiased average of a variable converges to the ensemble average. We routinely determine the ensemble mean of a stationary, time-varying signal at a point in space through time averaging.

2.4 The convergence of averages

The ensemble average has ideal properties: it commutes with linear operations, a second application has no effect (i.e., the average of an average is the average), and the average of a fluctuation is zero. But it can be very tedious to use with observational data, and as discussed in Section 2.3 experimentalists typically use a time average in stationary conditions instead.

Mechanically driven turbulent flows are often stationary by design. The atmospheric boundary layer is inherently nonstationary, however, because of the diurnal cycle and the changing synoptic conditions. But one can often find quasi-stationary periods of up to a few hours length, which could be long enough in some problems. This raises the question: How long must we average in time to get acceptably close to the ensemble average? We'll outline the general answer here using statistical concepts that we'll discuss more fully in Part III.

To answer this question† we begin with the time average of a stationary function of time $\tilde{u}(t)$ over an interval T,

$$\overline{u}^T = \frac{1}{T}\int_{t_0}^{t_0+T} \tilde{u}(t')\,dt', \tag{2.27}$$

the initial time t_0 being arbitrary. Here $\tilde{u}(t)$ could be the time series of the streamwise velocity component at a point in a turbulent flow. We shall call T the *averaging time.* Now let $\tilde{u}(t) = U + u(t)$, the sum of ensemble-mean and fluctuating parts. By our stationarity assumption U does not depend on time. The difference between the time and ensemble means of $\tilde{u}$ is then

† This discussion is adapted from Lumley and Panofsky (1964).

$$\overline{u}^T - U = \frac{1}{T}\int_{t_0}^{t_0+T} [U + u(t')]\, dt' - U = \frac{1}{T}\int_{t_0}^{t_0+T} u(t')\, dt', \tag{2.28}$$

which is a random variable with an ensemble mean of zero. One measure of $\overline{u}^T - U$ is its variance σ^2, which can be written as

$$\sigma^2 \equiv \overline{(\overline{u}^T - U)^2} = \frac{1}{T^2}\int_{t_0}^{t_0+T}\int_{t_0}^{t_0+T} \overline{u(t')u(t'')}\, dt'\, dt''. \tag{2.29}$$

$\overline{u(t')u(t'')}$ is called the *autocovariance* of $u(t)$. For a stationary process it is a function only of the time difference $t' - t''$, so we write it as

$$\overline{u(t')u(t'')} = \overline{u^2}\rho(t' - t''), \tag{2.30}$$

with $\overline{u^2} = \overline{u(t)u(t)}$ the variance of $u(t)$ and ρ its *autocorrelation function*. Equation (2.30) shows that ρ is an even function; i.e., $\rho(t' - t'') = \rho(t'' - t')$. Thus, we can write Eq. (2.29) as

$$\sigma^2 = \frac{\overline{u^2}}{T^2}\int_{t_0}^{t_0+T}\int_{t_0}^{t_0+T} \rho(t' - t'')\, dt'\, dt''. \tag{2.31}$$

One can reduce Eq. (2.31) to a single integral by changing variables to $\eta = t'' - t'$ and $\zeta = t'' + t'$ and doing the integration over ζ, noting that the limits of that integral then depend on η (Problem 2.4.) The result is

$$\sigma^2 = \frac{2\overline{u^2}}{T}\int_0^T \left(1 - \frac{t}{T}\right)\rho(t)\, dt. \tag{2.32}$$

The Eulerian *integral time scale* τ,† defined as

$$\int_0^\infty \rho(t)\, dt = \tau, \tag{2.33}$$

is a measure of the "memory time" of the Eulerian velocity fluctuation $u(t)$. For averaging time T much longer than the integral scale τ we can approximate Eq. (2.32) as

$$\sigma^2 \simeq \frac{2\overline{u^2}}{T}\int_0^T \rho(t)\, dt = \frac{2\overline{u^2}\tau}{T}. \tag{2.34}$$

This is quite an important result, for with Eq. (2.29) it quantifies the statistical difference between the ensemble average, which we can seldom use in applications, and the time average that we usually use instead.

† We'll introduce the Lagrangian integral time scale in Chapter 4.

To help interpret Eq. (2.34) we define the rms uncertainty e of the time mean through Eq. (2.29):

$$e \equiv \frac{\left[\overline{(\overline{u}^T - U)^2}\right]^{1/2}}{U} = \frac{\sigma}{U}. \tag{2.35}$$

e is a measure of the fractional error incurred in taking the finite-time mean as the ensemble mean; a small e means that $\overline{u}^T$ is a good approximation to U. From Eqs. (2.34) and (2.35) the averaging time required for determining to an rms fractional uncertainty e the mean of a signal $\tilde{u}(t)$ whose ensemble mean is U and integral scale is τ is therefore

$$T = \frac{2\tau}{e^2}\left[\frac{\overline{u^2}}{U^2}\right]. \tag{2.36}$$

Equation (2.36) says that the required averaging time is

- proportional to τ, the integral scale of the time series;
- proportional to $\overline{u^2}$, the variance of the time series; and
- inversely proportional to e^2, the square of the rms fractional uncertainty in the time mean.

As it turns out, obtaining atmospheric turbulence statistics of low rms uncertainty from stationary sensors can require unreasonably long averaging times; for this reason poor estimates of statistics plague atmospheric turbulence research. This makes it difficult to use atmospheric observations to develop and test models of atmospheric turbulence (Part II). Measurements made in a laboratory model of the atmospheric boundary layer converge very much faster than the corresponding atmospheric measurements (Problem 2.8).

For an average of a homogeneous turbulence signal of integral scale ℓ over a line of length L, the one-dimensional result (2.36) can be written as

$$e^2(\text{line average}) \simeq \left(\frac{\ell}{L}\right)\frac{\overline{u^2}}{U^2}. \tag{2.37}$$

The corresponding expressions for averaging homogeneous turbulence over a square of area L^2 and a cube of volume L^3 are

$$e^2(\text{plane average}) \simeq \left(\frac{\ell}{L}\right)^2\frac{\overline{u^2}}{U^2}, \quad e^2(\text{volume average}) \simeq \left(\frac{\ell}{L}\right)^3\frac{\overline{u^2}}{U^2}, \tag{2.38}$$

where for simplicity we have taken ℓ to be the same in each direction. Since in general ℓ/L is small, this shows the advantages of area and volume averaging in reducing squared uncertainty e^2. In large-eddy simulation of atmospheric boundary layers, for example, we can average the calculated fields over homogeneous horizontal planes. In this way one "snapshot" of a simulated field on a sufficiently large plane can yield a good estimate of the ensemble average.

2.5 The turbulence spectrum and the eddy velocity scale

In Chapter 1 we introduced the velocity scales u, v and length scales ℓ, η of the energy-containing and dissipative eddies, respectively, and showed that $\ell/\eta \sim R_t^{3/4}$, with R_t the large-eddy Reynolds number $u\ell/\nu$. R_t varies from less than 10^3 in some engineering flows to about 10^8 in the convective atmospheric boundary layer and perhaps 10^{10} in a supercell thunderstorm; over that range ℓ/η varies from about 10^2 to 10^7.

The *power spectral density* of the turbulent velocity field (loosely called "the turbulence spectrum") allows the velocity scale u of the energy-containing eddies to be generalized to $u(r)$, the velocity scale of an eddy of size r, with $\ell \geq r \geq \eta$. We shall present here an informal derivation of the turbulence spectrum, beginning for simplicity with a homogeneous scalar function of a single variable. Part III contains a more formal presentation.

2.5.1 The spectrum of a one-dimensional, real, random, homogeneous scalar function

Let $\tilde{f}(x)$ be a real, homogeneous function, the sum of an ensemble-mean part F and a fluctuation $f(x)$, defined over record of length L. It could be a spatial record of temperature or a velocity component in a turbulent flow, for example. We can approximate $\tilde{f}(x)$ through a Fourier series, a sum of sines and cosines of wavelengths $L/n,\ n = 0, \ldots, N$:

$$\tilde{f}(x) \simeq \frac{a_0}{2} + \sum_{n=1}^{N} a_n \cos\left(\frac{2\pi n x}{L}\right) + \sum_{n=1}^{N} b_n \sin\left(\frac{2\pi n x}{L}\right). \tag{2.39}$$

The coefficients a_n and b_n are real numbers called *Fourier coefficients*.

Since each of the sine and cosine terms in Eq. (2.39) integrates to zero over the record length, it follows that the average of $\tilde{f}(x)$ over the record length is

$$\frac{1}{L}\int_0^L \tilde{f}(x)\,dx = \frac{a_0}{2}. \tag{2.40}$$

The other Fourier components together represent the variation of $\tilde{f}(x)$ with x.

In turbulence we usually write Fourier series in terms of the wavenumber $\kappa_n = 2\pi n/L$, so we write Eq. (2.39) as

$$\tilde{f}(x) \simeq \frac{a_0}{2} + \sum_{n=1}^{N} a_n \cos(\kappa_n x) + \sum_{n=1}^{N} b_n \sin(\kappa_n x), \quad \kappa_n = 2\pi n/L. \tag{2.41}$$

It is a property of Fourier series that by increasing N we can approximate $f(x)$ as well as we like. Thus, we can formally write

$$\tilde{f}(x) = \frac{a_0}{2} + \sum_{n=1}^{\infty} a_n \cos(\kappa_n x) + \sum_{n=1}^{\infty} b_n \sin(\kappa_n x), \quad \kappa_n = 2\pi n/L. \tag{2.42}$$

Furthermore, in turbulence $\tilde{f}(x)$ is a random function, different in every realization α, so its Fourier coefficients a_n and b_n are also random. Thus we generalize Eq. (2.42) to include this randomness through the realization index α:

$$\tilde{f}(x,\alpha) = \frac{a_0(\alpha)}{2} + \sum_{n=1}^{\infty} a_n(\alpha) \cos(\kappa_n x) + \sum_{n=1}^{\infty} b_n(\alpha) \sin(\kappa_n x), \quad \kappa_n = 2\pi n/L. \tag{2.43}$$

Today we use a computer to calculate Fourier coefficients, and so it is convenient to express the series in exponential notation. By using the identities

$$\cos\theta = \frac{e^{i\theta} + e^{-i\theta}}{2}, \qquad \sin\theta = \frac{e^{i\theta} - e^{-i\theta}}{2i} \tag{2.44}$$

we can write (2.43) as

$$\tilde{f}(x;\alpha) = \frac{a_0}{2} + \sum_{n=1}^{\infty} \left(\frac{a_n - ib_n}{2}\right) e^{i\kappa_n x} + \sum_{n=1}^{\infty} \left(\frac{a_n + ib_n}{2}\right) e^{-i\kappa_n x}. \tag{2.45}$$

We can rewrite this as

$$\tilde{f}(x;\alpha) = \frac{a_0}{2} + \sum_{n=1}^{\infty} \left(\frac{a_n - ib_n}{2}\right) e^{i\kappa_n x} + \sum_{n=-1}^{-\infty} \left(\frac{a_{-n} + ib_{-n}}{2}\right) e^{-i\kappa_{-n} x}. \tag{2.46}$$

If we define $\kappa_{-n} = -\kappa_n$ we can express this more compactly as

$$\tilde{f}(x;\alpha) = \sum_{n=-\infty}^{\infty} \hat{f}(\kappa_n;\alpha) e^{i\kappa_n x};$$

$$\hat{f}(\kappa_n;\alpha) = \frac{a_n - ib_n}{2}, \; n+; \qquad \hat{f}(\kappa_n;\alpha) = \frac{a_n + ib_n}{2}, \; n-. \tag{2.47}$$

The variance of f is determined by squaring and ensemble averaging Eq. (2.47). Since f is real, f is equal to its complex conjugate f^* and we can write

$$\overline{f^2} = \overline{ff^*} = \sum_{n=-\infty}^{\infty} \sum_{m=-\infty}^{\infty} \overline{\hat{f}(\kappa_n)\hat{f}^*(\kappa_m)}\, e^{i(\kappa_n - \kappa_m)x}. \tag{2.48}$$

It is a property of the Fourier-series representation of a homogeneous function that Fourier coefficients of different wavenumbers are uncorrelated – that is,

$$\overline{\hat{f}(\kappa_n)\hat{f}^*(\kappa_m)} = 0, \quad \kappa_n \neq \kappa_m. \tag{2.49}$$

With Eq. (2.48) we can make this plausible as follows. The homogeneity of $f(x)$ implies that $\overline{f^2} \neq \overline{f^2}(x)$, and therefore the lhs of Eq. (2.48) is independent of x. On the rhs, for $\kappa_n \neq \kappa_m$ the exponential is a nonzero function of x, so $\overline{\hat{f}(\kappa_n)\hat{f}^*(\kappa_m)}$ must vanish to make the rhs independent of x. This is Eq. (2.49).

Given the constraint (2.49), the variance in Eq. (2.48) is given by a single sum:

$$\overline{f^2} = \sum_{n=-\infty}^{\infty} \overline{\hat{f}(\kappa_n)\hat{f}^*(\kappa_n)}. \tag{2.50}$$

We define $\phi(\kappa_n)$, the *power spectral density* of f, as the contribution to the variance $\overline{f^2}$ per unit interval of wavenumber,

$$\phi(\kappa_n) = \frac{\overline{\hat{f}(\kappa_n)\hat{f}^*(\kappa_n)}}{\Delta\kappa}, \quad \Delta\kappa = \frac{2\pi}{L}, \tag{2.51}$$

so that

$$\overline{f^2} = \sum_{n=-\infty}^{\infty} \phi(\kappa_n)\Delta\kappa. \tag{2.52}$$

Thus, $\phi(\kappa)$ is the density of contributions to the variance. In the limit as L and N approach infinity Eq. (2.52) becomes an integral:

$$\overline{f^2} = \int_{-\infty}^{\infty} \phi(\kappa)\, d\kappa. \tag{2.53}$$

2.5.2 Extension to three dimensions

Next we generalize Eq. (2.47) to represent a real, homogeneous, three-dimensional, random, conserved scalar field $c(x_1, x_2, x_3; \alpha) = c(\mathbf{x}; \alpha)$ in a cube of side L. Now the wavenumber is a vector $\boldsymbol{\kappa} = (\kappa_1, \kappa_2, \kappa_3)$ so we write

$$c(\mathbf{x}; \alpha) = \sum_{\boldsymbol{\kappa}} \hat{c}(\boldsymbol{\kappa}; \alpha) e^{i(\boldsymbol{\kappa}\cdot\mathbf{x})}. \tag{2.54}$$

The variance is given by

$$\overline{cc*} = \overline{c^2} = \sum_{\kappa}\sum_{\kappa'} \overline{\hat{c}(\kappa, \alpha)\hat{c}^*(\kappa', \alpha)} e^{i(\kappa - \kappa')\cdot \mathbf{x}}$$
$$= \sum_{\kappa} \overline{\hat{c}(\kappa, \alpha)\hat{c}^*(\kappa, \alpha)} = \sum_{\kappa} \phi(\kappa)(\Delta\kappa)^3. \tag{2.55}$$

In the limit as both L and N approach infinity Eq. (2.55) becomes an integral:

$$\overline{c^2} = \int\int\int_{-\infty}^{\infty} \phi(\kappa)\, d\kappa_1\, d\kappa_2\, d\kappa_3. \tag{2.56}$$

The integral in Eq. (2.56) can be done first over a sphere of radius $\kappa = (\kappa_1^2 + \kappa_2^2 + \kappa_3^2)^{1/2}$ and then over κ. The *three-dimensional spectrum* $E_c(\kappa)$† is defined as the integral of ϕ over the sphere,

$$E_c(\kappa) = \int\int_{\kappa_i\kappa_i=\kappa^2} \phi(\kappa_1, \kappa_2, \kappa_3)\, d\sigma, \tag{2.57}$$

so the variance is

$$\overline{c^2} = \int_0^{\infty} E_c(\kappa)\, d\kappa. \tag{2.58}$$

2.5.3 Application to a homogeneous velocity field

For a random, zero-mean, homogeneous velocity field $u_i(\mathbf{x}; \alpha)$ defined in a cube of side L the Fourier coefficients are vectors:

$$u_i(\mathbf{x}; \alpha) = \sum_{\kappa} \hat{u}_i(\kappa; \alpha) e^{i(\kappa\cdot\mathbf{x})}. \tag{2.59}$$

The covariance is

$$\overline{u_i u_j} = \overline{u_i u_j^*} = \sum_{\kappa} \overline{\hat{u}_i(\kappa)\hat{u}_j^*(\kappa)}$$
$$= \sum_{\kappa} \phi_{ij}(\kappa)(\Delta\kappa)^3. \tag{2.60}$$

In the limit Eq. (2.60) becomes an integral,

$$\overline{u_i u_j} = \int\int\int_{-\infty}^{\infty} \phi_{ij}(\kappa_1, \kappa_2, \kappa_3)\, d\kappa_1\, d\kappa_2\, d\kappa_3. \tag{2.61}$$

† This is so-named because it is a function of the magnitude of a three-dimensional wavenumber.

Here it is traditional to define a *three-dimensional energy spectrum* $E(\kappa)$ as the integral of $\phi_{ii}/2$ over a sphere of radius κ,

$$E(\kappa) = \int\int_{\kappa_i\kappa_i=\kappa^2} \frac{\phi_{ii}(\kappa_1, \kappa_2, \kappa_3)}{2} d\sigma. \tag{2.62}$$

The factor of 2 is used so that $E(\kappa)$ integrates to the kinetic energy per unit mass:

$$\frac{\overline{u_i u_i}}{2} = \int_0^\infty E(\kappa)\, d\kappa. \tag{2.63}$$

2.5.4 The eddy velocity scale u(r)

We can now use $E(\kappa)$ to estimate $u(r)$, the *velocity scale* or *characteristic velocity* (the typical root-mean-square velocity, say) of eddies of spatial scale r, or, equivalently, spatial wavenumber magnitude $\kappa \sim 1/r$. Following Tennekes and Lumley (1972), we define "an eddy of scale r" as one of spatial scale between roughly $r/2$ and $3r/2$, so that it lies in a band of width $\Delta r \sim r$ about scale r. In wavenumber terms, we take "an eddy of wavenumber magnitude $\kappa \sim 1/r$" to lie in a wavenumber band $\Delta\kappa \sim \kappa$ about wavenumber magnitude κ. Then we have

$$[u(r)]^2 \sim \kappa E(\kappa), \qquad \kappa \sim 1/r. \tag{2.64}$$

To proceed further we need to know $E(\kappa)$. As we shall discuss in Chapter 7, Kolmogorov (1941) argued that for wavenumbers in the *inertial subrange*, $1/\ell \ll \kappa \ll 1/\eta$, E depends only on ϵ and κ and so on dimensional grounds has the form

$$E(\kappa) \sim \epsilon^{2/3}\kappa^{-5/3}. \tag{2.65}$$

Through the interpretation $r \sim 1/\kappa$, the inertial subrange corresponds to scales $\ell \gg r \gg \eta$. Using (2.65) in (2.64) gives

$$u(r) \sim \left(\frac{E(1/r)}{r}\right)^{1/2} \sim \left(\frac{\epsilon^{2/3} r^{5/3}}{r}\right)^{1/2} \sim (\epsilon r)^{1/3}. \tag{2.66}$$

Equation (2.66) holds for $\ell \geq r \geq \eta$; that is, it yields $u(\ell) = u$, $u(\eta) = \upsilon$ (Problem 2.11). This is a much wider range of applicability than we might have expected.

The *turnover time* of an eddy of size r is defined as $r/u(r)$; it is often taken as a rough estimate of the lifetime of an eddy of size r.

2.5.5 The restriction to homogeneous fields

You have now met some of the classical concepts in turbulence:

- the power spectral density, or spectrum;
- Kolmogorov's inertial subrange;
- the velocity scale $u(r)$ of an eddy of size r.

We quantified these concepts through a lightly mathematical model of homogeneous turbulence, but engineering and geophysical flows are seldom homogeneous. The atmospheric boundary layer, for example, is inhomogeneous in the vertical. So perhaps you wonder: do these classical concepts apply to real flows?

The answer has two parts. First, these concepts apply only in homogeneous *directions*. In the atmospheric boundary layer, for example, one applies spectral analysis only in the homogeneous horizontal plane (with data from scanning radar or lidar, or from numerical simulation) or along a homogeneous horizontal line (with aircraft data). Second, at wavenumbers such that $\kappa L \gg 1$, with L the scale of the inhomogeneity (i.e., at turbulence scales small compared to the scale of the inhomogeneity), it appears that spectra can be interpreted in the fully homogeneous context (Part III).

2.6 Turbulent vorticity

As with velocity, we can define a characteristic amplitude ω of vorticity fluctuations through $\omega^2 = \overline{\omega_i \omega_i}$. We showed in Chapter 1 that at large R_t this vorticity is contained in the smallest eddies, whose velocity and length scales are the Kolmogorov scales υ and η. Thus we can express the characteristic vorticity ω as

$$\omega \sim \frac{\upsilon}{\eta}. \tag{2.67}$$

The vorticity characteristic of eddies of size r, $\omega(r)$, is of order $u(r)/r$, which from Eq. (2.66) and $\epsilon \sim u^3/\ell$ can be expressed as

$$\omega(r) \sim \frac{u(r)}{r} \sim \frac{\epsilon^{1/3}}{r^{2/3}}. \tag{2.68}$$

Thus, as eddy size r decreases its characteristic vorticity $\omega(r)$ increases; at the smallest scales it is

$$\omega(\eta) \sim \frac{u(\eta)}{\eta} \sim \frac{\upsilon}{\eta} \sim \omega, \tag{2.69}$$

as stated in Eq. (2.67).

The contrast between the characteristic velocity and vorticity fluctuations u and ω at large R_t is striking: $u \sim u(\ell)$, since the velocity fluctuations are dominated by the larger eddies; but $\omega \sim \omega(\eta)$, since virtually all the vorticity lies in the smallest eddies.

The turbulent vorticity can be surprisingly large. In an atmospheric boundary layer with energy-containing scales $u = 1$ m s^{-1} and $\ell = 1000$ m, so that the Kolmogorov scales are $\upsilon = 10^{-2}$ m s^{-1} and $\eta = 10^{-3}$ m, ω is of order 10 s^{-1}. This is equal to the vorticity in a tornado funnel of core diameter 60 m and winds of 300 m s^{-1}!

Following Eq. (2.64), the vorticity $\omega(r)$ of eddies of size r is related to the three-dimensional vorticity spectrum $\psi(\kappa)$ by

$$[\omega(r)]^2 \sim \kappa\psi(\kappa). \tag{2.70}$$

Since $\omega(r) \sim u(r)/r$, it follows that $\psi(\kappa) \sim \kappa^2 E(\kappa)$, so that the vorticity spectrum in the inertial subrange behaves as

$$\psi(\kappa) \sim \kappa^2 E(\kappa) \sim \epsilon^{2/3}\kappa^{1/3}, \tag{2.71}$$

which increases with increasing wavenumber. The increase is ultimately damped by viscosity at $\kappa \sim 1/\eta$.

2.7 Turbulent pressure

Taking the divergence of the Navier–Stokes equation (1.26) gives a Poisson equation for the pressure field:

$$-\frac{1}{\rho}\nabla^2 p = \frac{\partial u_j}{\partial x_i}\frac{\partial u_i}{\partial x_j}. \tag{2.72}$$

We can express a velocity gradient as a sum of strain-rate and rotation-rate tensors s_{ij} and r_{ij}:

$$\frac{\partial u_i}{\partial x_j} = \frac{1}{2}\left(\frac{\partial u_i}{\partial x_j} + \frac{\partial u_j}{\partial x_i}\right) + \frac{1}{2}\left(\frac{\partial u_i}{\partial x_j} - \frac{\partial u_j}{\partial x_i}\right) = s_{ij} + r_{ij}. \tag{2.73}$$

s_{ij} and r_{ij} are symmetric and antisymmetric tensors, respectively. Bradshaw and Koh (1981) pointed out that with Eq. (2.73) one can rewrite the Poisson equation (2.72) in terms of s_{ij} and r_{ij}:

$$-\frac{1}{\rho}\nabla^2 p = (s_{ij} + r_{ij})(s_{ij} - r_{ij}) = s_{ij}s_{ij} - r_{ij}r_{ij}. \tag{2.74}$$

They offered a physical interpretation of Eq. (2.74):

The rate-of-strain contribution comes from near saddle points in the streamline pattern (streamlines approaching a given point from north and south, and leaving it from east to west, as seen by an observer moving with the fluid at the point). The contribution to the right-hand side is positive, so that $\nabla^2 p$ tends to become negative near the point considered, and p tends to be a maximum there. Crudely, the rate-of-strain contribution to fluctuating pressure results from eddy collisions. The vorticity contribution is negative, implying pressure minima, and results from eddy rotation. The two may be succinctly labeled the "splat" and "spin" contributions, respectively.

We'll see in Chapter 5 that turbulent pressure fluctuations play a critical role in the maintenance of turbulent fluxes, and we'll revisit the form (2.74) of the pressure equation in Chapter 7.

2.8 Eddy diffusivity

Molecular diffusion, a macroscopic effect of molecular collisions in a medium, produces a flux of a property (e.g., thermal energy, momentum, constituent concentration) in the direction opposite to that of its gradient. This *down-gradient* diffusion occurs in both fluids and solids. A familiar form is conduction heat transfer, the molecular diffusion of thermal energy down the temperature gradient.

Although *turbulent diffusion* differs in many ways from molecular diffusion, it is convenient and tempting to treat it like molecular diffusion but with a much larger, "eddy" diffusivity. We can explore this eddy-diffusivity representation through a simple thought problem.†

Figure 2.4 shows a tank filled to depth d, halfway with dyed water (with dye concentration $c = c_{\text{initial}}$) and the rest with clear water ($c = 0$). The horizontal dimensions of the tank are much greater than d, making the problem one-dimensional. The fluids are motionless and separated by a thin membrane so the boundary between them is initially perfectly flat and distinct. We imagine that at time zero the membrane is dissolved and the dye begins to diffuse into the clear fluid.

Here the dye conservation equation (1.31) reduces to

$$\frac{\partial c}{\partial t} = \gamma \frac{\partial^2 c}{\partial z^2}. \tag{2.75}$$

We can remove the dependence on c_{initial} by defining a new dependent variable $c^* = c/c_{\text{initial}}$ which still satisfies Eq. (2.75). c^* is initially 1 in the lower half of the tank and 0 above.

One expects that for large time $c^*(z, t) \to$ constant $= 0.5$. We can infer the time required to approach this final state as follows.‡ The only important physical parameters in the problem are the tank depth d and the molecular diffusivity γ, so

† This problem is motivated by a discussion in Tennekes and Lumley (1972).

‡ In this simple example of *dimensional analysis* the answer follows immediately on dimensional grounds. In problems with more physical parameters and dimensions a systematic approach is necessary. We'll discuss it in Chapter 10.

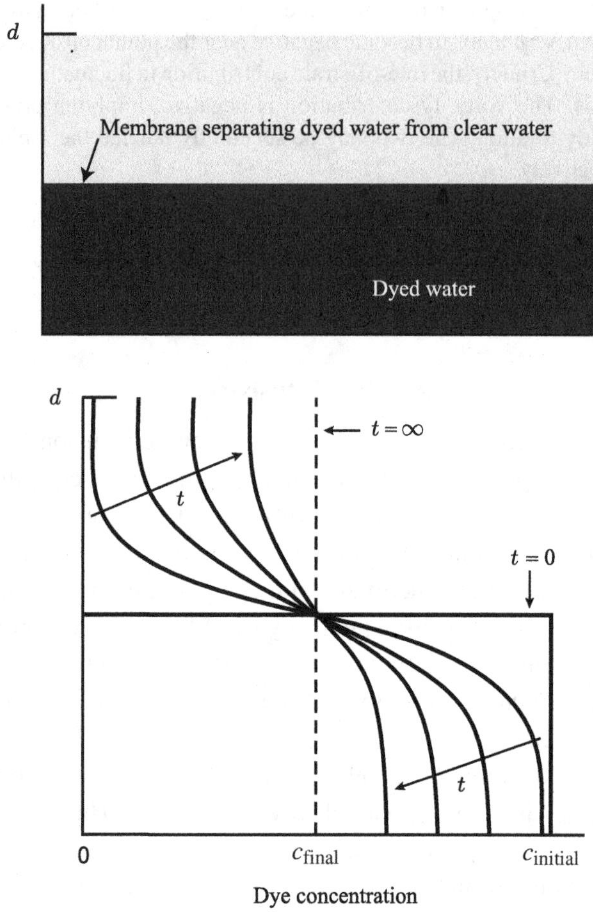

Figure 2.4 Molecular diffusion of dye in a tank of motionless fluid. The top panel shows the initial state; the bottom panel shows the evolving vertical profile of dye concentration during the diffusion process.

the response time τ_m must depend on them alone: $\tau_m = \tau_m(d, \gamma)$. It follows on dimensional grounds that the only possible form is $\tau_m \sim d^2/\gamma$. If $d = 1$ m and $\gamma = 10^{-5}$ m^2 s^{-1}, for example, then the time scale for molecular diffusion is 10^5 s, about one day. Molecular diffusion can be very slow.

Figure 2.5 shows a turbulent version of this problem, with the turbulence driven by bottom heating, say. The concentration field is now very complicated, every term in the conserved scalar equation (1.31) being active. How can we deduce from that equation what we know from experience – that turbulence can mix much faster than molecular diffusion?

As sketched in Figure 2.5, we expect that the ensemble-averaged concentration $C(z, t)$ – the average of $\tilde{c}(z, t)$ over many realizations of the same

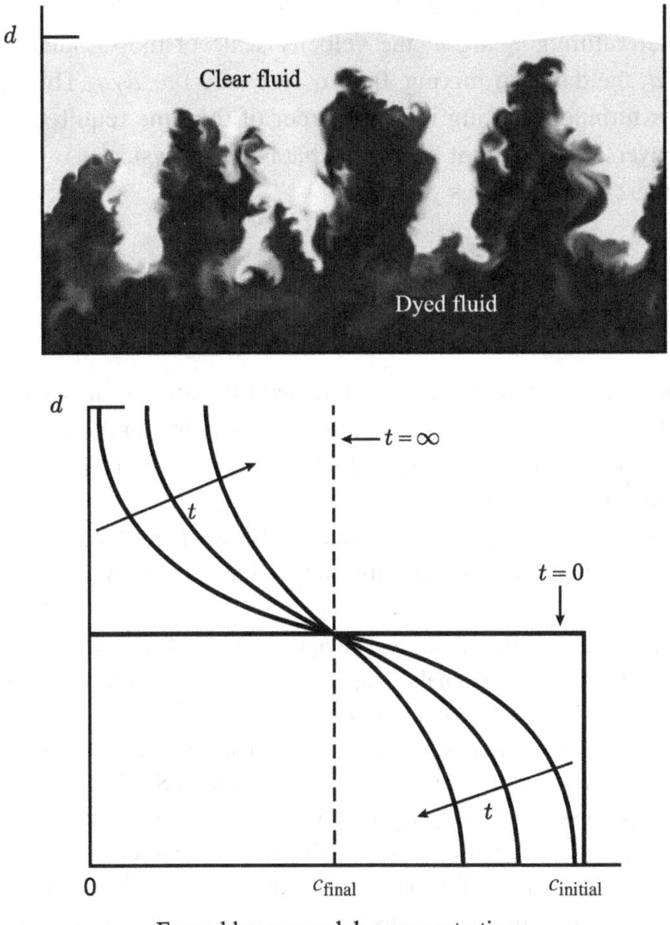

Figure 2.5 Diffusion of dye in a tank of turbulent fluid. The initial state is the motionless one in Figure 2.4. The top panel is an artist's sketch (motivated by Van Dyke's (1982) photographs) of an instantaneous cross section shortly after the onset of turbulence. The bottom panel shows the time evolution of the ensemble-averaged dye concentration profile.

experiment – evolves qualitatively like the concentration in the molecular diffusion problem. But it does so much more rapidly, so we'll use a much larger *eddy diffusivity* K in the diffusion equation (2.75),

$$\frac{\partial C}{\partial t} = K\frac{\partial^2 C}{\partial z^2}. \tag{2.76}$$

The time τ_t required for turbulent mixing is then $\tau_t \sim d^2/K$.

We can make another estimate of τ_t as follows. If the only important physical parameters determining τ_t are u, the velocity scale of the turbulence doing the mixing, and d, the depth of mixing, then $\tau_t = \tau_t(d, u) \sim d/u$. This says that the time scale for turbulent mixing is of the order of the time required for turbulent motions to traverse the depth of the tank. That is very physical.

Equating these two estimates yields

$$\tau_t \sim \frac{d^2}{K} \sim \frac{d}{u}, \quad \text{so that } K \sim ud. \tag{2.77}$$

Since we expect the dominant eddy size ℓ to be of the order of d, we can write this as $K \sim u\ell$. Thus we have inferred on plausible grounds that the turbulence in our "thought problem" has an eddy diffusivity $K \sim u\ell$, the product of the velocity and length scales of its dominant eddies.

Does this mean there is a formal analogy between molecular and turbulent diffusion? No, for there are several important differences between the two:

- In molecular diffusion the time and length scales of the diffusing motion, a microscopic process, are very small compared to those of the diffusion problem. Such scale separation tends not to exist in turbulent diffusion.
- The most general linear relation between the flux f_i of a scalar constituent and its gradient g_i is $f_i = -\gamma_{ij} g_j$, with γ_{ij} a second-rank diffusivity tensor. Since molecular diffusion is a microscopic process one expects γ_{ij} to be isotropic – having no preferred direction – so that (Part III) $\gamma_{ij} = \gamma \delta_{ij}$ with γ a scalar. Thus in molecular diffusion $f_i = -\gamma_{ij} g_j = -\gamma \delta_{ij} g_j = -\gamma g_i$. Turbulent diffusion is done by the dominant eddies, which can span the flow and are inherently anisotropic; thus there is no reason to expect the eddy diffusivity tensor K_{ij} to be isotropic (Chapter 5).
- The eddy diffusivity is a property of the turbulent flow, but the molecular diffusivity is a property of the fluid and the constituent being diffused.
- A diffusivity relates a mean flux and a mean gradient; it has no meaning without an averaging process. The averaging involved in the molecular diffusivity has been done for us in going from molecular-scale interactions to continuum fluid mechanics. Thus, instantaneous turbulence fields have a molecular diffusivity but not an eddy diffusivity.
- As we shall discuss in Chapter 4 and also in Part II, there are situations in which the eddy diffusivity is poorly behaved (e.g., has a singularity and becomes negative).

In summary, there is a tempting, but shallow, analogy between molecular and turbulent diffusion. Turbulent fluxes and mean gradients can show an eddy-diffusivity-like relationship (Chapter 4), but the eddy diffusivity is typically a spatially variable flow property that can be a tensor, rather than a scalar. In a given flow it can also depend on the diffusion geometry (Chapter 11).

2.9 Reynolds-number similarity

The statistics of the energy-containing eddies in a turbulent flow of a given type, when nondimensionalized with the scales of the energy-containing range, u and ℓ, are often observed not to depend significantly on Reynolds number when it is larger than a threshold value. A physical interpretation is that since the turbulent Reynolds number R_t is large in turbulent flow, the energy-containing eddies do not experience significant viscous forces and their statistics eventually become independent of Reynolds number. This is called *Reynolds-number similarity*.

Reynolds-number similarity, while only approximate, is useful because it allows us to make geometrically identical but smaller-scale laboratory models of geophysical flows with some confidence that their energy-containing structure will be representative of the much-larger-Reynolds-number geophysical flow. Thus, it provides the basis for the physical modeling we discussed in Chapter 1. Laboratory flows can provide less expensive measurements and have much smaller required averaging times (Problem 2.6) than atmospheric flows. Much of our knowledge of turbulent diffusion from sources in the convective atmospheric boundary layer was actually gained through measurements in laboratory convection tanks at vastly smaller Reynolds number (Willis and Deardorff, 1974).

Reynolds-number similarity also provides some justification for carrying out direct numerical simulation (DNS) of geophysical turbulent flows. Coleman *et al.* (1990) used DNS to simulate an atmospheric boundary layer at Reynolds numbers far lower than in nature. Their simulation corresponded to a layer with wind speed on the order of 1 cm s^{-1}.

2.10 Coherent structures

Among the largest eddies in a turbulent flow are apt to be "coherent structures." These are quasi-steady, significant-amplitude circulations that exist in essentially the same form, intensity, and location in each realization of the flow. With their Reynolds stresses they extract kinetic energy from the mean flow, and they lose it through the Reynolds stresses exerted on them by smaller eddies. Their form and intensity depend on the structure of the parent flow.

According to Holmes *et al.* (1996), Liu (1988) identified the first appearance of the idea of coherent structures in turbulent flows as being in the late 1930s. Townsend's (1956) book on turbulent shear flow presents analyses of the velocity and length scales of the coherent structures in various shear flows. The Holmes *et al.* monograph has extensive discussions of analytical methods for isolating them and predicting their form.

In boundary-layer meteorology coherent structures are apt to be called *secondary flows*. Examples include "convective rolls," large, counter-rotating, horizontal vortices that can be visible through the "cloud streets" they induce at their top (LeMone, 1973). Because of their relatively large size and intensity they can be important in applications such as atmospheric dispersion. Morrison *et al.* (2005) point to their prevalence in the hurricane boundary layer. We'll discuss these atmospheric boundary layer examples in Chapter 11.

Questions on key concepts

2.1 Since turbulence varies continuously in time and in the three spatial directions, how can it ever be steady and homogeneous?

2.2 Describe the physical nature of the top of the atmospheric boundary layer.

2.3 Under what conditions does a time average converge to an ensemble average? Explain in physical terms why this convergence occurs.

2.4 If $\overline{u}^T$ is the time average (over time interval T) of a stationary random function $\tilde{u}(t)$ whose ensemble average is U, then

$$\overline{(\overline{u}^T - U)^2} = \frac{2\overline{u^2}\tau}{T}, \tag{1}$$

where $\overline{u^2} = \overline{(\tilde{u} - U)^2}$ is the variance of $\tilde{u}(t)$. Explain physically the dependence on $\overline{u^2}$, τ, and T.

2.5 Explain on physical grounds why we can roughly estimate τ in Eq. (1) as ℓ/U.

2.6 Explain why atmospheric turbulent flows tend to have much larger required averaging times than engineering flows. What does this say about the relative scatter in measurements in the two types of flows?

2.7 Why is averaging time T apt to be limited in atmospheric flows but not in engineering flows?

2.8 Explain why we use the ensemble average in formal developments, even though the time average can converge to it.

2.9 Explain some of the key differences between turbulent and molecular diffusion. Explain their interplay.

2.10 Explain the concept of $u(r)$. How might one measure it?

2.11 Explain the concept of Reynolds-number similarity. Why is it important?

2.12 The eddy-diffusivity model has been criticized on the grounds that visual observation of turbulence gives no support to the notion that it diffuses properties down their local mean gradients. Is this a legitimate point of view? Discuss.

2.13 Explain why we say that derivatives of velocity and conserved scalars in turbulent flow are small-scale quantities, whereas velocity and conserved scalars are large-scale quantities.

2.14 Explain how Reynolds-number similarity allows one to model the atmospheric boundary layer in the laboratory.

Problems

2.1 Prove the ensemble-averaging rules (2.5)–(2.8).

2.2 The updrafts in a convective boundary layer occupy less than one-half of the area in a horizontal plane; the larger fraction of the area is occupied by downdrafts. Updrafts and downdrafts are most often detected through analysis of the time series measured from aircraft or from sensors on towers. Assume these records are linear spatial traverses. Explain the relationship between the fractions of a linear record and the horizontal cross-sectional area occupied by updrafts. (Hint: use ergodicity.)

2.3 Near the earth's surface in the daytime the temperature and horizontal velocity fluctuations are negatively correlated. Warmer air associated with plumes tends to have a lower than average horizontal velocity, and cooler air tends to be moving faster than average. It might seem that sampling a temperature signal at a constant rate in time will therefore give relatively more samples in updraft air, and less samples in downdraft air, compared to samples taken at uniform spatial intervals. Will the temporal record therefore give biased statistics? (Hint: use ergodicity.)

2.4 Derive Eq. (2.32).

2.5 Show that determining the variance of a stationary random signal through time averaging typically requires averaging over more integral scales than does determining the mean.

2.6 Explain why ensemble averaging commutes with differentiation and integration (Eq. (2.4)) but not with the substantial derivative.

2.7 It has been proposed that the eddy diffusivity in a thin, broad (stratiform) cloud be measured by creating a circular hole in the cloud with a cloud-seeding agent, observing the time required for turbulent diffusion to fill in the hole, and then using the analytical solution for this problem to infer the eddy diffusivity. Comment on this technique.

2.8 Compare required averaging times for a plume-dispersion problem in the atmospheric boundary layer and in a laboratory model of it. (Hint: assume the flows are geometrically similar, with their velocity and length scales in the ratio f_u and f_ℓ, respectively.)

2.9 Use the Kolmogorov scaling of Subsection 2.5.4 to show that viscous stresses are largest in the dissipative eddies.

2.10 A thin, sinuous plume with instantaneous concentration c occasionally "hits" a concentration sensor. If the average duration of the plume in the sensor is d and the average duration between hits is D, estimate the mean concentration and the concentration variance. What are the implications for measuring mean concentration at the edge of a plume?

2.11 Show that Eq. (2.66) holds even for $r = \ell$ and for $r = \eta$.

2.12 Using the inequalities

$$\overline{\left(\frac{u}{\sigma_u} - \frac{v}{\sigma_v}\right)^2} \geq 0, \qquad \overline{\left(\frac{u}{\sigma_u} + \frac{v}{\sigma_v}\right)^2} \geq 0,$$

prove that the correlation coefficient is bounded by 1.0 in magnitude.

2.13 Discuss the relative rates of convergence of line, area, and volume averages to the ensemble average under homogeneous conditions.

2.14 A direct numerical simulation of a turbulent boundary layer is done at a Reynolds number $R_h = U_\infty h/\nu = 10^4$. It is claimed to be representative of the atmospheric boundary layer. Calculate R_h for an atmospheric boundary layer with $h = 10^3$ m and $U_\infty = 10$ m s^{-1}. Given the difference in R_h values, how could the claim be justifiable? What is the kinematic viscosity needed if the flow had the atmospheric h and U_∞ values but $R_h = 10^4$? To what fluid would that roughly correspond? Does it seem plausible that a boundary layer in that fluid could be a good model of the atmospheric boundary layer?

2.15 A layer of fluid of depth h in the vertical is subject to the sudden application of a horizontal velocity U at its lower surface. Write the equation of motion for the fluid, assuming the problem has no horizontal variations. Estimate the time required for the vertical profile of fluid velocity to reach a constant value. Under what conditions would the fluid become turbulent? Estimate the time in that case.

2.16 Before Kolmogorov published his 1941 hypotheses about the dissipative scales in turbulence, Taylor (1935) had shown that $\epsilon \sim \nu u^2/\lambda^2$, with λ a length scale defined through the behavior of the autocorrelation function at the origin (Part III). (In his honor λ is now called the Taylor microscale.) He interpreted λ as the size of the dissipative eddies. Using $\epsilon \sim \nu u^2/\lambda^2$, determine the Reynolds number $u(\lambda)\lambda/\nu$ of λ-sized eddies, $u(\lambda)$ being the velocity characteristic of λ-sized eddies. Would you say they are directly influenced by viscosity? Develop an expression for λ/η. Was Taylor correct in interpreting λ as the size of the dissipative eddies?

2.17 Derive Eq. (2.72) from the Navier–Stokes equation. Then show that it can be rewritten as Eq. (2.74).

2.18 Sketch a sequence of stochastic functions of time having decreasing integral scale τ. Use it to explain why Eq. (2.36) says that the required averaging time approaches zero as τ does.

References

Bradshaw, P., and Y. M. Koh, 1981: A note on Poisson's equation for pressure in a turbulent flow. *Phys. Fluids*, **24**, 777.

Coleman, G. N., J. H. Ferziger, and P. R. Spalart, 1990: A numerical study of the turbulent Ekman layer. *J. Fluid Mech.*, **213**, 313–348.

Holmes, P., J. L. Lumley, and G. Berkooz, 1996: *Turbulence, Coherent Structures, Dynamical Systems and Symmetry.* Cambridge University Press.

Kolmogorov, A. N., 1941: The local structure of turbulence in incompressible viscous fluid for very large Reynolds numbers. *Doklady ANSSSR*, **30**, 301–305.

LeMone, M. A., 1973: The structure and dynamics of horizontal roll vortices in the planetary boundary layer. *J. Atmos. Sci.*, **30**, 1077–1091.

Lenschow, D. H., V. Patel, and A. Isbell, 1988: Measurements of fine-scale structure at the top of marine stratocumulus. Preprint Volume, *Eighth Symposium on Turbulence and Diffusion of the American Meteorological Society*, pp. 29–32.

Liu, J. T. C., 1988: Contributions to the understanding of large-scale coherent structures in developing free turbulent shear flows. *Adv. Appl. Mech.*, **26**, 183–309.

Lumley, J. L., and H. A. Panofsky, 1964: *The Structure of Atmospheric Turbulence.* New York: Interscience.

Monin, A. S., and A. M. Yaglom, 1971: *Statistical Fluid Mechanics*, Part 1. Cambridge, MA: MIT Press.

Morrison, I., S. Businger, F. Marks, P. Dodge, and J. A. Businger, 2005: An observational case for the prevalence of roll vortices in the hurricane boundary layer. *J. Atmos. Sci.*, **62**, 2662–2673.

Reynolds, O., 1895: On the dynamical theory of incompressible viscous fluids and the determination of the criterion. *Philos. Trans. R. Soc. London, Ser. A*, **186**, 123–164.

Tavoularis, S., and S. Corrsin, 1981: Experiments in nearly homogeneous turbulent shear flow with a uniform mean temperature gradient. *J. Fluid Mech.*, **104**, 311–347.

Taylor, G. I., 1935: Statistical theory of turbulence. Parts I–IV. *Proc. R. Soc.* **A151**, 421–478.

Tennekes, H., and J. L. Lumley, 1972: *A First Course in Turbulence.* Cambridge, MA: MIT Press.

Townsend, A. A., 1956: *The Structure of Turbulent Shear Flow.* Cambridge University Press.

Van Dyke, M., 1982: *An Album of Fluid Motion.* Stanford: Parabolic Press.

Willis, G. E., and J. W. Deardorff, 1974: A laboratory model of the unstable planetary boundary layer. *J. Atmos. Sci.*, **31**, 1297–1307.

3

Equations for averaged variables

3.1 Introduction

We saw in Chapter 1 that turbulent flows dissipate their kinetic energy into internal energy at an average rate (per unit mass) $\epsilon \sim u^3/\ell$, where u and ℓ are the velocity and length scales of the energy-containing eddies. Since the turbulent kinetic energy per unit mass is of order u^2, this implies that if its production mechanism were shut off turbulence would decay in a time of order $u^2/\epsilon \sim \ell/u$, about one large-eddy turnover time. That is surprisingly fast. You wouldn't easily ride a bicycle or drive a car having that much friction.

The decay of the energy-containing eddies in this scenario is caused not by their viscous friction (their large Reynolds number $R_\mathrm{t} = u\ell/\nu$ makes that negligible), but by the *energy cascade*. As we'll discuss in Chapter 6, this involves the full range of eddies, energy-containing to dissipative. The cascade begins in the large eddies, which drain kinetic energy from the mean flow and transfer it to smaller ones through eddy–eddy interactions; it terminates in the smallest eddies, which convert kinetic energy into internal energy through viscous friction.

Any direct numerical solution of the turbulent fluid equations must therefore resolve the entire eddy scale range. Because the length scales at the extremes of this range are in the ratio $\ell/\eta \sim R_\mathrm{t}^{3/4}$, this requires of order $(\ell/\eta)^3 \sim R_\mathrm{t}^{9/4}$ computational grid points. Since R_t values are on the order of 10^8 in the atmospheric boundary layer and perhaps 10^{10} in supercell thunderstorms, they would require on the order of 10^{18} and 10^{22} grid points, respectively. Today's computers allow only 10^{10}–10^{11} grid points. The R_t values in large-vehicle aerodynamics, while smaller than these geophysical values, are still too large to allow direct numerical simulation of the drag-producing turbulent flow around them.

Ensemble or space averaging can drastically reduce these computational requirements. With some exceptions, such as very near surfaces, the ensemble-mean field is resolvable on a relatively coarse numerical grid. We'll see that space averaging can be designed to remove eddies smaller than a *cutoff scale* Δ, say, chosen such that $\ell \gg \Delta \gg \eta$.

The reduction in computational requirements that results from averaging comes at a price. As we saw in Chapter 2, the averaged Navier–Stokes equation has a *Reynolds stress* that must be approximated, or *modeled*, before the equation can be solved. The resulting flow models fall into two broad classes. *Reynolds-averaged Navier–Stokes* (RANS), also called *second-order closure*, discussed in Chapter 5, is based on ensemble averaging; *large-eddy simulation* (LES), discussed in Chapter 6, uses space averaging.

3.2 Ensemble-averaged equations

We begin with the continuity equation, which for a constant-density fluid reduces to a zero-divergence requirement for the full velocity field $\tilde{u}_i$:

$$\frac{\partial \tilde{u}_i}{\partial x_i} = \frac{\partial (U_i + u_i)}{\partial x_i} = 0. \tag{3.1}$$

Ensemble averaging this equation, using the rules in Chapter 2, yields

$$\overline{\frac{\partial \tilde{u}_i}{\partial x_i}} = \frac{\partial \overline{\tilde{u}_i}}{\partial x_i} = \frac{\partial U_i}{\partial x_i} = 0, \tag{3.2}$$

so the mean field has zero divergence. Subtracting this from Eq. (3.1) then yields

$$\frac{\partial u_i}{\partial x_i} = 0, \tag{3.3}$$

so the fluctuating field also has zero divergence.

We consider next the Navier–Stokes equation (1.26), which by incompressibility we can write as

$$\frac{\partial \tilde{u}_i}{\partial t} + \frac{\partial \tilde{u}_i \tilde{u}_j}{\partial x_j} = -\frac{1}{\rho}\frac{\partial \tilde{p}}{\partial x_i} + \nu \frac{\partial^2 \tilde{u}_i}{\partial x_j \partial x_j}. \tag{3.4}$$

Ensemble averaging yields

$$\frac{\partial \overline{\tilde{u}_i}}{\partial t} + \frac{\partial \overline{\tilde{u}_i \tilde{u}_j}}{\partial x_j} = -\frac{1}{\rho}\frac{\partial \overline{\tilde{p}}}{\partial x_i} + \nu \frac{\partial^2 \overline{\tilde{u}_i}}{\partial x_j \partial x_j}. \tag{3.5}$$

Using $\overline{\tilde{u}_i} = U_i$, $\overline{\tilde{p}} = P$ and applying the ensemble-averaging rules to $\overline{\tilde{u}_i \tilde{u}_j}$ yields

$$\frac{\partial U_i}{\partial t} + \frac{\partial}{\partial x_j}\left(U_i U_j + \overline{u_i u_j}\right) = -\frac{1}{\rho}\frac{\partial P}{\partial x_i} + \nu \frac{\partial^2 U_i}{\partial x_j \partial x_j}. \tag{3.6}$$

The symmetric tensor $\overline{u_i u_j}$, which is shorthand for $\overline{u_i(\mathbf{x}, t) u_j(\mathbf{x}, t)}$, involves only the fluctuating part of the velocity field. As discussed in Chapter 1, it can be interpreted in terms of stress or momentum flux. $-\rho\overline{u_i u_j}$ is the mean force (in the i-direction) per unit area (whose normal is in the j-direction) due to the fluctuating turbulent motion. $\rho\overline{u_i u_j}$ can be interpreted as the mean i-direction turbulent flux of j-direction turbulent momentum.

G. I. Taylor called $-\rho\overline{u_i u_j}$ a *virtual mean stress* because it exists only as an ensemble mean. Today it is more commonly called the *Reynolds stress* after Osborne Reynolds, who first identified it in 1895 in the space-averaged equations.

We can also write the ensemble-averaged Navier–Stokes equation (3.6) in *flux form*:

$$\frac{\partial U_i}{\partial t} + \frac{\partial}{\partial x_j}\left[U_i U_j + \overline{u_i u_j} - \nu\left(\frac{\partial U_i}{\partial x_j} + \frac{\partial U_j}{\partial x_i}\right)\right] = -\frac{1}{\rho}\frac{\partial P}{\partial x_i}. \tag{3.7}$$

$U_i U_j + \overline{u_i u_j} - \nu(\partial U_i/\partial x_j + \partial U_j/\partial x_i)$ is the mean kinematic momentum flux tensor. In the simplest situation, where the length and velocity scales of both the mean flow and the turbulence are ℓ and u, the magnitudes of its turbulence and viscous contributions are in the ratio

$$\frac{\text{turbulence contribution}}{\text{viscous contribution}} \sim \frac{u^2}{\nu u/\ell} \sim \frac{u\ell}{\nu} = R_\mathrm{t} \gg 1. \tag{3.8}$$

This says that away from solid boundaries (where velocity fluctuations vanish) the viscous stress in the ensemble-averaged Navier–Stokes equation can be neglected in comparison to the Reynolds stress.

In turbulent flow the Reynolds-stress term in the averaged Navier–Stokes equation is generally as important as any other term. As a result, the fidelity of any numerical solution of (3.6) can depend strongly on the fidelity of the model used for the Reynolds stress.

3.2.1 Example: Steady turbulent flow in a channel

Figure 3.1 shows turbulent flow in a channel, the first flow studied through LES (Deardorff, 1970). The channel length in the streamwise (x_1 or x)† direction is much greater than its depth $2D$, so the flow is homogeneous in x and $\partial U/\partial x = 0$. The channel width in the y-direction is also much greater than $2D$, so that away from the lateral walls the flow is homogeneous in y as well. That implies $\partial V/\partial y = 0$, and from the symmetry of the mean flow about the $y = 0$ plane that implies $V = 0$. The mean continuity equation (3.2) then reduces to $\partial W/\partial z = 0$. From the symmetry of the mean flow about the $z = 0$ plane we conclude that $W = 0$

† As is common in the literature, we'll tend to write velocity and position components as u, v, w and x, y, z.

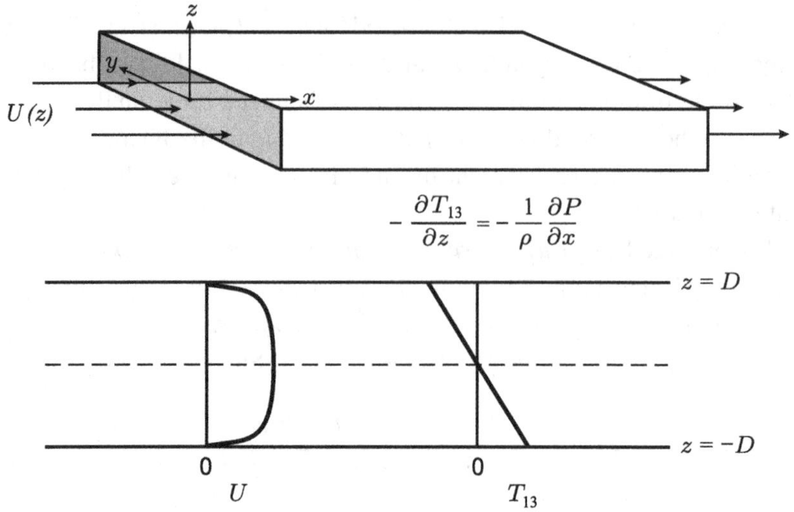

Figure 3.1 Steady turbulent channel flow and its profiles of mean velocity U and mean kinematic shear stress T_{13}.

as well. We can then write $U_i = [U(z), 0, 0]$, and in steady conditions the ensemble-averaged momentum equation (3.6) reduces to

$$\frac{\partial}{\partial x_j}\left(\overline{u_i u_j}\right) = -\frac{1}{\rho}\frac{\partial P}{\partial x_i} + \nu\frac{\partial^2 U_i}{\partial x_j \partial x_j}. \tag{3.9}$$

This expresses a balance among the turbulent stress divergence, the mean pressure gradient, and the viscous stress divergence. The homogeneity in y makes that component equation vanish, so we have

$$i = 1: \quad \frac{\partial}{\partial z}\left(\overline{uw} - \nu\frac{\partial U}{\partial z}\right) = -\frac{1}{\rho}\frac{\partial P}{\partial x}. \tag{3.10}$$

$$i = 3: \quad \frac{\partial \overline{w^2}}{\partial z} = -\frac{1}{\rho}\frac{\partial P}{\partial z}. \tag{3.11}$$

By differentiating Eq. (3.11) with respect to x, using homogeneity, and changing the order of differentiation we conclude that

$$\frac{\partial}{\partial x}\frac{\partial \overline{w^2}}{\partial z} = 0 = -\frac{1}{\rho}\frac{\partial}{\partial x}\frac{\partial P}{\partial z} = -\frac{1}{\rho}\frac{\partial}{\partial z}\frac{\partial P}{\partial x}, \tag{3.12}$$

which implies that $\partial P/\partial x$ does not depend on z. Equation (3.10) then implies that the mean kinematic shear stress $T_{13} = -\overline{uw} + \nu \partial U/\partial z$ varies linearly with z, as sketched in Figure 3.1. As we showed through Eq. (3.8), the viscous part of the shear stress is negligible except very near the walls.

3.2.2 Conserved scalars

Substituting the mean-turbulent decomposition into Eq. (1.29) for a scalar constituent $\tilde{c}$ in a turbulent flow and ensemble averaging yields

$$\frac{\partial C}{\partial t} + \frac{\partial}{\partial x_i}\left(U_i C + \overline{u_i c} - \gamma \frac{\partial C}{\partial x_i}\right) = 0. \tag{3.13}$$

$U_i C + \overline{u_i c} - \gamma \partial C/\partial x_i$ is the total mean flux of c-stuff, a vector. It has a contribution due to turbulence, $\overline{u_i c}$, the *turbulent flux of* c. In a flow with velocity and length scales u and ℓ, the turbulent and molecular fluxes are in the ratio $u\ell/\gamma$. If this ratio is large, then we can again neglect the molecular flux except very near solid boundaries.

In Chapter 2 we discussed the turbulent diffusion of a conserved dye that was initially concentrated in the lower half of a tank of water heated from below. This problem is horizontally homogeneous with $U = V = W = 0$, so from (3.13) the ensemble-mean concentration of dye follows

$$\frac{\partial C}{\partial t} + \frac{\partial}{\partial z}\left(\overline{wc} - \gamma \frac{\partial C}{\partial z}\right) = 0. \tag{3.14}$$

We assume the dye does not penetrate the tank surface, so the molecular flux vanishes there. Thus, if $u\ell/\gamma$ is large the molecular flux is everywhere negligible and Eq. (3.14) simplifies to

$$\frac{\partial C}{\partial t} + \frac{\partial \overline{wc}}{\partial z} = 0. \tag{3.15}$$

This says that the mean concentration evolves solely due to the divergence of the turbulent flux of c. Figure 3.2 sketches the evolution of this turbulent mass flux profile. We commonly call this process *turbulent diffusion.*

3.3 Interpreting the ensemble-averaged equations

The physical processes underlying the conservation equations of Chapter 1 are straightfoward. But the ensemble-averaged equations describe what Taylor might have called "virtual physics," which can be more difficult to understand.

3.3.1 Example: A conserved scalar diffusing from a point source

A conserved scalar released from a point source in a turbulent flow creates a highly irregular, filament-like instantaneous plume downstream. Figure 3.3 shows an artist's sketch of such an instantaneous plume.

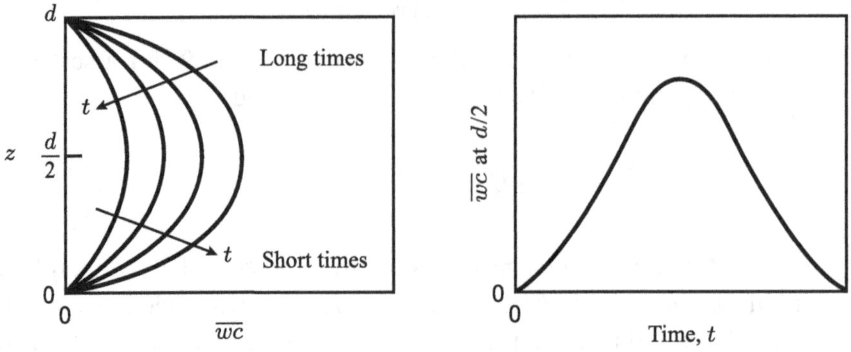

Figure 3.2 The evolution of the profile of mass flux $\overline{wc}$ in the turbulent diffusion problem of Figure 2.5.

During its travel the diameter d of the filament increases due to molecular diffusion of the scalar $\tilde{c}$ into the fluid, causing the concentration averaged across its cross section, $\tilde{c}^{\text{ave}}$, to decrease with time. From the conserved scalar equation (1.31) we estimate the magnitude of this rate of change of $\tilde{c}^{\text{ave}}$ as

$$\frac{D}{Dt}\tilde{c}^{\text{ave}} \sim \gamma \nabla^2 \tilde{c}^{\text{ave}} \sim \gamma \frac{\tilde{c}^{\text{ave}}}{d^2} \sim \frac{\tilde{c}^{\text{ave}}}{\tau_{\text{molec}}}, \tag{3.16}$$

with $\tau_{\text{molec}} \sim d^2/\gamma$ the time scale of this molecular diffusion process. The proportionality of τ_{molec} to d^2 indicates that only concentration anomalies of small spatial scale are removed quickly by molecular diffusion. If $d = 10^{-3}$ m, for example, and $\gamma = 10^{-5}$ m^2 s^{-1} (roughly the value for temperature and water vapor in air), $\tau_{\text{molec}} \sim 10^{-1}$ s. But if $d = 1$ m then $\tau_{\text{molec}} \sim 10^5$ s, about one day.

Under very stably stratified conditions (Part II) flow in the lower atmosphere can be laminar not far above the surface. In such cases an effluent plume from a stack of 1 m diameter in a wind of 10 m s^{-1} could extend hundreds of kilometers downwind in nonturbulent air!

In turbulent flow the ensemble-averaged concentration field $C(x, y, z, t)$ downstream of the source is formally defined by an average over many instantaneous plumes:

$$C(x, y, z, t) = \lim_{N\to\infty} \frac{1}{N} \sum_{\alpha=1}^{N} \tilde{c}(x, y, z, t; \alpha). \tag{3.17}$$

We'll refer to "ensemble-averaged" simply as "mean." Figure 3.3 shows a sketch of the downstream development of the mean plume in the x–z plane. Its contrast with the instantaneous plume is striking.

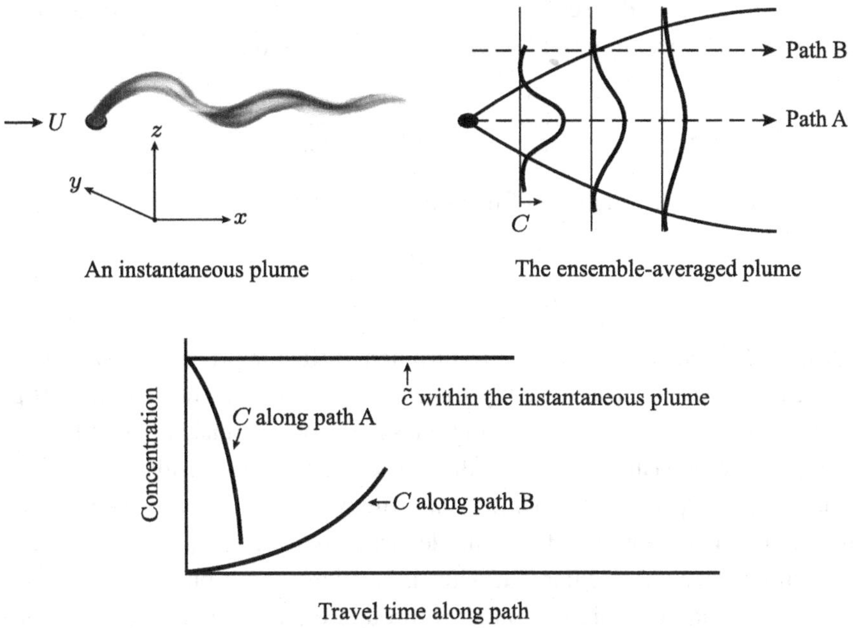

Figure 3.3 Instantaneous (top left) and ensemble-averaged (top right) plumes downstream of a continuous point source in turbulent flow. Bottom: concentrations observed along various paths in those plumes. On a path within the instantaneous plume the concentration $\tilde{c}$ changes only through molecular diffusion, which can be slow (Section 3.3.1). On path A along the centerline of the ensemble-averaged plume, the downstream broadening of that plume causes C to decrease; on path B, away from the centerline, this broadening causes C to initially increase as the path becomes closer (in plume widths) to the centerline.

In the steady case Eq. (3.13) for the mean concentration reduces to

$$U\frac{\partial C}{\partial x} + \frac{\partial \overline{cu_i}}{\partial x_i} = 0, \tag{3.18}$$

the effects of molecular diffusion on C being negligible. We can write this as

$$\frac{D^{\mathrm{m}}C}{Dt^{\mathrm{m}}} = U\frac{\partial C}{\partial x} = -\frac{\partial \overline{u_1 c}}{\partial x_1} - \frac{\partial \overline{u_2 c}}{\partial x_2} - \frac{\partial \overline{u_3 c}}{\partial x_3}, \tag{3.19}$$

where $D^{\mathrm{m}}/Dt^{\mathrm{m}}$ is the time derivative following the mean motion. Equation (3.19) says that on this "virtual mean trajectory" C changes solely due to the divergence of the turbulent flux of c-stuff.

If the mean plume is thin compared with distance from the source, the turbulent flux divergence in Eq. (3.19) is dominated by its cross-plume contributions. Taking the width of the mean plume (Figure 3.3) as L_{p}, the turbulent velocity scale as

u, and the turbulent scalar scale as c, an estimate for the magnitude of the rate of change of mean concentration following the mean motion is

$$\frac{D^{\mathrm{m}}}{Dt^{\mathrm{m}}}C = -\frac{\partial \overline{cu_i}}{\partial x_i} \sim \frac{cu}{L_{\mathrm{p}}}. \tag{3.20}$$

Here we expect $c \sim C$, in order of magnitude, so that

$$\frac{D^{\mathrm{m}}}{Dt^{\mathrm{m}}}C \sim \frac{Cu}{L_{\mathrm{p}}} \sim \frac{C}{\tau_{\mathrm{turb}}}, \tag{3.21}$$

with $\tau_{\mathrm{turb}} \sim L_{\mathrm{p}}/u$ the time scale of this turbulent diffusion process. Equation (3.21) says we can interpret τ_{turb} as the eddy-traverse time across the mean plume. If L_{p} is 1 m and u is 1 m s^{-1}, τ_{turb} is 1 s, strikingly less than our estimate of 10^5 s for τ_{molec}.

Put another way, if (in analogy with the molecular diffusion result $\tau_{\mathrm{molec}} \sim d^2/\gamma$) we write $\tau_{\mathrm{turb}} \sim L_{\mathrm{p}}^2/K$, with K a *turbulent* diffusivity, we find that $K \sim uL_{\mathrm{p}}$, which in our example is five orders of magnitude larger than γ.

Let's summarize. Molecular diffusion is a microscopic physical process that acts to remove macroscopic $\tilde{c}$-anomalies through the collective effect of molecular collisions. It seems well described by *intermingling*, a dictionary definition of diffusion. Turbulent diffusion, viewed most broadly, involves three processes. Two are physical – turbulent mixing by the chaotic, random deformation and advection of $\tilde{c}$-anomalies by turbulence; and molecular diffusion, which acts most strongly on the smallest of these $\tilde{c}$-anomalies. The all-important third, and new, process is ensemble averaging, which produces a virtual concentration field C that is much smoother, broader, and has smaller maximum concentrations than the turbulent $\tilde{c}$ field in any realization. As a result the concentrations observed along paths in the instantaneous and ensemble-average plumes can differ strikingly, as shown in the lower panel of Figure 3.3.

3.3.2 Enhancement of molecular diffusion by turbulence

Figure 3.4 depicts the distortion of a blob of conserved scalar in a turbulent flow in the process we call *turbulent mixing*. If we think of the distorted blob as a sheet having a surface area and a thickness, then as the distortion increases the surface area it decreases the thickness. The increased area and the larger scalar gradients on it greatly increase the total rate of molecular diffusion of the scalar out of the blob.

This suggests that we examine the equation for the evolution of the gradient $\tilde{g}_i$ of a conserved scalar (Problem 1.11):

$$\frac{D\tilde{g}_i}{Dt} = \frac{\partial \tilde{g}_i}{\partial t} + \tilde{u}_j\frac{\partial \tilde{g}_i}{\partial x_j} = -\frac{\partial \tilde{u}_j}{\partial x_i}\tilde{g}_j + \gamma\frac{\partial^2 \tilde{g}_i}{\partial x_j \partial x_j}. \tag{3.22}$$

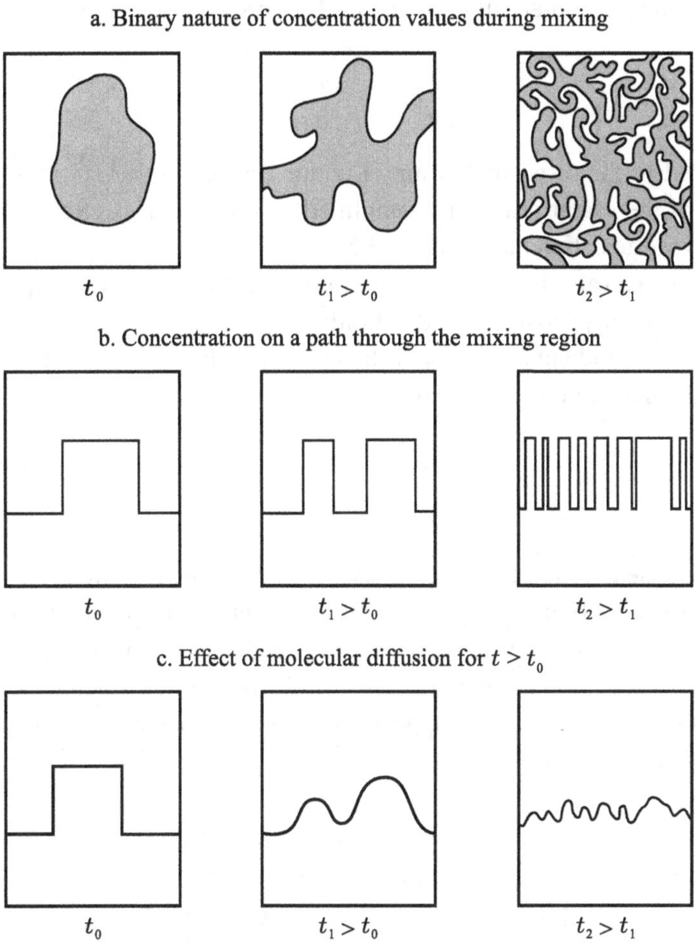

Figure 3.4 Schematic sketch of turbulent mixing of a contaminant, without molecular diffusion (a and b) and with it (c). A key process is the deformation of the blob by the turbulent strain field, which increases its surface area and the concentration gradients on it, thereby greatly enhancing the molecular diffusion. Adapted from Corrsin (1961).

The first term on the far right is the rate of change of scalar gradient due to deformation by the velocity field. This mechanism can either amplify or attenuate gradients; the gradient of a conserved scalar is not conserved. To illustrate, let the scalar isosurfaces be planes oriented such that their normal and the scalar gradient point in the x_1-direction, say:

$$\tilde{g}_j = (\tilde{g}_1,\ 0,\ 0)\,. \tag{3.23}$$

Then neglecting molecular diffusion Eq. (3.22) says the gradient in the x_1-direction evolves as

$$\frac{D}{Dt}\tilde{g}_1 \simeq -\frac{\partial \tilde{u}_1}{\partial x_1}\tilde{g}_1. \tag{3.24}$$

This is called linear or normal strain (Kundu, 1990). If $\partial \tilde{u}_1/\partial x_1$ is negative the magnitude of the scalar gradient is amplified; if positive, it is attenuated. This is analogous to vortex stretching (Figure 3.5).

Deformation can also reorient a scalar gradient. In the example of Eq. (3.23) with a scalar field initially having a gradient only in the x_1-direction, if the $\tilde{u}_1$ velocity component has a gradient in the $\alpha \neq 1$ direction, say, then components of the scalar gradient are induced in that direction:

$$\frac{D\tilde{g}_\alpha}{Dt} \simeq -\frac{\partial \tilde{u}_1}{\partial x_\alpha}\tilde{g}_1. \tag{3.25}$$

This is called shear strain. It is analogous to vortex tilting (Figure 3.5).

In summary, a blob of $\tilde{c}$ in a realization of a turbulent flow follows an irregular trajectory and is contorted by turbulent velocity gradients as it travels; this contortion increases the magnitude of scalar gradients within the blob, enhancing the molecular diffusion. As a result, the blob disappears more quickly than it would in the absence of turbulence (Figure 3.4).

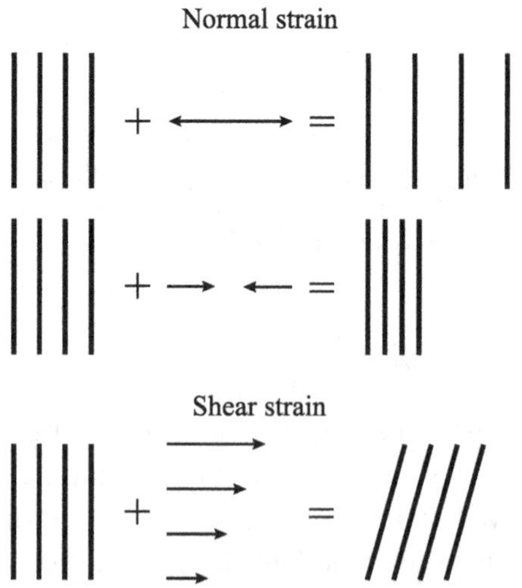

Figure 3.5 A schematic of the distortion of a scalar-gradient field by a strain field. The lines are iso-concentration contours. The upper two panels depict linear or normal strain, Eq. (3.24); the lower panel depicts shear strain, Eq. (3.25).

This prompts two final questions. The first concerns the substantial-derivative form of a conservation equation, which we often interpret as the equation *following a parcel*. How can we follow parcels in a turbulent flow if they are rapidly distorted and annihilated?

The answer is that we need "follow" a parcel only long enough to define this time derivative. Since the derivative involves the limit process $\Delta t \to 0$, this is a vanishingly small time.

A second question concerns mixing: if an advected constituent in a flow is conserved, how can it mix?

We said that a scalar constituent following Eq. (1.31) changes only *through the effects of molecular diffusion*. Molecular diffusion is the final stage of turbulent mixing. Without molecular diffusion a scalar constituent cannot truly mix; it will ultimately be finely but nonuniformly distributed by the turbulence, the spatial scale of the nonuniformities being that of the smallest turbulent eddies (Figure 3.4b).

3.4 Space-averaged equations

We stated the "Reynolds averaging rules" in Eqs. (2.5)–(2.8). Any linear average has the distributive property (2.5), and (2.7) is a consequence of (2.6). Rules (2.6) and (2.8) are

- The average of an average is the average:

$$\overline{\overline{\tilde{a}}} = \overline{\tilde{a}}. \tag{2.6}$$

- The average of a derivative is the derivative of the average (the *commutative* property):

$$\overline{\frac{\partial \tilde{a}}{\partial x_i}} = \frac{\partial \overline{\tilde{a}}}{\partial x_i}; \qquad \overline{\frac{\partial \tilde{a}}{\partial t}} = \frac{\partial \overline{\tilde{a}}}{\partial t}. \tag{2.8}$$

The ensemble average satisfies both of these rules. Now let's consider their applicability to two other commonly used averages.

The "record average," the average of a recorded measurement of $f(x,t)$, say, over an x-record of length L, is

$$\overline{f}^{\mathrm{rec}} = \frac{1}{L}\int_0^L f(x,t)\,dx. \tag{3.26}$$

Since $\overline{f}^{\mathrm{rec}}$ does not depend on x, averaging it over the record yields

$$\overline{\left(\overline{f}^{\mathrm{rec}}\right)}^{\mathrm{rec}} = \frac{1}{L}\int_0^L \overline{f}^{\mathrm{rec}}\,dx = \frac{\overline{f}^{\mathrm{rec}}}{L}\int_0^L dx = \overline{f}^{\mathrm{rec}}, \tag{3.27}$$

so the record average satisfies rule (2.6).

The record average commutes with differentiation with respect to independent variables except that in which it is recorded. Thus for $f(x, t)$ (Problem 3.10)

$$\frac{\partial}{\partial t}\left(\overline{f}^{\text{rec}}\right) = \overline{\left(\frac{\partial f}{\partial t}\right)}^{\text{rec}}, \quad \text{but} \quad \frac{\partial}{\partial x}\left(\overline{f}^{\text{rec}}\right) \neq \overline{\left(\frac{\partial f}{\partial x}\right)}^{\text{rec}}. \tag{3.28}$$

The "local average" at any point is an average over a neighborhood of that point:

$$\overline{f}^{\text{loc}}(x, t, \Delta) = \frac{1}{\Delta}\int_{-\Delta/2}^{\Delta/2} f(x + x', t)\, dx'. \tag{3.29}$$

This does not satisfy rule (2.6) (Problem 3.11), but it commutes with both derivatives (Problem 3.12).

Reynolds (1895) averaged the equation of motion "over a small region of the flow." This could be, for example, the local average

$$\overline{f}^{\text{loc}}(\mathbf{x}, t, \Delta) = \frac{1}{\Delta^3}\int_{-\Delta/2}^{\Delta/2}\int_{-\Delta/2}^{\Delta/2}\int_{-\Delta/2}^{\Delta/2} \tilde{f}(\mathbf{x} + \mathbf{x}', t)\, dx_1'\, dx_2'\, dx_3'. \tag{3.30}$$

This essentially removes eddies that are much smaller than the cube side Δ and minimally affects those that are much larger than Δ. By contrast, ensemble averaging removes all eddies. Thus the two become more alike when $\Delta \gg \ell$ – i.e., in coarse-resolution models. Perhaps this is why the type of averaging used in coarse-resolution numerical models is not always explicitly discussed.

3.4.1 The generalization to spatial filtering

By the early 1970s increases in computer size and speed made it possible to resolve the energy-containing eddies in an evolving turbulent flow. A conceptual turning point was Leonard's (1974) generalization of Reynolds' (1895) notion of local averaging in space, as in Eq. (3.30), to spatial *filtering*:

$$\overline{f}^{\text{filt}}(\mathbf{x}, t) = \int_{-\infty}^{\infty}\int_{-\infty}^{\infty}\int_{-\infty}^{\infty} \tilde{f}(\mathbf{x} + \mathbf{x}', t)\, G(\mathbf{x} - \mathbf{x}')\, dx_1'\, dx_2'\, dx_3'. \tag{3.31}$$

G is called the *filter function*. This is the local average of Eq. (3.30) if G is taken as $1/\Delta^3$ within a cube of side Δ centered at the origin and 0 outside it. As we shall discuss in Chapter 6, other filter functions are now also used, and we now interpret "local averaging" more broadly as "spatial filtering." In the usual (*low-pass*) form of this filtering Fourier components of small wavenumber ($\kappa \ll 1/\Delta$) are retained and those of larger wavenumber are removed.

Today the principal purpose of averaging the turbulent-flow equations is to enable their numerical solution, so we shall denote a spatially filtered variable through a superscript "r", this part of the variable being computationally *resolvable*. We write

$$\tilde{f} = \overline{\tilde{f}}^{\text{filt}} + (\tilde{f} - \overline{\tilde{f}}^{\text{filt}}) = \tilde{f}^{\text{r}} + \tilde{f}^{\text{s}}, \tag{3.32}$$

so that spatial filtering decomposes a turbulent variable into *resolvable* (r) and *subfilter-scale* (s) parts. In general subsequent applications of a filter also have effects;[†] that is,

$$(\tilde{f}^{\text{r}})^{\text{r}} \neq \tilde{f}^{\text{r}}. \tag{3.33}$$

From Eq. (3.32) it follows that in general the spatially filtered subfilter-scale field does not vanish:

$$(\tilde{f}^{\text{s}})^{\text{r}} \neq 0. \tag{3.34}$$

Unlike ensemble averaging, spatial filtering can preserve the turbulent character of the flow variable if the spatial scale of the filter function G is much less than the spatial scale of the turbulence.[‡] It can profoundly reduce the computational requirements for numerical solution of the equations, as we shall see.

3.4.2 Spatially filtered governing equations

We can now apply the local average (interpreted as Leonard's more general spatial filter, Eq. (3.31)) to the governing equations. With the decomposition of Eq. (3.32) the continuity equation (3.1) for the full velocity implies

$$\frac{\partial \tilde{u}_i}{\partial x_i} = \frac{\partial}{\partial x_i}(\tilde{u}_i^{\text{r}} + \tilde{u}_i^{\text{s}}) = \frac{\partial \tilde{u}_i^{\text{r}}}{\partial x_i} + \frac{\partial \tilde{u}_i^{\text{s}}}{\partial x_i} = 0. \tag{3.35}$$

Assuming that spatial filtering commutes with differentiation,[§] applying it to the full continuity equation (3.1) yields

$$\left(\frac{\partial \tilde{u}_i}{\partial x_i}\right)^{\text{r}} = \frac{\partial \tilde{u}_i^{\text{r}}}{\partial x_i} = 0. \tag{3.36}$$

Subtracting Eq. (3.36) from Eq. (3.35) then gives

$$\frac{\partial \tilde{u}_i^{\text{s}}}{\partial x_i} = 0. \tag{3.37}$$

Thus, both $\tilde{u}_i^{\text{r}}$ and $\tilde{u}_i^{\text{s}}$ are divergence free if $\tilde{u}_i$ is.

† The exception is the "top-hat" filter, Chapter 6.
‡ Equation (3.31) shows that when G is vanishingly narrow – a "delta function" – it leaves the function unchanged.
§ Leonard (1974) shows this is true if the function vanishes on the boundaries.

Consider next the Navier–Stokes equation (1.26), which we shall write in the form

$$\frac{\partial \tilde{u}_i}{\partial t} + \frac{\partial \tilde{u}_i \tilde{u}_j}{\partial x_j} = -\frac{1}{\rho}\frac{\partial \tilde{p}}{\partial x_i} + \nu \frac{\partial^2 \tilde{u}_i}{\partial x_j \partial x_j}. \tag{3.38}$$

Applying the spatial filter and using its commutativity with differentiation yields

$$\frac{\partial \tilde{u}_i^{\mathrm{r}}}{\partial t} + \frac{\partial (\tilde{u}_i \tilde{u}_j)^{\mathrm{r}}}{\partial x_j} = -\frac{1}{\rho}\frac{\partial \tilde{p}^{\mathrm{r}}}{\partial x_i} + \nu \frac{\partial^2 \tilde{u}_i^{\mathrm{r}}}{\partial x_j \partial x_j}. \tag{3.39}$$

As in Eq. (1.39), we write

$$(\tilde{u}_i \tilde{u}_j)^{\mathrm{r}} = \tilde{u}_i^{\mathrm{r}} \tilde{u}_j^{\mathrm{r}} + \left[(\tilde{u}_i \tilde{u}_j)^{\mathrm{r}} - \tilde{u}_i^{\mathrm{r}} \tilde{u}_j^{\mathrm{r}}\right] = \tilde{u}_i^{\mathrm{r}} \tilde{u}_j^{\mathrm{r}} - \frac{\tau_{ij}}{\rho}, \tag{3.40}$$

with τ_{ij} a Reynolds stress due to spatial filtering. From its definition in Eq. (3.40) we can write this generalized Reynolds stress as

$$\tau_{ij} \equiv \rho\left[\tilde{u}_i^{\mathrm{r}} \tilde{u}_j^{\mathrm{r}} - (\tilde{u}_i \tilde{u}_j)^{\mathrm{r}}\right] = \rho\left[(\tilde{u}_i^{\mathrm{r}} \tilde{u}_j^{\mathrm{r}})^{\mathrm{s}} - (\tilde{u}_i^{\mathrm{r}} \tilde{u}_j^{\mathrm{s}} + \tilde{u}_i^{\mathrm{s}} \tilde{u}_j^{\mathrm{r}} + \tilde{u}_i^{\mathrm{s}} \tilde{u}_j^{\mathrm{s}})^{\mathrm{r}}\right]. \tag{3.41}$$

τ_{ij} is neither a resolved nor a subfilter-scale quantity, for it has both a subfilter-scale part, $\rho(\tilde{u}_i^{\mathrm{r}} \tilde{u}_j^{\mathrm{r}})^{\mathrm{s}}$, and a resolved part, $-\rho(\tilde{u}_i^{\mathrm{r}} \tilde{u}_j^{\mathrm{s}} + \tilde{u}_i^{\mathrm{s}} \tilde{u}_j^{\mathrm{r}} + \tilde{u}_i^{\mathrm{s}} \tilde{u}_j^{\mathrm{s}})^{\mathrm{r}}$. But it does vanish in the high-resolution limit, so it is called the subfilter-scale (sfs) Reynolds stress.

Substituting Eq. (3.40) into Eq. (3.39) and reintroducing the viscous stress yields

$$\frac{\partial \tilde{u}_i^{\mathrm{r}}}{\partial t} + \frac{\partial}{\partial x_j}\left[\tilde{u}_i^{\mathrm{r}} \tilde{u}_j^{\mathrm{r}} - \frac{\tau_{ij}}{\rho} - \nu\left(\frac{\partial \tilde{u}_i^{\mathrm{r}}}{\partial x_j} + \frac{\partial \tilde{u}_j^{\mathrm{r}}}{\partial x_i}\right)\right] = -\frac{1}{\rho}\frac{\partial \tilde{p}^{\mathrm{r}}}{\partial x_i}. \tag{3.42}$$

Equation (3.42) and the resolved continuity equation (3.36) comprise four equations, but because of the structure of the sfs Reynolds stress the unknowns exceed four in number. Thus one needs a model for τ_{ij} in order to solve Eqs. (3.42) and (3.36) numerically.

As discussed in Chapter 16, with discrete spatial filtering techniques τ_{ij} can be measured (Tong *et al.*, 1999), and it can also be computed (at relatively small R_{t} values) from numerical-simulation fields. Thus there is also some observational and computational guidance that can be brought to such *subgrid* or *subfilter-scale* modeling, as it is called.

We shall see in Chapter 6 that even if τ_{ij}/ρ is much smaller than $\tilde{u}_i^{\mathrm{r}} \tilde{u}_j^{\mathrm{r}}$, which is the case with very high spatial resolution (small filter width), we still need to include it in the filtered equation (3.42). It is responsible for the extraction of kinetic energy from the resolved scales, the manifestation here of the energy cascade that is a key property of turbulence.

Applying the same spatial filtering process to the scalar equation (1.30) yields

$$\frac{\partial \tilde{c}^{\mathrm{r}}}{\partial t} + \frac{\partial}{\partial x_i}\left(\tilde{u}_i^{\mathrm{r}}\tilde{c}^{\mathrm{r}} + f_i - \gamma\frac{\partial \tilde{c}^{\mathrm{r}}}{\partial x_i}\right) = 0, \tag{3.43}$$

where $f_i = (\tilde{u}_i\tilde{c})^{\mathrm{r}} - \tilde{u}_i^{\mathrm{r}}\tilde{c}^{\mathrm{r}}$. If the resolved velocity field $\tilde{u}_i^{\mathrm{r}}$ is known this is one equation in two unknowns, $\tilde{c}^{\mathrm{r}}$ and f_i, and so one needs a model for the sfs scalar flux f_i.

3.4.3 Resolved and subfilter-scale turbulent fluxes

Ensemble averaging Eq. (3.42) produces

$$\frac{\partial \overline{\tilde{u}_i^{\mathrm{r}}}}{\partial t} + \frac{\partial}{\partial x_j}\left[\overline{\tilde{u}_i^{\mathrm{r}}\tilde{u}_j^{\mathrm{r}}} - \frac{\overline{\tau_{ij}}}{\rho} - \nu\left(\frac{\partial \overline{\tilde{u}_i^{\mathrm{r}}}}{\partial x_j} + \frac{\partial \overline{\tilde{u}_j^{\mathrm{r}}}}{\partial x_i}\right)\right] = -\frac{1}{\rho}\frac{\partial \overline{\tilde{p}^{\mathrm{r}}}}{\partial x_i}. \tag{3.44}$$

Let us assume that the spatial filter resolves the ensemble-average fields perfectly, so that $U_i^{\mathrm{r}} = U_i$, $P^{\mathrm{r}} = P$. This should be the case if the filter scale is small compared with ℓ. Then if we use the commutativity of spatial filtering and ensemble averaging and the expressions

$$\tilde{u}_i^{\mathrm{r}} = (U_i + u_i)^{\mathrm{r}} = U_i + u_i^{\mathrm{r}}, \qquad \tilde{p}^{\mathrm{r}} = P + p^{\mathrm{r}}, \tag{3.45}$$

we can write Eq. (3.44) as

$$\frac{\partial U_i}{\partial t} + \frac{\partial}{\partial x_j}\left[\overline{(U_i + u_i^{\mathrm{r}})(U_j + u_j^{\mathrm{r}})} - \frac{\overline{\tau_{ij}}}{\rho} - \nu\left(\frac{\partial U_i}{\partial x_j} + \frac{\partial U_j}{\partial x_i}\right)\right] = -\frac{1}{\rho}\frac{\partial P}{\partial x_i}. \tag{3.46}$$

Using the ensemble-averaging rules this becomes

$$\frac{\partial U_i}{\partial t} + \frac{\partial}{\partial x_j}\left[U_iU_j + \overline{u_i^{\mathrm{r}}u_j^{\mathrm{r}}} - \frac{\overline{\tau_{ij}}}{\rho} - \nu\left(\frac{\partial U_i}{\partial x_j} + \frac{\partial U_j}{\partial x_i}\right)\right] = -\frac{1}{\rho}\frac{\partial P}{\partial x_i}. \tag{3.47}$$

Comparing Eq. (3.47) and the ensemble-averaged Navier–Stokes equation (3.7) shows that

$$\overline{u_i^{\mathrm{r}}u_j^{\mathrm{r}}} - \frac{\overline{\tau_{ij}}}{\rho} = \overline{u_iu_j}. \tag{3.48}$$

This says that of the total kinematic momentum flux carried by the turbulence, $\overline{u_iu_j}$, the spatially filtered fluid equations directly resolve a portion, $\overline{u_i^{\mathrm{r}}u_j^{\mathrm{r}}}$, and require that the remainder, $-\overline{\tau_{ij}}/\rho$, be modeled. Since the contributions to $\overline{u_iu_j}$ come from the energy-containing eddies, if the scale of the spatial filtering lies in the inertial range the resolved portion is generally much larger than the portion that must be modeled. An ensemble-averaged model, by contrast, resolves none of the turbulent flux; all of it must be modeled. Because turbulence models are not perfect, we expect that (other things being equal) an ensemble-averaged model is inherently less reliable than a spatially filtered model. Thus the two approaches have two important differences:

- A spatially filtered model requires three-dimensional, time-dependent calculations. The ensemble-averaged model requires only as many spatial dimensions as there are inhomogeneous directions, and can be time independent if the flow is steady. Thus the computational requirements for a space-averaged model can be much larger.
- Since a spatially filtered model resolves part of the turbulent flux, and since the reliability of turbulent-flux models is uneven, other things being equal it tends to be the more reliable.

3.5 Summary

The equations for the ensemble-averaged fields of velocity and a conserved scalar in an incompressible turbulent flow are

$$\frac{\partial U_i}{\partial x_i} = 0, \tag{3.2}$$

$$\frac{\partial U_i}{\partial t} + \frac{\partial}{\partial x_j}\left[U_i U_j + \overline{u_i u_j} - \nu\left(\frac{\partial U_i}{\partial x_j} + \frac{\partial U_j}{\partial x_i}\right)\right] = -\frac{1}{\rho}\frac{\partial P}{\partial x_i}, \tag{3.7}$$

$$\frac{\partial C}{\partial t} + \frac{\partial}{\partial x_i}\left(U_i C + \overline{u_i c} - \gamma\frac{\partial C}{\partial x_i}\right) = 0. \tag{3.13}$$

Without the new terms involving $\overline{u_i u_j}$ and $\overline{u_i c}$, Eqs. (3.7) and (3.13) are the Navier–Stokes and scalar conservation equations (1.26) and (1.30), which have turbulent solutions at large Reynolds number. But here the turbulent-flux terms, being of leading order, ensure smooth, nonturbulent solutions that reveal little, if anything, about the instantaneous structure of a turbulent flow.

The spatially filtered equations are

$$\frac{\partial \tilde{u}_i^{\mathrm{r}}}{\partial x_i} = 0, \tag{3.36}$$

$$\frac{\partial \tilde{u}_i^{\mathrm{r}}}{\partial t} + \frac{\partial}{\partial x_j}\left[\tilde{u}_i^{\mathrm{r}}\tilde{u}_j^{\mathrm{r}} - \frac{\tau_{ij}}{\rho} - \nu\left(\frac{\partial \tilde{u}_i^{\mathrm{r}}}{\partial x_j} + \frac{\partial \tilde{u}_j^{\mathrm{r}}}{\partial x_i}\right)\right] = -\frac{1}{\rho}\frac{\partial \tilde{p}^{\mathrm{r}}}{\partial x_i}, \tag{3.42}$$

$$\frac{\partial \tilde{c}^{\mathrm{r}}}{\partial t} + \frac{\partial}{\partial x_i}\left(\tilde{u}_i^{\mathrm{r}}\tilde{c}^{\mathrm{r}} + f_i - \gamma\frac{\partial \tilde{c}^{\mathrm{r}}}{\partial x_i}\right) = 0. \tag{3.43}$$

Here the spatial filtering has produced the new terms $\tau_{ij}/\rho = \tilde{u}_i^{\mathrm{r}}\tilde{u}_j^{\mathrm{r}} - (\tilde{u}_i\tilde{u}_j)^{\mathrm{r}}$ and $f_i = (\tilde{u}_i\tilde{c})^{\mathrm{r}} - \tilde{u}_i^{\mathrm{r}}\tilde{c}^{\mathrm{r}}$ involving the subfilter-scale turbulence.

Both equation sets (3.2), (3.7), (3.13) and (3.36), (3.42), and (3.43) have more unknowns than equations, giving what is called a *closure problem*. We shall explore it in the following chapters.

Questions on key concepts

3.1 What is a flux? What three types of fluxes can emerge in an ensemble-mean equation? Illustrate with the momentum or conserved scalar equations.

3.2 What is a Reynolds stress? A Reynolds flux? Explain G. I. Taylor's term "virtual mean flux."

3.3 Explain how symmetry arguments can be used to evaluate Reynolds fluxes.

3.4 Explain the statement "turbulent mixing need not be mixing at all."

3.5 Explain the interplay of turbulent and molecular diffusion.

3.6 Give an example of an ensemble-mean model in boundary-layer meteorology.

3.7 Explain why it is unlikely that an ensemble-mean profile – for example, the mean-wind profile in a boundary layer – would ever appear in a realization of a turbulent flow.

3.8 If an advected constituent in a turbulent flow is conserved, how can it ever mix?

3.9 Explain why a spatially filtered turbulent flow model can be more reliable than an ensemble-averaged model.

3.10 How could a spatially filtered model be made to have the randomness property – i.e., to be able to generate an ensemble of predictions?

3.11 How could a spatially filtered model of the type described in Question 3.10 be used to predict both an ensemble mean and variances about that mean?

3.12 What is the *closure problem*? How does it originate?

3.13 With an eddy-viscosity closure the ensemble-mean equation of motion (3.7) can be written in the same form as the Navier–Stokes equation. Then why does it not have turbulent solutions?

3.14 Why must we exclude mean pressure from homogeneity considerations? What other mean variables fall into this category?

Problems

3.1 Starting from Eq. (3.7), form the equation governing the evolution of the kinetic energy of a steady, ensemble-averaged flow.

(a) Is the kinetic energy of the ensemble-averaged flow the same as the ensemble-averaged kinetic energy? Discuss.

(b) Write the term involving the viscous stress as the sum of a divergence and a dissipative term.
(c) Write the remaining terms as the sum of a divergence and a remainder, and integrate the equation over a volume. Interpret the result.
(d) Interpret the volume-integrated equation for steady flow in a pipe. What can you conclude about the role of the turbulent flux term?
(e) Repeat (a)–(d) for a conserved scalar.

3.2 Derive the equation for the evolution of mean vorticity and interpret the terms.

3.3 Derive the equation for the evolution of the mean gradient of a conserved scalar and interpret the terms.

3.4 Interpret the kinematic Reynolds stress $\overline{uw}$ in channel flow (Figure 3.1) formally in terms of an average over realizations. Indicate the arguments of u and w. Then explain how we determine $\overline{uw}$ from time series measured at a point, again indicating arguments. Explain physically when and why this is justifiable.

3.5 Write the Poisson equation for mean pressure by taking the divergence of the mean momentum equation (3.6).

3.6 Write the vertical mean momentum equation for steady flow in a channel (Subsection 3.2.1).

3.7 Sketch instantaneous and ensemble-averaged effluent plumes from a stack. What are their important differences?

3.8 Explain why we say that "turbulent diffusion" is partly real, partly virtual.

3.9 Evaluate the lateral equation of mean motion in turbulent channel flow (Figure 3.1).

3.10 Prove the averaging rules in Eqs. (3.28).

3.11 Show with an example that the local average, Eq. (3.29), does not satisfy averaging rule (2.6).

3.12 Show that the local average, Eq. (3.29), commutes with both derivatives.

3.13 Explain why the space-averaging operation defined in Eq. (3.30) removes eddies that are much smaller than the cube side h and minimally affects those that are much larger than h.

3.14 Do the expansion in Eq. (3.41) and show that it produces the rhs of the equation.

References

Corrsin, S., 1961: Turbulent flow. *Am. Sci.*, **49**, 300–325.
Deardorff, J. W., 1970: A numerical study of three-dimensional turbulent channel flow at large Reynolds numbers. *J. Fluid Mech.*, **41**, 453–480.
Kundu, Pijush K., 1990: *Fluid Mechanics*. San Diego: Academic Press.

Leonard, A., 1974: Energy cascade in large-eddy simulations of turbulent fluid flows. *Adv. Geophys.*, **18A**, 237–248.

Reynolds, O., 1895: On the dynamical theory of incompressible viscous fluids and the determination of the criterion. *Philos. Trans. R. Soc. London, Series A*, **186**, 123–164.

Tong, C., J. Wyngaard, and J. Brasseur, 1999: Experimental study of subgrid-scale stresses in the atmospheric surface layer. *J. Atmos. Sci.*, **56**, 2277–2292.

4

Turbulent fluxes

4.1 Introduction

We saw in Chapter 3 that averaging produces new types of fluxes in the momentum and scalar conservation equations. In this chapter we'll discuss those produced by ensemble averaging; we'll follow the usual convention in referring to these as *turbulent fluxes*. We'll use turbulent heat flux† to illustrate the concepts.

To obtain some insight into the nature of heat flux in turbulent flow, we multiply the fluid temperature Eq. (1.32) by $\rho c_p = \kappa/\alpha$, here a constant, and write it in "flux form":

$$\frac{\partial \rho c_p \tilde{T}}{\partial t} = -\frac{\partial}{\partial x_i}\left(\rho c_p \tilde{T}\tilde{u}_i - k\frac{\partial \tilde{T}}{\partial x_i}\right) = -\frac{\partial \tilde{H}_i}{\partial x_i}. \tag{4.1}$$

The heat flux $\tilde{H}_i$ is a turbulent vector variable: it fluctuates in space and time and is different in every realization. Equation (4.1) says it has a contribution from the fluid motion, $\rho c_p \tilde{T}\tilde{u}_i$, with $\rho c_p \tilde{T}$ the enthalpy of the fluid (joules m^{-3}); and a contribution from thermal conduction, $-k\partial\tilde{T}/\partial x_i$, with k the fluid thermal conductivity (watts m^{-1} K^{-1}).

In our pipe-flow example of Chapter 1 the radial component of heat flux is the most important. At the wall it is due solely to conduction, since the fluid velocity vanishes there; above the wall it is dominated by the turbulent radial velocity.

Hereafter in this chapter we shall deal with the temperature equation (1.32) rather than with (4.1), its product with ρc_p. Although by the usual convention a flux refers to an extensive property, in using Eq. (1.32) we shall make an exception by referring to the flux of temperature.

† Thermodynamicists (e.g., Bohren and Albrecht, 1998) have long railed against this use of "heat," but here it is the conventional term.

4.2 Temperature flux in a boundary layer

With the decomposition $\tilde{T} = \Theta + \theta$, ensemble averaging the temperature equation (1.32) in a turbulent boundary layer yields

$$\frac{\partial \Theta}{\partial t} + U_i \frac{\partial \Theta}{\partial x_i} + \frac{\partial \overline{u_i \theta}}{\partial x_i} = \alpha \frac{\partial^2 \Theta}{\partial x_i \partial x_i}. \tag{4.2}$$

The molecular-diffusion term is important only in the diffusive sublayer on the surface. If that surface is very large in extent and horizontally homogeneous, the flow approaches horizontal homogeneity in which $U_i = [U(z,t), 0, 0]$ and $\Theta = \Theta(z,t)$. This is a common model of the atmospheric boundary layer. Then the time rate of change of the mean fluid temperature above the surface is due to the divergence of the turbulent flux,

$$\frac{\partial \Theta}{\partial t} = -\frac{\partial \overline{u_i \theta}}{\partial x_i} = -\frac{\partial \overline{w\theta}}{\partial z}. \tag{4.3}$$

If the surface is warmer than the fluid above, rising (positive w) fluid tends to be warmer than its local environment (positive θ), since it came from the warmer region nearer the surface; likewise, sinking (negative w) fluid tends to be cooler than its environment (negative θ), since it came from the cooler region farther away. Thus, we expect $\overline{w\theta}$ to be positive. Correspondingly, $\overline{w\theta}$ is negative over a cooler surface. $\overline{w\theta}$ vanishes at the boundary-layer top because the turbulence vanishes there. Thus the divergence of the turbulent temperature flux is nonzero within the boundary layer, which through Eq. (4.3) causes Θ to change with time.

The familiar near-surface warming of the atmosphere on a sunny morning is well described by Eq. (4.3) (interpreted for potential temperature, defined in Part II). Warming typically occurs at nearly the same rate over much or most of the deepening boundary layer. If so, Eq. (4.3) says that the temperature flux divergence is nearly constant with height.

Typical daytime, clear-weather $\overline{w\theta}$ and Θ profiles are sketched in Figure 4.1. There is a pronounced negative mean temperature gradient near the surface, where $\overline{w\theta}$ is positive. This is consistent with the existence of an eddy diffusivity K such that $\overline{w\theta} = -K\,\partial\Theta/\partial z$. Above the surface the magnitude of $\partial\Theta/\partial z$ decreases with height more sharply than $\overline{w\theta}$ does, implying that K increases away from the surface. The top of the daytime boundary layer is typically determined by a "capping" inversion – a stably stratified layer (Θ increasing with height) that damps turbulence and acts as a "lid" for rising convective elements. The entrainment of this inversion by turbulence makes $\overline{w\theta}$ negative in this interfacial region, as sketched in Figure 4.1.

As Figure 4.1 indicates, this region of positive $\partial\Theta/\partial z$ aloft can extend down into the boundary layer. As a result the eddy diffusivity for temperature can be poorly

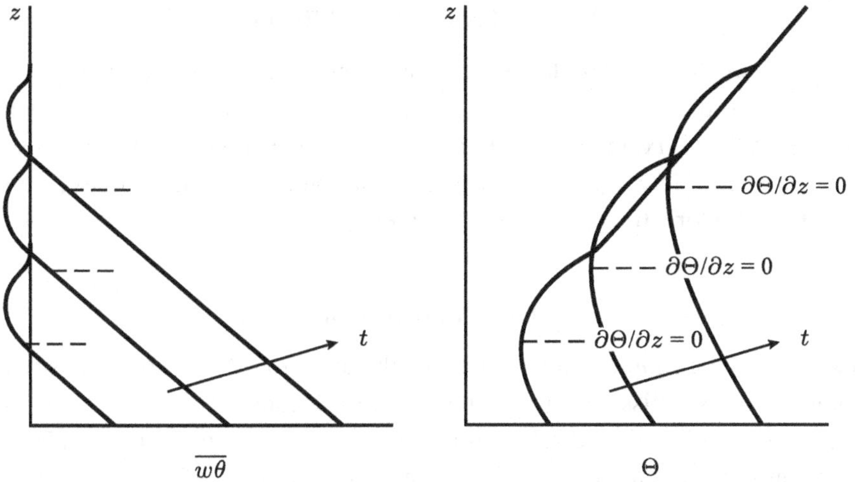

Figure 4.1 A sketch of the profiles of vertical turbulent temperature flux $\overline{w\theta}$ (left) and mean temperature Θ (right) in a growing convective boundary layer. In midday $\partial\Theta/\partial z$ can vanish in midlayer; those points are indicated by dashed lines. If $\overline{w\theta}$ is nonzero at those points (as at the later times, left) the eddy diffusivity K has a singularity there; for some distance above that K is negative. Here "temperature" is potential temperature (Part II).

behaved; it has a singularity if $\partial\Theta/\partial z$ and $\overline{w\theta}$ change sign at different heights, as at the later times in Figure 4.1.

The complete mean-temperature equation (4.2) for the one-dimensional problem reduces to

$$\frac{\partial\Theta}{\partial t} + \frac{\partial}{\partial z}\left(\overline{w\theta} - \alpha\frac{\partial\Theta}{\partial z}\right) = 0. \tag{4.4}$$

A typical value of the surface temperature flux Q_0 over land on a sunny day is $0.1\ \mathrm{m\ s^{-1}\ K}$, which requires a mean temperature gradient at the surface of

$$\frac{\partial\Theta}{\partial z} = -\frac{Q_0}{\alpha} \sim -10^4\ \mathrm{K\ m^{-1}}, \tag{4.5}$$

using $\alpha = 10^{-5}\ \mathrm{m^2\ s^{-1}}$.

Above the thin diffusive sublayer the temperature flux is carried almost entirely by the turbulence. The eddy diffusivity is of order $u\ell$, much larger than α, so the magnitude of $\partial\Theta/\partial z$ is much less than at the surface. At a height $z \ll h$ where the eddy diffusivity is $\sim 10^5\alpha$, for example, which we shall see in Part II occurs in the surface layer, turbulence reduces $\partial\Theta/\partial z$ to $\sim 0.1\ \mathrm{K\ m^{-1}}$, five orders of magnitude smaller than at the surface.

4.3 Mass flux in scalar diffusion

In Chapter 3 we discussed molecular and turbulent diffusion of an effluent released from a point source into a flow. We showed that the turbulent problem, which involves turbulent mixing, molecular diffusion, and ensemble averaging, has some "virtual" aspects. Here, after discussing the solution for the laminar case, we will present G. I. Taylor's treatment of the turbulent problem.

4.3.1 *Laminar flow*

Figure 4.2 shows a simple problem in molecular diffusion. A line source of strength Q (units of mass emission rate per unit length), located at $x = 0,\ z = 0$ and oriented in the y-direction, discharges a conserved effluent into a uniform laminar flow of constant velocity $u_i = (u, 0, 0)$. Under steady conditions the conservation equation (1.31) for the effluent reduces to a balance between streamwise advection and molecular diffusion in the streamwise and vertical directions:

$$u\frac{\partial c}{\partial x} = \gamma\left(\frac{\partial^2 c}{\partial x^2} + \frac{\partial^2 c}{\partial z^2}\right). \tag{4.6}$$

We will make the "thin-plume" approximation that the streamwise length scale of the plume is much larger than the plume thickness. This allows us to neglect streamwise diffusion, and Eq. (4.6) reduces to

$$u\frac{\partial c}{\partial x} = \gamma\frac{\partial^2 c}{\partial z^2}, \tag{4.7}$$

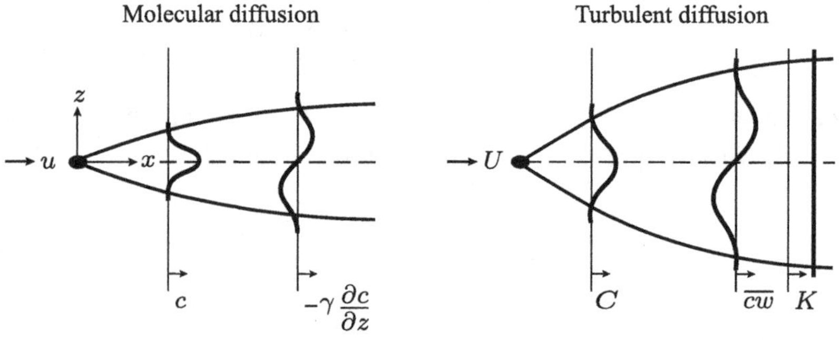

Figure 4.2 An artist's sketch, exaggerated in the vertical (more so for the left panel), of diffusion of a conserved scalar from a crosswind (y-direction) line source in a uniform flow. Left: The profiles of concentration c and lateral flux $-\gamma\partial c/\partial z$ in laminar flow. The plume width grows as $x^{1/2}$, Eq. (4.9). Right: Profiles of mean concentration C, lateral turbulent flux $\overline{wc}$, and eddy diffusivity K in the long-time limit in homogeneous turbulent flow. Here the plume width grows initially as x and then as $x^{1/2}$, Eq. (4.29).

whose solution (Problem 4.5) is the *Gaussian plume*,

$$c(x, z) = \frac{Q}{\sqrt{2\pi}u\sigma} e^{\frac{-z^2}{2\sigma^2}}. \tag{4.8}$$

The "plume width" $\sigma(x)$ is

$$\sigma = \left(\frac{2\gamma x}{u}\right)^{1/2}, \quad \frac{\sigma}{x} = \left(\frac{2\gamma}{ux}\right)^{1/2}. \tag{4.9}$$

Equation (4.9) shows that σ grows more slowly than streamwise distance x. If $u = 5$ m s^{-1} and $\gamma = 10^{-5}$ m^2 s^{-1}, at 4 cm downstream of the source σ is only 1% of that. The thin-plume approximation only improves farther downstream; at a point 4 m downstream, for example, it is 0.1%.

The vertical flux of c here is $-\gamma \partial c/\partial z$. From Eqs. (4.8) and (4.9) this is

$$-\gamma \frac{\partial c}{\partial z} = \frac{\gamma z}{\sigma^2} c(z). \tag{4.10}$$

It is sketched on Figure 4.2 along with the c profile. Because of the symmetry of the problem the c flux vanishes on the centerplane; elsewhere the flux is directed away from the c-rich fluid there.

4.3.2 Turbulent flow

Now let's consider the same problem in a homogeneous turbulent flow. An instantaneous turbulent plume has an irregular shape; the ensemble-averaged (mean) plume (the result of averaging many instantaneous, sharp-edged plumes, each different in detail) is smooth (Figure 2.3). The mean concentration profile across it (which has been measured in homogeneous laboratory turbulent flows) follows the Gaussian solution (4.8) closely.

The ensemble-averaged concentration equation in the homogeneous-turbulence case is

$$U\frac{\partial C}{\partial x} = -\frac{\partial \overline{uc}}{\partial x} - \frac{\partial \overline{wc}}{\partial z}, \tag{4.11}$$

where the mean velocity $U_i = (U, 0, 0)$ and U is constant. As in the laminar case (4.6) this is a balance between streamwise advection and turbulent diffusion in the streamwise and vertical directions. The molecular diffusion terms are negligible everywhere, there being no solid boundaries in this problem.

If we make the *thin-plume* assumption, which implies here that $\partial\overline{uc}/\partial x \ll \partial\overline{wc}/\partial z$, Eq. (4.11) reduces to

$$U\frac{\partial C}{\partial x} = -\frac{\partial \overline{wc}}{\partial z}. \tag{4.12}$$

We can diagnose the behavior of $\overline{wc}$, the vertical component of the turbulent flux of effluent. In the upper half of the plume we expect upward motion (positive w) to be associated with concentrated effluent (positive c) that originates nearer the centerplane, and downward motion (negative w) associated with diluted effluent (negative c) that originates farther from the centerplane. ($\tilde{c}$ cannot be negative, but its fluctuating part c can be.) Each leads to positive $\overline{wc}$. In the same way, we expect negative $\overline{wc}$ in the bottom half of the plume. Since the mean flow is symmetric about the centerplane, $\overline{wc}$ must vanish there, just as the molecular flux does in the laminar problem. Finally, the vanishing of c in the effluent-free flow far away from the centerplane makes $\overline{wc} = 0$ there.

These deduced C and $\overline{wc}$ profiles in the turbulent-diffusion problem are also sketched in Figure 4.2. Their similarity to those in molecular diffusion suggests a similarity between the concentration equation (4.7) in laminar flow and the mean concentration equation (4.12) in homogeneous turbulent flow. If there were a z-independent eddy diffusivity $K(x)$ such that

$$\overline{wc}(x, z) = -K(x)\frac{\partial C}{\partial z}, \tag{4.13}$$

then at any x the C equation (4.12) would have the same form as Eq. (4.7) for laminar flow,

$$U\frac{\partial C}{\partial x} = K(x)\frac{\partial^2 C}{\partial z^2}, \tag{4.14}$$

and the mean concentration $C(z)$ at any x would also be Gaussian, as observed:

$$C(x, z) = \frac{Q}{\sqrt{2\pi}U\sigma_t}e^{\frac{-z^2}{2\sigma_t^2}}. \tag{4.15}$$

This is the molecular diffusion form (4.8) but with the mean velocity U and the width parameter of the mean concentration distribution, $\sigma_t(x)$.

The behavior of the eddy diffusivity $K(x)$ in this problem can be deduced through a Lagrangian analysis as follows.[†] If we substitute the Gaussian solution (4.15) into Eq. (4.14) for the mean concentration and solve for $K(x)$ we find

$$K(x) = \frac{U}{2}\frac{d\sigma_t^2}{dx}. \tag{4.16}$$

At any $x = Ut$ the mean concentration $C(x, z)$ is proportional to the probability that $z_p(t)$, the vertical displacement of an effluent particle at time t after release, is $\simeq z$. Thus we can interpret σ_t^2 as the Lagrangian quantity $\overline{z_p^2}(t)$, the mean-squared

† This analysis of G. I. Taylor's 1921 solution is adapted from Csanady (1973).

value of the vertical displacement of diffusing effluent particles. $z_{\mathrm{p}}(t)$ is related to another Lagrangian quantity, the particle vertical velocity $w_{\mathrm{p}}(t)$, by

$$z_{\mathrm{p}}(t) = \int_0^t w_{\mathrm{p}}(t')\,dt'. \tag{4.17}$$

Multiplying by $2w_{\mathrm{p}} = 2dz_{\mathrm{p}}/dt$, ensemble averaging, and using the commutativity of time differentiation and ensemble averaging yields

$$\overline{2z_{\mathrm{p}}\frac{dz_{\mathrm{p}}}{dt}} = \overline{\frac{dz_{\mathrm{p}}^2}{dt}} = \frac{d\overline{z_{\mathrm{p}}^2}}{dt} = 2\int_0^t \overline{w_{\mathrm{p}}(t)w_{\mathrm{p}}(t')}\,dt'. \tag{4.18}$$

This can be rewritten as

$$\frac{d\overline{z_{\mathrm{p}}^2}}{dt} = 2\overline{w_{\mathrm{p}}^2}\int_0^t R(\tau)\,d\tau, \tag{4.19}$$

with R the vertical velocity autocorrelation function of a particle,

$$R(\tau) = \frac{\overline{w_{\mathrm{p}}(t)w_{\mathrm{p}}(t+\tau)}}{\overline{w_{\mathrm{p}}^2}}. \tag{4.20}$$

We can diagnose the behavior of R in the short-time and long-time limits. The first is simple: $R(0) = 1$. At long times we would expect a moving particle to "forget" its initial vertical velocity, so that $R \to 0$.

In homogeneous, stationary turbulence in a constant-density fluid the Lagrangian variance $\overline{w_{\mathrm{p}}^2}$ is equal to the Eulerian variance $\overline{w^2}$ (Lumley, 1962; Corrsin, 1963). Thus, Eq. (4.19) can be written

$$\frac{d\overline{z_{\mathrm{p}}^2}}{dt} = 2\overline{w^2}\int_0^t R(\tau)d\tau. \tag{4.21}$$

This result is due to G. I. Taylor (1921) and is known as "Taylor's theorem." The short-time limit $R = 1$ implies

$$\frac{d\overline{z_{\mathrm{p}}^2}}{dt} \to 2\overline{w^2}t, \quad t \to 0. \tag{4.22}$$

The long-time limit $R = 0$ yields

$$\frac{d\overline{z_{\mathrm{p}}^2}}{dt} \to 2\overline{w^2}\tau_{\mathrm{L}}, \quad t \to \infty, \tag{4.23}$$

where τ_L is the Lagrangian integral time scale defined by

$$\tau_L = \int_0^\infty R(t)\, dt. \tag{4.24}$$

τ_L is of the order of the Eulerian time integral scale τ (Chapter 2), but there is no simple relation between them (Corrsin, 1963).

We can also interpret these results in spatial terms. We can write

$$x_p(t) = \int_0^t u_p(t')\, dt', \tag{4.25}$$

and ensemble averaging this gives, using the equivalence of Eulerian and Lagrangian statistics in homogeneous turbulence,

$$\overline{x_p}(t) = \int_0^t \overline{u_p}(t')\, dt' = Ut. \tag{4.26}$$

Equation (4.26) says that the ensemble of particles moves downstream at velocity U. Thus, we can make the conversion $x = Ut$, $d/dt = U d/dx$ and write Eq. (4.16) for K as

$$K = \frac{U}{2}\frac{d\sigma_t^2}{dx} = \frac{U}{2}\frac{d\overline{z_p^2}}{dx} = \frac{1}{2}\frac{d\overline{z_p^2}}{dt}. \tag{4.27}$$

In the short-time and long-time limits (4.22) and (4.23), which we can now interpret as $t \ll \tau_L$ and $t \gg \tau_L$, respectively, we then have

$$K(x) \sim \overline{w^2}t = \frac{\overline{w^2}}{U}x, \quad t \ll \tau_L; \quad K(x) \sim \overline{w^2}\tau_L = \text{constant}, \quad t \gg \tau_L. \tag{4.28}$$

The behavior of the mean plume width σ_t is, from (4.27),

$$\sigma_t \sim \frac{(\overline{w^2})^{1/2}}{U}x, \quad x \ll U\tau_L; \quad \sigma_t \sim \left(\frac{\overline{w^2}\tau_L}{U}\right)^{1/2} x^{1/2}, \quad x \gg U\tau_L. \tag{4.29}$$

This indicates the mean plume grows linearly with distance downstream in the initial stages, and slows to parabolic growth (as sketched in Figure 4.2) after a distance of order $U\tau_L$.

In summary, the mean concentration profile $C(x, z)$ in a slender plume diffusing from a continuous, crosswind line source in homogeneous turbulence of mean velocity U has the same form as the laminar solution (4.8),

$$C(x, z) = \frac{Q}{\sqrt{2\pi}U\sigma_t} e^{\frac{-z^2}{2\sigma_t^2}},$$

but the mean-plume parameters depend on turbulence statistics:

$$\frac{d\sigma_t^2}{dt} = 2K = 2\overline{w^2}\int_0^t R(t')\,dt', \quad R(t') = \frac{\overline{w_p(t)w_p(t+t')}}{\overline{w_p^2}}, \quad x = Ut.$$

As Csanady (1973) points out, the behavior of the eddy diffusivity K is odd when there is more than one effluent source. If there are two sources, one upstream from the other, then at a point where both contribute to the total mean concentration the two effluents have different K values. This prompted Taylor (1959) to label the eddy diffusivity "an illogical conception."

4.4 Momentum flux in channel flow

In Figure 3.1 we sketched steady turbulent flow down a channel of rectangular cross section. We showed that the streamwise component of its mean momentum equation is

$$\frac{\partial}{\partial z}\left(\overline{uw} - \nu\frac{\partial U}{\partial z}\right) = -\frac{1}{\rho}\frac{\partial P}{\partial x}, \tag{3.10}$$

and that the profile of the 1-3 component of its mean kinematic stress is

$$T_{13} = -\overline{uw} + \nu\partial U/\partial z = -\frac{u_*^2 z}{D}, \tag{4.30}$$

with $2D$ the distance between the channel walls and u_*^2 the magnitude of the kinematic wall stress. u_* is called the *friction velocity*. On smooth walls that stress is a viscous one, but just above the diffusive sublayer it is carried by the turbulence. For this reason u_* is an important velocity scale for near-wall turbulence.

We sketch the viscous and turbulent components of this kinematic mean stress profile in Figure 4.3. The turbulent component is consistent with the existence of a

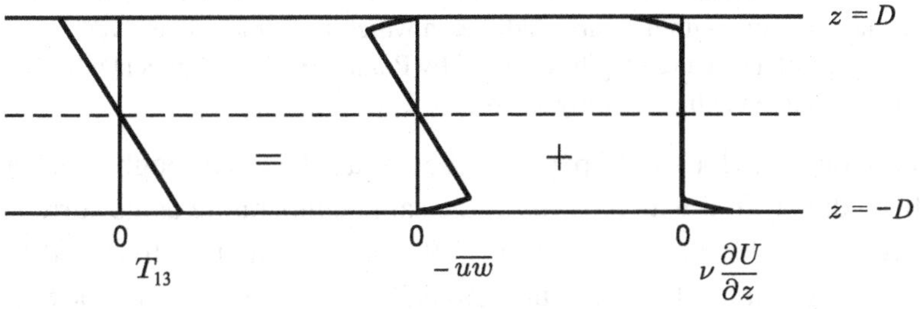

Figure 4.3 A sketch of the mean kinematic stress profile T_{13} and its turbulence and viscous components in the turbulent channel flow of Figure 3.1.

positive eddy diffusivity K such that $\overline{uw} = -K\,\partial U/\partial z$. The shape of the U profile, Figure 3.1, also suggests that K is smaller near the channel walls and larger near the center.

4.5 The "mixture length"

In this chapter we've seen that the turbulent flux of a quantity – the covariance of turbulent velocity and the turbulent part of the quantity being transported – has a central role in turbulent flows. In the example of Section 4.2 the vertical flux of temperature $\overline{w\theta}$ dictates the evolution of the daytime temperature profile.

It is natural to try to relate the turbulent part of the quantity being transported to the corresponding mean gradient, e.g.,

$$\theta \sim -d\frac{\partial \Theta}{\partial z}, \tag{4.31}$$

with d a length scale related to the turbulent fluid displacements in z. Then the flux is

$$\overline{\theta w} \sim -\overline{dw}\frac{\partial \Theta}{\partial z} \sim -K\frac{\partial \Theta}{\partial z}, \tag{4.32}$$

with K an "eddy diffusivity."

The notion of a length scale d related to turbulent mixing was evidently developed independently by Taylor and Prandtl. In a 1913 experiment Taylor measured the vertical profiles of temperature and humidity above the cold water of the Grand Banks of Newfoundland, finding that the rate at which the cooling wave penetrated into the atmosphere could be accounted for "by a process like molecular conduction but much more vigorous." In reflecting on his early attempts to understand the "effective conductivity due to turbulence," Taylor (1970) wrote:

> To complete a theoretical analogy between molecular and turbulent transfer it is necessary to think up some length connected with turbulence which is analogous to the mean free path of molecules. I was driven to imagine a purely hypothetical process to represent the collisions which terminate each molecular free path, and in 1915 I put out the idea that coherent fluid masses move a certain distance up or down vertically carrying all their transferable properties and then mix with the surroundings in which they find themselves.
>
> The … idea … of a mixture length was used by Prandtl [in a 1925 paper] who afterwards told me that he had never heard of my 1915 paper.

A key notion in Taylor's 1915 paper is "the average height through which an eddy moves from the layer at which it was at the same temperature as its surroundings, to the layer with which it mixes." In his 1925 paper Prandtl introduced the term "Mischungsweg," in English "mixture length," by which it became known. Taylor (1970) made clear that he had considered these notions about what had come to be known as the mixture length to be quite crude:

I was not satisfied with the mixture-length theory, because the idea that a fluid mass would go a certain distance unchanged and then deliver up its transferable property, and become identical with the mean condition at that point, is not a realistic picture of a physical process.

He credited his misgivings about mixture-length theory to his development of an exact solution for the growth of the mean plume from a continuous point source (Section 4.3.2):

While thinking of [analytical "fall out" from attempts to rationalize a mixture-length theory] I became interested in the form which a smoke trail takes after leaving a chimney. Any theory of diffusion which is based on a virtual coefficient of diffusion must predict a mean shape for a smoke plume which is paraboloidal, and it was quite clear to me that near the emitting source the mean outline of a smoke plume is pointed. This led me to think of other ways than mixture-length theory to describe turbulent diffusion. The result was my paper "Diffusion by continuous movements," (1921) in which the idea of correlation was introduced into the subject, I think for the first time.

Taylor's "pointed mean outline of a smoke plume" is the linear limit of Eq. (4.29).

Perhaps Taylor's (1921) paper, which presented a formal solution to dispersion from a continuous point source in steady, homogeneous turbulence, dampened interest in the much less elegant mixing-length approaches. Indeed Lumley (1989), in commenting on the mixing-length (as it is now usually called) assumption, wrote:

When I was a graduate student, I did not know that the mixing length assumption existed.... It was only later that I learned about it.

Taylor subsequently published a four-part series on a statistical approach to turbulence (Taylor, 1935), which has heavily influenced our present perspectives.

4.6 Summary

We have deduced the behavior of turbulent fluxes in three problems. In each case they were much larger than the corresponding molecular fluxes, except at a solid surface. We also found evidence that the turbulent flux, mean-gradient relation need not be as simple and direct as that in molecular diffusion; we'll see that it rarely is.

We discussed Taylor's early experience with observations of turbulent flows, his evolving physical notions about turbulent fluxes and their maintenance, and his attempts to develop a simple "mixture-length" theory for turbulence. We discussed his misgivings about that theory and his continuing efforts to find better alternatives. We'll take that path further in Chapter 5 by examining the evolution equations for turbulent fluxes.

Questions on key concepts

4.1 Show that the use of the mean-plus-fluctuating decomposition produces several types of conserved-scalar fluxes in turbulent flow. Which one is the *turbulent flux*?

4.2 Explain how the averaging process allows us to write the mean flux in the averaged form of the scalar conservation equation (1.29) as the sum of mean-mean, turbulent, and molecular diffusion parts.

4.3 In equation terms, why does the mean temperature outside in the morning increase predictably on sunny days? What is the physical mechanism causing this increase?

4.4 What is the physical mechanism underlying the diffusion of a plume from a point source in laminar flow? In turbulent flow? By what ratio do the effective diffusivities differ?

4.5 Explain how we can often diagnose the turbulent flux profiles in turbulent flow. Illustrate with an example.

4.6 Explain the essence of the "mixing-length" idea.

4.7 Interpret the physical origins of the horizontal turbulent flux term in Eq. (4.11).

4.8 Interpret Eq. (4.11) through a mean mass balance for a control volume downstream of a line source in the cross-stream direction. Explain physically why each of its terms is nonzero.

4.9 Explain why the entrainment flux at the top of the convective boundary layer, Figure 4.1, is negative.

4.10 Explain physically why we expect the Lagrangian autcorrelation function to be nonzero for some initial time period.

4.11 In an Eulerian frame the Taylor problem of Subsection 4.3.2 is stationary. Why is it nonstationary in a Lagrangian frame? In what problem would it also be stationary in a Lagrangian frame?

Problems

4.1 A reactive trace constituent having the evolution equation

$$\frac{D\tilde{c}}{Dt} = \gamma \nabla^2 \tilde{c} - \frac{\tilde{c}}{\tau},$$

with τ a decay time scale, diffuses in a turbulent flow. Assume that in this flow a conserved constituent has an eddy diffusivity. Using the mixing-length ideas, discuss under what conditions you would and would not expect this reactive constituent to have the eddy diffusivity of a conserved constituent.

4.2 Why might the horizontal turbulent temperature flux be of little practical consequence in a boundary layer, even though it can be larger than the vertical flux?

4.3 Have you seen chimney plumes that do not diffuse? Under what atmospheric conditions have you seen them? Sketch what you have seen. What causes this behavior?

4.4 What is the physical origin of the flux-gradient relation in molecular diffusion? What averaging process is involved? Such a flux-gradient relation need not apply to turbulent diffusion. What difference in the two processes could account for this?

4.5 Prove that Eq. (4.8), with its parameters defined as in Eq. (4.9), is a solution of Eq. (4.7).

4.6 Flow between oppositely moving parallel planes is called Couette flow. When is it laminar? Turbulent? Calculate the stress profile and the velocity profile in laminar Couette flow. Why is the pressure gradient zero?

4.7 Calculate the stress profile in turbulent Couette flow. Sketch the mean velocity profile, contrasting it with the laminar case. Sketch the corresponding profile of eddy diffusivity.

4.8 In a turbulent Couette flow one plane is held at temperature T_1 and the other at T_2. Write the mean temperature equation for this problem. Sketch the vertical profile of turbulent temperature flux.

4.9 Water vapor enters a horizontally homogeneous atmospheric boundary layer by evaporation from the underlying surface. The boundary layer is deepening with time by entrainment of the overlying air, which is drier than that in the boundary layer. The mean water vapor concentration does not change with time.

(a) Write the mean water vapor concentration equation for this problem.

(b) Sketch and explain the profile of the vertical turbulent flux of water vapor.

References

Bohren, C. F., and B. A. Albrecht, 1998: *Atmospheric Thermodynamics.* New York: Oxford University Press.

Corrsin, S., 1963: Estimates of the relations between Eulerian and Lagrangian scales in large Reynolds number turbulence. *J. Atmos. Sci.*, **20**, 115–119.

Csanady, G. T., 1973: *Turbulent Diffusion in the Environment.* Dordrecht: Reidel.

Lumley, J. L., 1962: The mathematical nature of the problem of relating Lagrangian and Eulerian statistical functions in turbulence. *Mécanique de la Turbulence*, Paris: CNRS, pp. 17–26.

Lumley, J. L., 1989: The state of turbulence research. *Advances in Turbulence*, W. K. George and R. Arndt, Eds., New York: Hemisphere, pp. 1–10.

Taylor, G. I., 1915: Eddy motion in the atmosphere. *Philos. Trans. R. Soc. London, Sec. A*, **215**, 1–26.

Taylor, G. I., 1921: Diffusion by continuous movements. *Proc. London Math. Soc., Sec. 2*, **20**, 196–211.

Taylor, G. I., 1935: Statistical theory of turbulence. Parts I–IV. *Proc. Roy. Soc. London Sec. A*, **151**, 421–478.
Taylor, G. I., 1959: The present position in the theory of turbulent diffusion. *Adv. Geophys.*, **6**, 101–112.
Taylor, G. I., 1970: Some early ideas about turbulence. *J. Fluid Mech.*, **41**, 3–11.

5

Conservation equations for covariances

5.1 Introduction and background

We saw in Chapter 1 that direct numerical solutions of the turbulence equations can be done only at relatively small values of the turbulence Reynolds number R_t. Calculations of the vastly larger R_t flows in engineering and geophysical applications use averaged forms of these equations. The averaging produces important *turbulent fluxes* that must be specified before the equations can be solved numerically. That specification allows today's calculations of averaged turbulence fields in applications ranging from flow in the atmospheric boundary layer to the general circulation of the earth's atmosphere and convection on the sun.

In this chapter we derive, interpret, and scale the conservation equations for several covariances, including turbulent fluxes, that arise from ensemble averaging. At very coarse resolution ergodicity (the property that any unbiased average converges to the ensemble average, Chapter 2) blurs the distinction between ensemble and space averaging, so these covariance equations are used also in traditional mesoscale-modeling and weather-forecasting applications.

Monin and Yaglom (1971) credit a 1924 paper by Keller and Friedmann as the first to present a method of deriving turbulence moment equations, of which the turbulent-flux budgets are an example. Because they contain unknown terms, the turbulence moment equations were little used until the late 1960s, when computers were large enough to solve approximate versions of them.

The literature on models of these flux budgets can be bewildering. The rationale for their closure approximations is not always discussed, and the models go under several names – e.g., "second-order closure," "Reynolds-averaged Navier–Stokes (RANS)," "single-point closure," "higher-order closure," and "invariant modeling." Occasionally the model developers' names are used, as with "Mellor–Yamada level 2.5."

5.2 The fluctuation equations

For two turbulent variables $\tilde{a} = A + a$ and $\tilde{b} = B + b$, the sum of ensemble-mean and fluctuating parts, we can derive the equation for their covariance $\overline{ab}$ as follows. Since ensemble averaging and differentiation commute, we can write

$$\frac{\partial \overline{ab}}{\partial t} = \overline{a\frac{\partial b}{\partial t}} + \overline{b\frac{\partial a}{\partial t}}. \tag{5.1}$$

Thus we form the conservation equation for $\overline{ab}$ by multiplying the b equation by a, multiplying the a equation by b, adding, and ensemble averaging.

To derive the conservation equation for a fluctuating variable we:

- Introduce the mean-fluctuating decomposition into the conservation equation for the full variable; we call this the *full* equation.
- Ensemble average the full equation to produce the mean equation.
- Subtract the mean equation from the full equation to find the fluctuating equation.

We'll take a conserved scalar through this process. With $\tilde{u}_i = U_i + u_i$, $\tilde{c} = C + c$, Eq. (1.31) for the full variable is

$$\frac{\partial (C+c)}{\partial t} + (U_j + u_j)\frac{\partial (C+c)}{\partial x_j} = \gamma \frac{\partial^2 (C+c)}{\partial x_j \partial x_j}. \tag{5.2}$$

Ensemble averaging and the averaging rules in Chapter 2 yield the mean equation:

$$\frac{\partial C}{\partial t} + U_j \frac{\partial C}{\partial x_j} + \frac{\partial \overline{u_j c}}{\partial x_j} = \gamma \frac{\partial^2 C}{\partial x_j \partial x_j}. \tag{5.3}$$

Subtracting the mean equation (5.3) from the full equation (5.2) produces the equation for the fluctuation:

$$\frac{\partial c}{\partial t} + U_j \frac{\partial c}{\partial x_j} + u_j \frac{\partial C}{\partial x_j} + \frac{\partial}{\partial x_j}\left(u_j c - \overline{u_j c}\right) = \gamma \frac{\partial^2 c}{\partial x_j \partial x_j}. \tag{5.4}$$

The full variable $\tilde{c}(\mathbf{x}, t)$ could be a constituent density, in which case it must be positive. But its fluctuation has no such restriction: $c = \tilde{c} - C$ is negative when the full variable $\tilde{c}(\mathbf{x}, t)$ is less than its ensemble mean $C(\mathbf{x}, t)$.

The same process yields the equation for the fluctuating velocity field:

$$\frac{\partial u_i}{\partial t} + U_j \frac{\partial u_i}{\partial x_j} + u_j \frac{\partial U_i}{\partial x_j} + \frac{\partial}{\partial x_j}\left(u_i u_j - \overline{u_i u_j}\right) = -\frac{1}{\rho}\frac{\partial p}{\partial x_i} + \nu \frac{\partial^2 u_i}{\partial x_j \partial x_j}. \tag{5.5}$$

The second, third, and fourth terms on the left sides of Eqs. (5.4) and (5.5) represent the three advection processes that produce fluctuations: advection of the fluctuating quantity by the mean velocity, advection of the mean quantity by the fluctuating velocity, and the fluctuating part of advection of the fluctuating quantity by the fluctuating velocity. The other terms in Eqs. (5.4) and (5.5) are the fluctuating counterparts of those in the full equations. Like their parent equations, Eqs. (5.4) and (5.5) are nonlinear since the fourth terms on their left sides involve products of the fluctuating quantities.

5.3 Example: The scalar variance equation

5.3.1 Derivation and interpretation

We'll demonstrate the derivation process with the scalar variance $\overline{c^2}$. Multiplying the c-equation (5.4) by $2c$, ensemble averaging, and using the averaging rules in Chapter 2 yields the following results, term by term:

$$\begin{aligned}
\overline{2c\frac{\partial c}{\partial t}} &= \frac{\partial \overline{c^2}}{\partial t}, \\
\overline{2cU_j\frac{\partial c}{\partial x_j}} &= U_j\frac{\partial \overline{c^2}}{\partial x_j}, \\
\overline{2cu_j\frac{\partial C}{\partial x_j}} &= 2\overline{cu_j}\frac{\partial C}{\partial x_j}, \\
\overline{2c\frac{\partial}{\partial x_j}\left(u_jc - \overline{u_jc}\right)} &= \frac{\partial \overline{c^2u_j}}{\partial x_j}, \\
\overline{2c\gamma\frac{\partial^2 c}{\partial x_j\partial x_j}} &= \gamma\frac{\partial^2 \overline{c^2}}{\partial x_j\partial x_j} - 2\gamma\overline{\frac{\partial c}{\partial x_j}\frac{\partial c}{\partial x_j}}.
\end{aligned} \tag{5.6}$$

The resulting conservation equation for scalar variance, with the meaning of its terms indicated, is

$$\begin{aligned}
\frac{\partial \overline{c^2}}{\partial t} = &- U_j\frac{\partial \overline{c^2}}{\partial x_j} \quad \text{(mean advection)} \\
&- 2\overline{u_jc}\frac{\partial C}{\partial x_j} \quad \text{(mean-gradient production)} \\
&- \frac{\partial \overline{c^2u_j}}{\partial x_j} \quad \text{(turbulent transport)}
\end{aligned}$$

$$+\gamma \frac{\partial^2 \overline{c^2}}{\partial x_j \partial x_j} \quad \text{(molecular diffusion)}$$

$$-2\gamma \overline{\frac{\partial c}{\partial x_j} \frac{\partial c}{\partial x_j}} \quad \text{(molecular destruction).} \tag{5.7}$$

As indicated, the molecular term represents two effects.

The combination of the local time change plus mean advection is the time derivative following the mean motion. Thus it differs from the substantial derivative, also called the total time derivative, which follows the instantaneous motion:

$$\text{time derivative following mean motion} \equiv \frac{\partial}{\partial t} + U_i \frac{\partial}{\partial x_i},$$

$$\text{time derivative following instantaneous motion} \equiv \frac{\partial}{\partial t} + \tilde{u}_i \frac{\partial}{\partial x_i}$$

$$= \frac{D}{Dt} = \frac{\partial}{\partial t} + (U_i + u_i) \frac{\partial}{\partial x_i}. \tag{5.8}$$

We can interpret the *mean-gradient-production* term as follows. A displacement d_j in the presence of a mean gradient $\partial C/\partial x_j$ produces a fluctuation $c = d_j \partial C/\partial x_j$, so $u_j \partial C/\partial x_j$ is the rate of production of c fluctuations. Multiplying by $2c$ and averaging then gives the rate of production of $\overline{c^2}$ as $2\overline{u_j c}\, \partial C/\partial x_j$.

The third term on the right side of Eq. (5.7) is the divergence of $\overline{c^2 u_j}$, the turbulent flux of squared scalar fluctuation. We call this *turbulent transport* of scalar variance. A term of this type appears in any second-moment equation.

The fourth term, molecular diffusion, is also a divergence. The sum of these three divergence terms is

$$\text{mean advection} + \text{turbulent transport} + \text{molecular diffusion}$$

$$= -\frac{\partial}{\partial x_j}\left(U_j \overline{c^2} + \overline{c^2 u_j} - \gamma \frac{\partial \overline{c^2}}{\partial x_j} \right). \tag{5.9}$$

With the divergence theorem we can express the volume integral of this divergence as an integral over the surface that bounds the volume:

$$\int_V \frac{\partial}{\partial x_j}\left(U_j \overline{c^2} + \overline{c^2 u_j} - \gamma \frac{\partial \overline{c^2}}{\partial x_j} \right) dV = \int_A \left(U_{\mathrm{n}} \overline{c^2} + \overline{c^2 u_{\mathrm{n}}} - \gamma \frac{\partial \overline{c^2}}{\partial x_{\mathrm{n}}} \right) dA, \tag{5.10}$$

where n is the outward normal to the surface element dA. If we integrate over the entire flow volume, so the velocity field vanishes on this bounding surface, the mean advection and turbulent transport contributions to the surface integral vanish. We conclude that they move scalar variance from one point in the flow to another.

We can rewrite the molecular diffusion term as

$$\gamma \frac{\partial^2 \overline{c^2}}{\partial x_j \partial x_j} = \frac{\partial}{\partial x_j}\left(\gamma \frac{\partial \overline{c^2}}{\partial x_j}\right) = -\frac{\partial}{\partial x_j}\left(-\gamma \frac{\partial \overline{c^2}}{\partial x_j}\right), \tag{5.11}$$

which is the negative divergence of the molecular flux of $\overline{c^2}$, or molecular diffusion. The second molecular term, being negative definite, represents the rate of molecular destruction of $\overline{c^2}$. We will label it χ_c:

$$\chi_c = 2\gamma \overline{\frac{\partial c}{\partial x_j} \frac{\partial c}{\partial x_j}}. \tag{5.12}$$

The factor 2 is not always used in its definition, which can be a source of confusion. We can rewrite χ_c as

$$\begin{aligned} \chi_c &= 2\,\overline{\left(\gamma \frac{\partial c}{\partial x_j}\right)\left(\frac{\partial c}{\partial x_j}\right)} \\ &= -2\,\overline{(\text{fluctuating molecular flux of } c) \cdot (\text{fluctuating gradient of } c)}, \end{aligned} \tag{5.13}$$

which has the same form as the mean-gradient production term, the scalar (dot) product of flux and gradient.

In steady conditions the integral of Eq. (5.7) over the flow volume therefore reduces to

$$\frac{\partial}{\partial t}\int_V \overline{c^2}\, dV = 0 = -\int_V 2\overline{u_j c}\frac{\partial C}{\partial x_j} dV - \int_V \chi_c\, dV, \tag{5.14}$$

so it follows that

$$-\int_V 2\overline{u_j c}\frac{\partial C}{\partial x_j} dV = \int_V \chi_c\, dV > 0. \tag{5.15}$$

Equation (5.14) says that in steady conditions the flow-integrated rates of production and molecular destruction of $\overline{c^2}$ are in balance. Equation (5.15) is the consequence: in steady conditions the flow-integrated rate of production is positive definite. The eddy-diffusivity closure

$$\overline{cu_j} = -K\frac{\partial C}{\partial x_j}, \quad K \geq 0, \tag{5.16}$$

which can be useful for simple estimates, satisfies this constraint by making the *local* mean-gradient production term positive definite:

$$-2\overline{u_j c}\frac{\partial C}{\partial x_j} = 2K\left(\frac{\partial C}{\partial x_j}\frac{\partial C}{\partial x_j}\right) \geq 0. \tag{5.17}$$

The scalar variance budget, Eq. (5.7), can be very useful in applications. For example, knowing only the mean concentration C of a diffusing hazardous species can be insufficient, for C gives no information on locally and instantaneously large values of $\tilde{c}$. In $\overline{c^2}$ we also have a measure of such fluctuations. One of the most difficult challenges in modeling Eq. (5.7) is formulating closures that maintain the positive-definite nature of $\overline{c^2}$.

5.3.2 Scaling guidelines

The following are guidelines for scaling terms in the evolution equations for covariances of energy-containing-range variables. We'll discuss the scaling of terms in covariance equations for dissipative-range quantities in Chapter 14.

1. Velocity fluctuations scale with $u = (\overline{u_i u_i})^{1/2}$; conserved scalar fluctuations scale with $s = (\overline{c^2})^{1/2}$; pressure fluctuations scale with $p = \rho u^2$.
2. The correlation coefficients of fluctuating velocity components and conserved scalar fluctuations are $O(1)$.[†]
3. Spatial variations in mean quantities scale with ℓ.
4. Mean scalar gradients and mean velocity gradients scale with s/ℓ and u/ℓ, respectively.
5. Constants can be ignored.
6. Mean-advection and local-time-change terms can involve externally imposed scales L and τ_e not directly related to the turbulence scales ℓ and u.
7. Avoid attempting to scale a "mixed-scale" covariance, one involving a small-scale (dissipative-range) property and a large-scale (variance-containing range) property, for its correlation coefficient is not $O(1)$.

An example[‡] of a "mixed-scale" covariance is $\partial\overline{c^2}/\partial x = 2\overline{c\partial c/\partial x}$. The left side is the spatial derivative of a mean quantity, so its order is s^2/ℓ. The right side is a mixed-scale covariance; $c \sim s$ is a large-scale property, while we shall see shortly that $\partial c/\partial x$ is a small-scale property. In this case we can rewrite it to determine its order (Problem 5.17), but it is not always this simple.

The scales u and s in Guideline 1 are the simplest, most direct measures of the fluctuation level of velocity and a conserved scalar in turbulent flow. Similarly, 3 is the simplest scaling of spatial variations in that it associates the same length scale with the energy-containing eddies and with variations in mean quantities. Guideline 4 incorporates the mixing-length notion that fluctuations are due to eddy motions in the presence of mean gradients.

† $O(1)$ is a mathematical term that in turbulence we can interpret as meaning "approaches a constant as $R_t \to \infty$." In general $O(1)$ implies nothing about the magnitude of this constant. Schwartz's inequality (Part III) limits correlation coefficients to 1 in magnitude.

‡ This is adapted from Tennekes and Lumley (1972).

Equation (5.15) indicates that the mean-gradient production and molecular destruction terms in Eq. (5.7) are of the same order, which by the scaling guidelines is

$$\overline{u_j c}\frac{\partial C}{\partial x_j} \sim us\frac{s}{\ell} = \frac{s^2 u}{\ell}. \tag{5.18}$$

The turbulent transport term is also of this order.

The molecular diffusion term scales as

$$\gamma\frac{\partial^2 \overline{c^2}}{\partial x_j \partial x_j} \sim \gamma\frac{s^2}{\ell^2} = \frac{s^2 u}{\ell}\left(\frac{\gamma}{u\ell}\right) \sim \frac{s^2 u}{\ell}\left(\frac{\gamma}{\nu}\right) R_{\mathrm{t}}^{-1} \ll \frac{s^2 u}{\ell}, \tag{5.19}$$

so it is negligible.

With these scaling results the scalar-variance conservation equation reduces to (retaining the time-change and mean-advection terms for now)

$$\frac{\partial \overline{c^2}}{\partial t} = -U_j\frac{\partial \overline{c^2}}{\partial x_j} - 2\overline{u_j c}\frac{\partial C}{\partial x_j} - \frac{\partial \overline{c^2 u_j}}{\partial x_j} - \chi_c. \tag{5.20}$$

Its terms (except for time change and mean advection) are of order $s^2 u/\ell$.

We have inferred through Eq. (5.15) that the molecular destruction term χ_c in the scalar variance budget is of leading order, $s^2 u/\ell$. Thus we can write

$$\chi_c = 2\gamma\overline{\frac{\partial c}{\partial x_j}\frac{\partial c}{\partial x_j}} \sim \frac{s^2 u}{\ell} = \frac{s^2}{\ell/u}, \tag{5.21}$$

which indicates that the time scale of the removal of $\overline{c^2}$ by molecular destruction is ℓ/u, the large-eddy turnover time. This says that if their production mechanism were suddenly shut off the scalar fluctuations would disappear within a time of the order of a large-eddy turnover time. It is difficult to imagine how this could happen more quickly. This reflects the strongly dissipative nature of turbulence.

5.3.3 Quasi-steadiness, local homogeneity

As Guideline 6 indicates, the mean-advection and time-change terms in a turbulence budget such as Eq. (5.20) can involve additional, externally imposed scales. For example, conditions in the atmospheric boundary layer change in the streamwise direction due to variations in the surface conditions (variable temperature or roughness, for example), and change in time due to evolving synoptic conditions and the diurnal cycle. Thus, for mean advection we introduce a scale U of the mean velocity and a scale L_x of streamwise variations in $\overline{c^2}$,

$$U_j\frac{\partial \overline{c^2}}{\partial x_j} \sim U\frac{s^2}{L_x}, \tag{5.22}$$

that need not be directly related to u and ℓ. Analogously, for the time-change term we allow a scale τ_e of time changes in $\overline{c^2}$ that can differ from ℓ/u:

$$\frac{\partial \overline{c^2}}{\partial t} \sim \frac{s^2}{\tau_e}. \tag{5.23}$$

The order of magnitude of the mean-advection term in Eq. (5.7) is then

$$\text{mean advection} \sim U\frac{s^2}{L_x} = \left(\frac{U\ell}{uL_x}\right)\frac{s^2 u}{\ell}. \tag{5.24}$$

U is typically of the order of but larger than u. If so, and if $L_x \gg \ell$, then the parameter $(U\ell/uL_x) \ll 1$ and mean advection is negligible, as in a homogeneous flow. We call this *local homogeneity.* Similarly, the time-change term scales as

$$\text{time change} \sim \frac{s^2}{\tau_e} = \left(\frac{\ell/u}{\tau_e}\right)\frac{s^2 u}{\ell}. \tag{5.25}$$

If $(\ell/u)/\tau_e \ll 1$, meaning that the large-eddy turnover time ℓ/u is much less than the time scale τ_e of the changing boundary conditions, then the time-change term is negligible. It is as if the mean flow is steady; we call it *quasi-steady.*

As we shall show in Part II, in homogeneous terrain and away from the morning and evening transitions in the surface heat flux this can allow a quasi-steady, locally homogeneous interpretation of the second-moment budgets. For Eq. (5.20) this is

$$\frac{\partial \overline{c^2}}{\partial t} \simeq 0 = -2\overline{wc}\frac{\partial C}{\partial z} - \frac{\partial \overline{c^2 w}}{\partial z} - \chi_c. \tag{5.26}$$

5.3.4 Interpreting the molecular destruction term

We can scale the derivative covariance in χ_c, Eq. (5.12), through the power spectral density that we introduced in Chapter 2. There we represented a real, statistically homogeneous, zero-mean, one-dimensional random scalar function $f(x;\alpha)$ on an interval $0 \le x \le L$ as a complex Fourier series. We'll express this as

$$f(x;\alpha) = \sum_{n=-N}^{N} \hat{f}(\kappa_n;\alpha)e^{i\kappa_n x}. \tag{5.27}$$

We also showed that

$$\overline{f^2} = \overline{ff^*} = \sum_{n=-N}^{N} \overline{\hat{f}(\kappa_n)\hat{f}^*(\kappa_n)} = \sum_{n=-N}^{N} \frac{\overline{\hat{f}(\kappa_n)\hat{f}^*(\kappa_n)}}{\Delta\kappa}\Delta\kappa$$
$$= \sum_{n=-N}^{N} \phi(\kappa_n)\Delta\kappa, \tag{5.28}$$

with ϕ the power spectral density of f. From Eq. (5.27) the derivative of f is

$$\frac{df}{dx} = \sum_{n=-N}^{N} i\kappa_n \hat{f}(\kappa_n; \alpha)e^{i\kappa_n x}, \tag{5.29}$$

so that the derivative variance is

$$\overline{\left(\frac{df}{dx}\right)^2} = \sum_{n=-N}^{N} \kappa_n^2\, \overline{\hat{f}(\kappa_n)\hat{f}^*(\kappa_n)} = \sum_{n=-N}^{N} \kappa_n^2\, \phi(\kappa_n)\Delta\kappa. \tag{5.30}$$

In the limit of large N and L this becomes

$$\overline{\left(\frac{df}{dx}\right)^2} = \int_{-\infty}^{\infty} \kappa^2\, \phi(\kappa)\, d\kappa. \tag{5.31}$$

In the three-dimensional case we showed that the variance of a conserved scalar fluctuation $c(\mathbf{x}, \alpha)$ is given by

$$\overline{c^2} = \int_0^{\infty} E_c(\kappa)\, d\kappa, \tag{2.58}$$

where E_c is the three-dimensional spectrum for the scalar,

$$E_c(\kappa) = \int\int_{\kappa_i\kappa_i=\kappa^2} \phi(\kappa_1, \kappa_2, \kappa_3)\, d\sigma. \tag{2.57}$$

The corresponding expression for the derivative variance is

$$\overline{\frac{\partial c}{\partial x_i}\frac{\partial c}{\partial x_i}} = \int_0^{\infty} \kappa^2\, E_c(\kappa)\, d\kappa. \tag{5.32}$$

In the inertial range of wavenumbers ($1/\ell \ll \kappa \ll 1/\eta$) it is found that $E_c(\kappa) \sim \kappa^{-5/3}$ (Chapter 7). Thus, Eq. (2.58) confirms that the principal contributions to the scalar variance come from smaller wavenumbers (larger scales), those in the energy-containing range. By contrast, Eq. (5.32) shows that the density of contributions to the scalar derivative variance grows as $\kappa^{1/3}$ in the inertial range, meaning that the contributions to the scalar derivative variance come mainly from wavenumbers in the dissipative range.

We'll denote the intensity scale of the dissipative-range c-fluctuations as s_{d} and their length scale as η, the Kolmogorov microscale. (We are assuming here that $\gamma \simeq \nu$. We discuss the general case when $\gamma \not\simeq \nu$ in Chapter 7.) Then χ_c scales as

$$\chi_c = 2\gamma \overline{\frac{\partial c}{\partial x_j}\frac{\partial c}{\partial x_j}} \sim \gamma \frac{s_{\mathrm{d}}^2}{\eta^2}. \tag{5.33}$$

Solving for s_{d} yields, using $\eta = (\nu^3/\epsilon)^{1/4}$,

$$s_{\mathrm{d}} \sim \frac{\chi_c^{1/2}\gamma^{1/4}}{\epsilon^{1/4}}. \tag{5.34}$$

We saw earlier that $\chi_c \sim s^2 u/\ell$. With $\epsilon \sim u^3/\ell$ this gives from (5.34)

$$\frac{s}{s_{\mathrm{d}}} \sim \left(\frac{u\ell}{\nu}\right)^{1/4} \sim R_{\mathrm{t}}^{1/4}. \tag{5.35}$$

Since R_{t} is large, this confirms that $s_{\mathrm{d}} \ll s$, meaning that the dissipative eddies of the scalar field are of low intensity. This is the same relation that connects the velocity scales of the dissipative and energy-containing eddies, Eq. (1.35).

If we write Eq. (5.33) for χ_c as

$$\chi_c \sim \frac{s_{\mathrm{d}}^2}{\eta^2/\gamma} = \frac{s_{\mathrm{d}}^2}{\tau_{\mathrm{r}}}, \tag{5.36}$$

then since the η-sized c fluctuations have amplitude s_{d}, we see that $\tau_{\mathrm{r}} \sim \eta^2/\gamma$ is the time scale for their removal by molecular diffusion. The time required for molecular diffusion through distance η should depend only on η and γ, which indeed yield a time of order η^2/γ. We conclude that χ_c does represent the diffusive removal of η-sized c fluctuations.

5.3.5 Simple limiting cases

As we shall discuss in Part II, near the surface in a quasi-steady, locally homogeneous turbulent boundary layer the turbulent transport term in the variance budget

(5.26) of a conserved scalar diffusing to or from the surface is observed to be negligible. In this case the budget reduces to a simple balance between mean-gradient production and molecular destruction,

$$2\overline{wc}\frac{\partial C}{\partial z} = -\chi_c = -2\gamma\overline{\frac{\partial c}{\partial x_j}\frac{\partial c}{\partial x_j}} \leq 0. \tag{5.37}$$

Thus here the global constraint of Eq. (5.15) becomes a local one that requires $\overline{wc}$ and $\partial C/\partial z$ to be of opposite sign.

In the steady flow downstream of a heated grid in a wind tunnel, the turbulent transport of temperature variance is also negligible and the budget reduces to a balance between streamwise mean advection and molecular destruction:

$$U\frac{\partial \overline{c^2}}{\partial x} = -\chi_c. \tag{5.38}$$

This provides a way of inferring χ_c from the rate of change of $\overline{c^2}$ with downstream distance, which can be simpler and more reliable than measuring it directly from its definition (5.12) (Part III).

5.4 The scalar flux and Reynolds stress budgets

5.4.1 The $\overline{cu_i}$ budget

The same derivation procedure, plus scaling arguments for the molecular terms (Appendix), yields the turbulent scalar flux budget:

$$\begin{aligned}
\frac{\partial \overline{cu_i}}{\partial t} = &- U_j\frac{\partial \overline{cu_i}}{\partial x_j} \quad \text{(mean advection)} \\
&- \overline{u_j u_i}\frac{\partial C}{\partial x_j} \quad \text{(mean-gradient production)} \\
&- \overline{cu_j}\frac{\partial U_i}{\partial x_j} \quad \text{(tilting production)} \\
&- \frac{\partial \overline{cu_i u_j}}{\partial x_j} \quad \text{(turbulent transport)} \\
&- \frac{1}{\rho}\left(\overline{c\frac{\partial p}{\partial x_i}}\right) \quad \text{(pressure-gradient interaction)} \\
&- (\gamma + \nu)\overline{\frac{\partial u_i}{\partial x_j}\frac{\partial c}{\partial x_j}} \quad \text{(molecular destruction).}
\end{aligned} \tag{5.39}$$

Since a variance is positive, the sign of a production term in a variance budget such as Eq. (5.7) is clear: when placed on the right side of the equation it is positive. But this does not hold for fluxes, which can be of either sign.

The *mean-gradient production* term is a contraction of kinematic Reynolds stress and the mean scalar gradient. The turbulent scalar flux so produced need not be aligned with the gradient of C producing it. In the vertical ($i = 3$) flux budget in the horizontally homogeneous surface layer, for example, this term is $-\overline{u_3 u_3}\partial C/\partial x_3$, and the flux and mean gradient are aligned. But in the horizontal ($i = 1$) budget the term is $-\overline{u_1 u_3}\partial C/\partial x_3$, which is the rate of production of *horizontal* scalar flux by interaction of the turbulence with the *vertical* gradient of C.

The *tilting production* term changes the magnitude and direction of the scalar flux through tilting by the mean velocity gradient. It is analogous to the stretching and tilting term in Eq. (1.28) for vorticity. Each of these two production terms is of order su^2/ℓ.

The pressure-covariance term in Eq. (5.39) has long been considered difficult, if not impossible, to measure reliably, and it was neglected in some early studies. The other terms in the budget have been measured in field programs in the atmospheric surface layer, allowing this pressure covariance to be inferred from their imbalance; it is now known to be important (Subsection 5.5.3). It has also been measured directly (Wilczak and Bedard, 2004).

As we explain in the Appendix, the molecular term in the scalar flux budget (5.39) is conventionally neglected on the grounds of local isotropy so it reduces to

$$\frac{\partial \overline{cu_i}}{\partial t} = -U_j \frac{\partial \overline{cu_i}}{\partial x_j} - \overline{u_j u_i}\frac{\partial C}{\partial x_j} - \overline{cu_j}\frac{\partial U_i}{\partial x_j} - \frac{\partial \overline{cu_i u_j}}{\partial x_j} - \frac{1}{\rho}\left(\overline{c\frac{\partial p}{\partial x_i}}\right). \tag{5.40}$$

In steady state this expresses a balance among, in order, mean advection; mean production through the interaction of Reynolds stress and the mean scalar gradient, and the interaction of scalar flux and mean velocity gradient; turbulent transport; and destruction through pressure effects. Its leading terms are of order su^2/ℓ.

5.4.2 The $\overline{u_i u_k}$ budget

Using the same scaling and local-isotropy arguments to simplify the molecular terms (Appendix), the budget of $\overline{u_i u_k}$ reduces to

$$\begin{aligned}\frac{\partial \overline{u_i u_k}}{\partial t} = &-U_j \frac{\partial \overline{u_i u_k}}{\partial x_j} \quad \text{(mean advection)}\\ &-\overline{u_j u_k}\frac{\partial U_i}{\partial x_j} - \overline{u_j u_i}\frac{\partial U_k}{\partial x_j} \quad \text{(mean-gradient production)}\end{aligned}$$

$$
\begin{aligned}
& - \frac{\partial \overline{u_i u_k u_j}}{\partial x_j} \quad \text{(turbulent transport)} \\
& - \frac{1}{\rho}\left(\overline{u_k \frac{\partial p}{\partial x_i}} + \overline{u_i \frac{\partial p}{\partial x_k}} \right) \quad \text{(pressure-gradient interaction)} \\
& - \frac{2\epsilon}{3}\delta_{ik} \quad \text{(viscous dissipation).}
\end{aligned} \tag{5.41}
$$

The interpretation of the terms here is analogous to that for the scalar flux budget. The leading terms are of order u^3/ℓ.

There was renewed interest in these equations in the late 1960s and early 1970s because of their potential as turbulence models (Daly and Harlow, 1970; Donaldson, 1971). The form of the mean-gradient and tilting production terms in the scalar flux equation (5.40), for example, suggests that the flux-gradient relation for scalars can be more complicated than the usual eddy-diffusivity expression.

5.5 Applications

The conservation equations for turbulent fluxes were first studied observationally in a comprehensive way in the 1968 Kansas experiment (Haugen *et al.*, 1971). The full suite of instrumentation and the quasi-steady, locally homogeneous conditions enabled analyses of the budgets of turbulence kinetic energy (TKE), Reynolds shear stress, and temperature flux in the surface layer. We'll focus here on their behavior when buoyancy effects are negligible; we cover the general case in Part II.

5.5.1 The TKE budget

Contracting i on k in Eq. (5.41) and dividing by 2 yields the equation for the evolution of the mean kinetic energy per unit mass of the turbulence, more commonly called the TKE budget:

$$
\begin{aligned}
\frac{1}{2}\frac{\partial}{\partial t}\overline{u_i u_i} = & - \frac{U_j}{2}\frac{\partial}{\partial x_j}\overline{u_i u_i} - \overline{u_i u_j}\frac{\partial U_i}{\partial x_j} - \frac{1}{2}\frac{\partial}{\partial x_j}\overline{u_i u_i u_j} \\
& - \frac{1}{\rho}\frac{\partial}{\partial x_i}\overline{p u_i} - \nu \overline{\frac{\partial u_i}{\partial x_j}\frac{\partial u_i}{\partial x_j}}.
\end{aligned} \tag{5.42}
$$

The viscous-dissipation term is typically labeled ϵ:

$$
\epsilon \equiv \nu \overline{\frac{\partial u_i}{\partial x_j}\frac{\partial u_i}{\partial x_j}}. \tag{5.43}
$$

The mean shear can be rewritten

$$\frac{\partial U_i}{\partial x_j} = \frac{1}{2}\left(\frac{\partial U_i}{\partial x_j} + \frac{\partial U_j}{\partial x_i}\right) + \frac{1}{2}\left(\frac{\partial U_i}{\partial x_j} - \frac{\partial U_j}{\partial x_i}\right) = S_{ij} + R_{ij}, \tag{5.44}$$

the sum of mean strain-rate and mean rotation-rate tensors. Since $\overline{u_i u_j}$ is symmetric, its contraction with the antisymmetric tensor R_{ij} vanishes (Problem 1.15) and the shear-production term becomes

$$\overline{u_i u_j}\frac{\partial U_i}{\partial x_j} = \overline{u_i u_j}\left(S_{ij} + R_{ij}\right) = \overline{u_i u_j}\, S_{ij}, \tag{5.45}$$

the contraction of kinematic stress and strain-rate tensors.

The viscous dissipation term can be interpreted similarly. By definition it is the average rate, per unit mass, of working against viscous stresses. Thus, it is the mean value of the contraction of the instantaneous viscous stress tensor σ_{ij} (Chapter 1) and the instantaneous strain-rate tensor s_{ij}, divided by density. Using the notation $\partial u_i/\partial x_j \equiv u_{i,j}$, this is

$$\begin{aligned}\epsilon \equiv \frac{\overline{\sigma_{ij} s_{ij}}}{\rho} &= \frac{\nu}{2}\overline{\left(u_{i,j} + u_{j,i}\right)\left(u_{i,j} + u_{j,i}\right)} = \nu\left(\overline{u_{i,j}u_{i,j}} + \overline{u_{i,j}u_{j,i}}\right) \\ &= \nu\left(\overline{u_{i,j}u_{i,j}} + \left(\overline{u_i u_j}\right)_{,ji}\right) \simeq \nu\, \overline{u_{i,j}u_{i,j}},\end{aligned} \tag{5.46}$$

the term involving the second derivative of $\overline{u_i u_j}$ being negligible by our scaling guidelines (Problem 5.13).

In a steady flow the right side of the TKE equation (5.42) sums to zero. The mean advection, turbulent transport, and pressure transport terms are divergences; if we integrate (5.42) over the entire flow volume, on the bounding surface of which the velocity vanishes, these divergence terms integrate to zero. That means these divergence terms only move TKE around in space. Then we have, using Eq. (5.45),

$$\frac{\partial}{\partial t}\int_{\text{volume}} \frac{\overline{u_i u_i}}{2} dV = 0 = -\int_{\text{volume}} \overline{u_i u_j}\, S_{ij}\, dV - \int_{\text{volume}} \epsilon\, dV. \tag{5.47}$$

The first term on the right side of (5.47) represents the volume-integrated mean rate of production of TKE through the interaction of the Reynolds stress and the mean strain rate. The second term is the mean rate at which the volume-integrated TKE is dissipated – irreversibly converted into internal energy through viscous forces.

Since ϵ is positive definite, we can write Eq. (5.47) as

$$-\int_{\text{volume}} \overline{u_i u_j}\, S_{ij}\, dV = \int_{\text{volume}} \epsilon\, dV > 0. \tag{5.48}$$

This is the analog of Eq. (5.15) for scalar variance. It says that a positive eddy diffusivity K, defined such that $\overline{u_i u_j} - \delta_{ij}\overline{u_k u_k}/3 = -K S_{ij}$, satsifies this constraint *locally* by making the integrand positive definite (Problem 5.21):

$$-\int_{\text{volume}} \overline{u_i u_j}\, S_{ij}\, dV = \int_{\text{volume}} K S_{ij} S_{ij}\, dV > 0. \tag{5.49}$$

The major contributions to the rate of dissipation of TKE come from motions at the smallest spatial scales. In Chapter 1 we identified these dissipative velocity and length scales as the Kolmogorov scales υ and η:

$$\upsilon = (\nu\epsilon)^{1/4}, \quad \eta = \left(\nu^3/\epsilon\right)^{1/4}, \quad \epsilon = \nu\, \overline{u_{i,j} u_{i,j}} \sim \nu \frac{\upsilon^2}{\eta^2}. \tag{5.50}$$

In Chapter 1 we interpreted the dependence of υ and η on ϵ as meaning that a turbulent flow adjusts the length and velocity scales of its smallest, viscous eddies in order to dissipate TKE at the correct mean rate. We shall see in Chapter 6 that this rate is set by the energy-containing turbulence, which establishes an "energy cascade" to smaller eddies that terminates in viscous dissipation.

5.5.2 The mean-flow kinetic energy equation

Multiplying the mean-momentum equation (3.7) by U_i and rearranging the terms yields the equation for the mean-flow kinetic energy per unit mass, or the MKE equation:

$$\begin{aligned}\frac{\partial}{\partial t}\frac{U_i U_i}{2} &= -\frac{\partial}{\partial x_j}\left(\frac{U_i U_i U_j}{2} + \overline{U_i u_i u_j} - \nu \frac{\partial}{\partial x_j}\frac{U_i U_i}{2}\right) - \frac{U_i}{\rho}\frac{\partial P}{\partial x_i} \\ &\quad - \nu \frac{\partial U_i}{\partial x_j}\frac{\partial U_i}{\partial x_j} + \overline{u_i u_j}\, S_{ij}.\end{aligned} \tag{5.51}$$

The first term on the right is a divergence of (in order) the flux of MKE by the mean flow, the turbulent flux of joint kinetic energy (including U_i under the overbar facilitates that interpretation), and the molecular flux of MKE. The next term, the rate of production of MKE by mean flow down the mean pressure gradient, can also be written as a divergence of PU_i. Next is the rate of viscous dissipation of MKE, and the final term is the negative of the rate of production of TKE. The viscous dissipation term is of order $\nu u^2/\ell^2$; the last is of order u^3/ℓ, a factor $u\ell/\nu = R_t$ larger. Thus, we can neglect the viscous dissipation term in Eq. (5.51) except very near solid boundaries.

If we now integrate Eq. (5.51) over a volume the divergence terms on the right become integrals over the bounding surface. In a homogeneous flow the first surface integral vanishes and we have

$$\frac{\partial}{\partial t}\int_{\text{volume}}\frac{U_iU_i}{2}dV = -\frac{1}{\rho}\int_{\text{surface}} PU_n\,d\sigma + \int_{\text{volume}}\overline{u_iu_j}\,S_{ij}\,dV. \qquad (5.52)$$

Equation (5.52) says that the time rate of change of volume-integrated MKE is due to an imbalance between its rate of gain through mean pressure differences and its rate of loss by working against turbulent stress in producing TKE (Problem 5.22). The latter is ultimately balanced by the rate of gain of internal energy of the fluid by viscous dissipation of TKE in the smallest eddies.

5.5.3 Insights into pressure covariances

5.5.3.1 Role in budgets of stress and scalar flux

In a steady, horizontally homogeneous boundary-layer flow the budgets (5.41) of $\overline{uw}$ and (5.40) of $\overline{cw}$ reduce to balances of mean-gradient production, turbulent transport, and pressure-gradient interaction:

$$\begin{aligned}\frac{\partial\overline{uw}}{\partial t} &= 0 = -\overline{w^2}\frac{\partial U}{\partial z} - \frac{\partial\overline{uw^2}}{\partial z} - \frac{1}{\rho}\left(\overline{u\frac{\partial p}{\partial z}} + \overline{w\frac{\partial p}{\partial x}}\right),\\ \frac{\partial\overline{cw}}{\partial t} &= 0 = -\overline{w^2}\frac{\partial C}{\partial z} - \frac{\partial\overline{cw^2}}{\partial z} - \frac{1}{\rho}\left(\overline{c\frac{\partial p}{\partial z}}\right).\end{aligned} \qquad (5.53)$$

Turbulent transport integrates to zero over the whole flow, so the steady, global balance is between mean-gradient production and pressure destruction. This was first confirmed observationally in the 1968 Kansas experiments with temperature as the scalar; the pressure covariances were inferred from the imbalance of the measured terms (Wyngaard *et al.*, 1971). It has since been verified by direct measurements of the pressure covariance (Wilczak and Bedard, 2004). Physically, this pressure covariance can be interpreted as a rate of production of opposite-signed flux (Question 5.12).

5.5.3.2 TKE budget

Equation (5.41) contains the component TKE equations for a horizontally homogeneous, quasi-steady boundary-layer flow. They are particularly revealing if we rewrite the pressure covariance as

$$\overline{u_\alpha\frac{\partial p}{\partial x_\alpha}} = \frac{\partial}{\partial x_\alpha}\overline{u_\alpha p} - \overline{p\frac{\partial u_\alpha}{\partial x_\alpha}}, \quad \alpha = 1, 2, 3, \quad \text{no sum on } \alpha, \qquad (5.54)$$

using horizontal homogeneity to eliminate the first term on the right for $\alpha = 1$ and 2. Then taking the x-axis along the mean-flow direction and assuming local isotropy for the dissipative term (Part III), the component TKE equations are

$$\frac{1}{2}\frac{\partial \overline{u^2}}{\partial t} = 0 = -\overline{uw}\frac{\partial U}{\partial z} - \frac{1}{2}\frac{\partial \overline{wu^2}}{\partial z} + \frac{1}{\rho}\overline{p\frac{\partial u}{\partial x}} - \frac{\epsilon}{3}, \tag{5.55}$$

$$\frac{1}{2}\frac{\partial \overline{v^2}}{\partial t} = 0 = -\frac{1}{2}\frac{\partial \overline{wv^2}}{\partial z} + \frac{1}{\rho}\overline{p\frac{\partial v}{\partial y}} - \frac{\epsilon}{3}, \tag{5.56}$$

$$\frac{1}{2}\frac{\partial \overline{w^2}}{\partial t} = 0 = -\frac{1}{2}\frac{\partial \overline{w^3}}{\partial z} - \frac{1}{\rho}\frac{\partial}{\partial z}\overline{pw} + \frac{1}{\rho}\overline{p\frac{\partial w}{\partial z}} - \frac{\epsilon}{3}. \tag{5.57}$$

The turbulent and pressure transport terms in Eqs. (5.55)–(5.57) are observed not to be large near the surface in the *neutral* case – i.e., in the absence of buoyancy effects. Then the steady balances of the TKE components near the surface reduce to

$$\frac{1}{2}\frac{\partial \overline{u^2}}{\partial t} = 0 = -\overline{uw}\frac{\partial U}{\partial z} + \frac{1}{\rho}\overline{p\frac{\partial u}{\partial x}} - \frac{\epsilon}{3}, \tag{5.58}$$

$$\frac{1}{2}\frac{\partial \overline{v^2}}{\partial t} = 0 = \frac{1}{\rho}\overline{p\frac{\partial v}{\partial y}} - \frac{\epsilon}{3}, \tag{5.59}$$

$$\frac{1}{2}\frac{\partial \overline{w^2}}{\partial t} = 0 = \frac{1}{\rho}\overline{p\frac{\partial w}{\partial z}} - \frac{\epsilon}{3}. \tag{5.60}$$

Incompressibility implies that the pressure covariances in Eqs. (5.58)–(5.60) sum to zero,

$$\overline{p\frac{\partial u}{\partial x}} + \overline{p\frac{\partial v}{\partial y}} + \overline{p\frac{\partial w}{\partial z}} = \overline{p\frac{\partial u_i}{\partial x_i}} = 0, \tag{5.61}$$

so we conclude that they represent *intercomponent TKE transfer*. Thus, we interpret Eqs. (5.58)–(5.60) as follows: the sole TKE source is the rate of shear production of $\overline{u^2}/2$; some of that production rate is balanced by the rate of dissipation in the $\overline{u^2}/2$ equation and the remainder is transferred by the pressure covariances at equal rates to the other two TKE components.

5.5.3.3 The role of turbulent pressure

Isotropic turbulence statistics are independent of translation, rotation, and reflection of the coordinate axes (Part III). Thus scalar fluxes vanish under isotropy, for a nonzero flux would imply a preferred direction and, hence, anisotropy. In the flux budget (5.40) the pressure term, as the principal loss mechanism for scalar flux, acts to drive it toward isotropy. Similarly, in an isotropic turbulence field the component velocity variances are equal. In Eqs. (5.59) and (5.60) pressure transfer is the principal source of $\overline{w^2}$ and $\overline{v^2}$, respectively, acting to drive them toward equality with $\overline{u^2}$.

In constant-density flows the rms value σ_p of pressure fluctuations is of the order of ρu^2. In the atmospheric boundary layer with $\rho \simeq 1$ kg m^{-3} and $u = 1$ m s^{-1} this gives $\sigma_p \simeq 1$ newton m^{-2}, or 10^{-5} atmospheres (10 microbars). These fluctuations are far smaller than the hydrostatic pressure changes that drive atmospheric motions.

Bradshaw (1994) has described turbulent pressure fluctuations as one of the "Great Unmeasurables," because in disturbing the flow near it a pressure probe generates spurious pressure fluctuations that can be as large as those being measured. It does appear, however, that in lower-atmospheric applications turbulent pressure fluctuations can now be reliably measured with a special "quad disc pressure port" (Nishiyama and Bedard, 1991; Wyngaard *et al.*, 1994). We'll discuss this more in Part II.

5.6 From the covariance equations to turbulence models

5.6.1 Background

The covariance equations (5.7), (5.39), and (5.41) describe exactly the spatial and temporal evolution of turbulent fluxes and other second moments. However, they cannot be directly solved because their turbulent-transport, pressure-covariance, and molecular-destruction terms are unknown. By the same technique one could derive equations for these unknowns, but those equations would involve yet further unknowns. This is a manifestation of the *closure problem* in turbulence.

Nonetheless these covariance equations provide important insights into turbulent fluxes. For example, the presence of the mean-gradient production and tilting-production terms in the scalar flux-conservation equation (5.39) suggests that the turbulent scalar flux need not be aligned with the mean scalar gradient, contrary to a common assumption.

Experience with these covariance equations has also shown (Launder, 1996) that the sign of a covariance is usually the sign of its principal production term. Thus if we write the conservation equation for a covariance C, say, in its lowest-order form

$$\frac{\partial C}{\partial t} \simeq \text{principal production term} - \text{principal destruction term}, \qquad (5.62)$$

and approximate the principal destruction term as $-C/T$, with T a time scale, we have

$$\frac{\partial C}{\partial t} \simeq P - \frac{C}{T}, \qquad (5.63)$$

whose quasi-steady solution is $C \simeq PT$. Using the analogy with monetary budgets, Launder (1996) calls this the WET (*Wealth* = *Earnings times Time*) model. It has useful diagnostic value (Problem 5.4).

Beginning in the late 1960s, when growth in the size and speed of computers made it feasible to solve such sets of partial differential equations numerically, the turbulence community has used approximate versions of these covariance equations as "second-moment" turbulence models. We now have nearly 40 years' experience with such models in both engineering and geophysical flows.

5.6.2 History, status, and outlook

The use of second-moment turbulence models grew rapidly in the early 1970s. In this period the modeling technique now called *large-eddy simulation*, or LES – which as we'll discuss in Chapter 6 is the time-dependent numerical calculation of turbulent flow on a three-dimensional grid fine enough to resolve the energy-containing eddies – also appeared (Deardorff, 1970). LES attracted great interest, but because it is vastly more demanding of computer resources it was not initially competitive with second-moment modeling.

In a review paper on turbulence Liepmann (1979) criticized much of this early second-moment modeling:

> Problems of technological importance are always approached by approximate methods, and a large body of turbulence modeling has been established under prodding from industrial users. The Reynolds-averaged equations are almost always applied in such work, and the hierarchy of equations is closed by semi-empirical arguments which range from very simple guesses ... to much more sophisticated hierarchies....
>
> I am convinced that much of this huge effort will be of passing interest only. Except for rare critical appraisals such as the 1968 Stanford contest for computation of turbulent boundary layers, much of this work is never subjected to any kind of critical or comparative judgment. The only encouraging prospect is that current progress in understanding turbulence will... guide these efforts to a more reliable discipline.

There were some early, in-depth assessments of the performance of second-moment turbulence models, mainly in engineering but in geophysical applications as well. The need was more pressing in engineering, and the requisite data were much more accessible there. In time the salient features of these models became evident. One is their lack of *universality* – their tendency to unreliability in flows different from those used to develop them. In acknowledging this attribute Lumley (1983) cautioned that one should not expect too much from these models, which he termed "calibrated surrogates for turbulence." He felt they should "work satisfactorily in situations not too far removed geometrically, or in parameter values, from the benchmark situations used to calibrate them." He went on to write:

Many of the initial successes of the models...have been in flows...where details of the models are irrelevant. Thus emboldened, the modelers have been over enthusiastic in promoting their models...often without considering in depth the difficult questions that arise. Consequently, there is some disillusionment with the models...This reaction is probably justified, but it would be a shame if it resulted in a cessation of efforts to put a little more physics and mathematics into the models.

By the mid-1980s LES was being widely used in research applications in both geophysical and engineering flows. It showed a universality lacking in second-moment turbulence models. A decade later, Bradshaw (1994) wrote:

...even if one makes generous estimates of required engineering accuracy and requires predictions only of the Reynolds stresses, the likelihood is that a simplified model of turbulence will be significantly less accurate, or significantly less widely applicable, than the Navier–Stokes equations themselves – i.e., it will not be "universal"....

Irrespective of the use to which a Reynolds-stress model will be put, lack of universality may interfere with its calibration. For example, it is customary to fix one of the coefficients...so that the model reproduces the decay of grid turbulence accurately. This involves the assumption that the model is valid for grid turbulence as well as in the flows for which it is intended – presumably shear layers, which have a very different structure from grid turbulence....It is becoming more and more probable that really reliable turbulence models are likely to be so long in development that large eddy simulations (from which, of course, all required statistics can be derived) will arrive at their maturity first.

In a later meeting Bradshaw (1999) summarized:

Perhaps the most important defect of current engineering turbulence models...is that their *non-universality* (the boundaries of their range of acceptable engineering accuracy) cannot be estimated at all usefully a priori....very few codes output warning messages when the model is leaving its region of proven reliability.

The 20 years spanned by these comments saw Liepmann's pessimism about turbulence modeling, then Lumley's appeal for broader understanding of its nature and more patience with the model-development process, next Bradshaw's tacit acceptance that before turbulence models become adequately reliable they may be replaced by LES, and finally Bradshaw's doubts that we know enough about turbulence-model reliability. The turn to LES is evident in geophysical applications, where it has been used since the 1970s to generate surrogate "databases" for research and more recently to evaluate turbulence models and parameterizations (Ayotte *et al.*, 1996).

As we shall see in Part II, because of their larger scales and smaller speeds geophysical flows tend to be much more strongly influenced by buoyancy. Because of the tendency of turbulence models to nonuniversality, those for geophysical applications need their own development and assessment process. The geophysical community has recently pointed to the need to recognize, accommodate, and foster "model parameterization science" to ensure continued progress in meteorological model development (NRC, 2005).

Appendix

Scaling the molecular-diffusivity terms in the budgets of scalar flux and Reynolds stress

The molecular-diffusivity terms in the scalar flux budget (5.39) involve one large-scale and one small-scale quantity, and so by Guideline 7 we try to avoid scaling them. Instead we rewrite them:

$$\begin{aligned} \gamma\overline{u_i c_{,jj}} &= \gamma(\overline{u_i c})_{,jj} - \gamma\left(\overline{c u_{i,j}}\right)_{,j} - \gamma\overline{u_{i,j} c_{,j}}\,, \\ \nu\overline{c u_{i,jj}} &= \nu(\overline{u_i c})_{,jj} - \nu\left(\overline{u_i c_{,j}}\right)_{,j} - \nu\overline{c_{,j} u_{i,j}}\,. \end{aligned} \tag{5.64}$$

The first terms on the right side of Eqs. (5.64) represent molecular diffusion. They scale as su^2/ℓ (the order of the leading terms in Eq. (5.39)) times $R_{\mathrm{t}}^{-1}\gamma/\nu$ and R_{t}^{-1}, respectively, so they are negligible.

Velocity and scalar derivatives scale as

$$u_{i,j} \sim \upsilon/\eta \sim (\epsilon/\nu)^{1/2} \sim \left(u^3/\ell\nu\right)^{1/2}, \quad c_{,j} \sim s_{\mathrm{d}}/\eta \sim \left(s^2 u/\ell\gamma\right)^{1/2}. \tag{5.65}$$

Thus the covariances in the second terms on the right side of Eqs. (5.64) scale as

$$\overline{c u_{i,j}} < s\left(u^3/\ell\nu\right)^{1/2} = \left(s^2 u^3/\ell\nu\right)^{1/2}, \quad \overline{u_i c_{,j}} < \left(s^2 u^3/\ell\gamma\right)^{1/2}, \tag{5.66}$$

the "<" indicating upper bounds since we are scaling covariances of large- and small-scale quantities, which are not well correlated (Problem 5.17). Thus, the second terms on the right side of Eqs. (5.64) scale as

$$\gamma\overline{\left(c u_{i,j}\right)_{,j}} < su^2/\ell\,(\gamma/\nu)\,R_{\mathrm{t}}^{-1/2}, \quad \nu\overline{\left(u_i c_{,j}\right)_{,j}} < su^2/\ell\,(\gamma/\nu)^{1/2}\,R_{\mathrm{t}}^{-1/2}. \tag{5.67}$$

Thus this pair of terms in Eqs. (5.64) is also negligible at large R_{t}.

We conclude that the molecular term in the budget of $\overline{u_i c}$ reduces to

$$\gamma\overline{u_i c_{,jj}} + \nu\overline{c u_{i,jj}} \simeq -(\gamma+\nu)\overline{u_{i,j} c_{,j}} \equiv -\chi_{u_i c}. \tag{5.68}$$

We label it $\chi_{u_i c}$ because it has the form of a molecular destruction term. But under local isotropy it vanishes (Part III), and the measurements of Mydlarski (2003) confirm that; thus we shall neglect it.

Parallel arguments hold for the molecular terms in the stress budgets, so we neglect them as well.

Questions on key concepts

5.1 Explain the physical meaning of the turbulent-flux conservation equations. Why are they needed? In what applications are they used?

5.2 How can the equations give indications of whether a given second moment is zero or nonzero?

5.3 How can the equations give indications that energy-containing-range fluctuating variables have an $O(1)$ correlation coefficient, as scaling guideline 2 indicates?

5.4 Explain scaling guideline 4 physically.

5.5 Explain the physical meaning of a "mixed-scale" covariance. Why is it difficult to scale?

5.6 Explain physically why a spatial derivative of a covariance of turbulent variables scales differently than a covariance of a spatial derivative of turbulent variables. What is their ratio?

5.7 Explain physically why the time-change and mean-advection terms in a second-moment budget can involve external scales. Explain how this can lead to a quasi-steady, locally homogeneous state. Give an example.

5.8 Which term in the velocity-velocity and velocity-scalar second-moment equations do you find most difficult to interpret physically? Explain.

5.9 Interpret physically each term in the conservation equation for fluctuation c. Which term can be interpreted as a rigorous statement of mixing-length ideas? Explain.

5.10 Discuss the range of scales involved in a variance conservation equation.

5.11 Discuss how the sign of a covariance can be diagnosed from its conservation equation.

5.12 Discuss how the conservation equation for the flux of a conserved scalar can reduce to an eddy-diffusivity model. Show that the vertical flux equation is consistent with a scalar eddy diffusivity, but the horizontal flux equation shows it to be a second-order tensor.

5.13 Explain why pressure fluctuations (scaling guideline 1, Subsection 5.3.2) do not scale with $\rho U u$, which is much larger than ρu^2. (Hint: consider a Galilean transformation of the coordinate system.)

5.14 Explain why the pressure covariance term in the turbulent flux budget can be interpreted as a rate of production of opposite-signed flux.

Problems

5.1 Derive and interpret the flow-integrated budget of full scalar variance $C^2 + \overline{c^2}$.

5.2 Derive, scale, and interpret the conservation equation for the covariance of two conserved scalars. Is there a complication if their molecular diffusivities differ? (Hint: use the form in Eq. (5.64).)

5.3 Explain how you would derive a conservation equation for ϵ.

5.4 One cannot mount w and c sensors at the same point. This raises the question: when measuring the scalar flux $\overline{wc}$ in the surface layer, is it better to mount the c sensor just above the w sensor, or vice-versa?† To answer the question, derive a rate equation for the difference of the displaced covariances, $\overline{w(z)c(z-d)} - \overline{w(z-d)c(z)}$, where d is the sensor separation in the vertical. Assume that you are in the "constant flux" layer where $\overline{wc}$ is independent of z and assume that each of these displaced covariances is smaller in magnitude than $\overline{wc}$. Use the WET model.

5.5 Derive the fluctuating vorticity equation.

5.6 Use the result of Problem 5.5 to derive the conservation equation for the variance of fluctuating vorticity. Scale it; what are the two leading terms?

5.7 Derive the conservation equation for mean helicity, the mean of the dot product of fluctuating vorticity and fluctuating velocity.

5.8 Write the TKE balance for *grid turbulence* – the flow downstream of a turbulence-producing grid in a wind tunnel. Assume homogeneity in the normal plane.

5.9 Write the conservation equation for horizontal scalar flux in a horizontally homogeneous boundary layer. It has been found that its turbulent transport and molecular terms are negligible. What then are its production and destruction terms? Can you explain why this budget provides a good test of a fluctuating pressure sensor (Wilczak and Bedard, 2004)?

5.10 Explain why the second term in the first group on the right side of (5.51) was interpreted as "the turbulent flux of joint kinetic energy."

5.11 Interpret the pressure transport term in Eq. (5.42) physically.

5.12 Show that the pressure covariance in the scalar flux budget (5.53) can be scaled as the product of a kinematic pressure gradient of order u^2/ℓ and a scalar fluctuation of order s.

5.13 Work out Eq. (5.46), showing that the second-derivative term can be neglected as claimed.

5.14 Derive and scale the conservation equation for scalar-gradient variance. What are its two leading terms?

† This problem is based on the paper by Kristensen *et al.* (1997).

5.15 Explain the claim that the tilting production term in Eq. (5.39) represents changes in the magnitude and direction of the flux through its interaction with the mean velocity gradient.

5.16 Using $\epsilon \sim u^3/\ell$, write the Kolmogorov microscale as an order-of-magnitude expression in ν, u, and ℓ.

5.17 Using the scaling of $\partial\overline{c^2}/\partial x$, find the dependence of the correlation coefficient of c and $\partial c/\partial x$ on R_t. Explain the dependence physically.

5.18 The horizontal turbulent heat flux $\rho c_p \overline{u\theta}$ has been found to be typically as large or larger than the vertical one. Why do we seldom (if ever) hear about it?

5.19 Derive the conservation equation for the covariance of the fluctuating gradient of a conserved scalar and fluctuating vorticity. Can you identify its production term(s)? Destruction term(s)?

5.20 Evaluate the TKE equation (5.42) on the stagnation streamline just upstream of a body of revolution. Can you interpret the terms?

5.21 Explain why in the development leading to Eq. (5.49) the eddy diffusivity K was written for the deviatoric stress $\overline{u_i u_j} - 2\delta_{ij}\overline{u_k u_k}/3$. (Hint: consider the *trace*, the form under contraction of the indices.)

5.22 Consider Eq. (5.51) for steady flow in a pipe. Integrate the equation over the region between two cross sections. Show that all but one of the divergence terms integrates to zero. Interpret your result.

5.23 Explain physically why the rate of molecular destruction of scalar flux vanishes, as argued in the Appendix.

5.24 The dependent variable $w(x,t)$ in a nonlinear system is governed on the interval $0 \le x \le L$ by

$$\frac{\partial w}{\partial t} + w\frac{\partial w}{\partial x} = \beta(x,t;\alpha) + \gamma\frac{\partial^2 w}{\partial x^2}, \quad w(0,t) = w(L,t) = 0, \quad \gamma \text{ a constant.}$$

β is a zero-mean, random, stochastic, forcing function.

(a) For a given realization sketch an example of $w(x)$ for fixed t and $w(t)$ for fixed x.

(b) Ensemble average the equation. Express the second term on the left side as a gradient. Write the steady equation. Integrate it once, using the boundary conditions.

(c) Derive the equation for $\overline{w^2}$, writing its advection term as a gradient and its molecular term as the sum of destruction and diffusion terms. Interpret its steady form. Integrate it from $x = 0$ to $x = L$, using the given boundary conditions, and interpret that.

(d) Use the equation and your inferences to sketch the profiles of $\overline{w}$ and $\overline{w^2}$ from 0 to L.

References

Ayotte, K., P. Sullivan, A. Andrén, *et al.*, 1996: An evaluation of neutral and convective planetary boundary-layer parameterizations relative to large-eddy simulations. *Bound.-Layer Meteor.*, **79**, 131–175.

Bradshaw, P., 1994: Turbulence: the chief outstanding difficulty of our subject. *Exp. Fluids*, **16**, 203–216.

Bradshaw, P., 1999: The best turbulence models for engineers. In *Modeling Complex Turbulent Flows*, M. D. Salas, J. N. Hefner, and L. Sakell, Eds., Dordrecht: Kluwer.

Daly, B. J., and F. H. Harlow, 1970: Transport equations in turbulence. *Phys. Fluids*, **13**, 2634–2649.

Deardorff, J. W., 1970: A numerical study of three-dimensional turbulent channel flow at large Reynolds numbers. *J. Fluid Mech.*, **41**, 453–480.

Donaldson, C. duP., 1971: Calculation of turbulent shear flows for atmospheric and vortex motions. *AIAA J.*, **7**, 272–278.

Haugen, D. A., J. C. Kaimal, and E. F. Bradley, 1971: An experimental study of Reynolds stress and heat flux in the atmospheric surface layer. *Quart. J. Roy. Meteor. Soc.*, **97**, 168–180.

Kristensen, L., J. Mann, S. P. Oncley, and J. C. Wyngaard, 1997: How close is close enough when measuring fluxes with displaced sensors? *J. Atmos. Ocean. Tech.*, **14**, 814–821.

Launder, B., 1996: An introduction to single-point closure methodology. In *Simulation and Modeling of Turbulent Flows*, T. Gatski, M. Hussani, and J. Lumley, Eds., Oxford University Press.

Liepmann, H., 1979: The rise and fall of ideas in turbulence. *Am. Sci.*, **69**, 221–228.

Lumley, J. L., 1983: Atmospheric modelling. *Mech. Eng. Trans. Inst. Eng. Australia*, **ME8**, 153–159.

Monin, A. S., and A. M. Yaglom, 1971: *Statistical Fluid Mechanics: Mechanics of Turbulence*. J. Lumley, Ed., Cambridge, MA: MIT Press.

Mydlarski, L., 2003: Mixed velocity-passive scalar statistics in high-Reynolds-number turbulence. *J. Fluid Mech.*, **475**, 173–203.

NRC, 2005: *Improving the Scientific Foundation for Atmosphere-Land-Ocean Simulations*. Washington, D.C.: Board on Atmospheric Sciences and Climate, The National Academies Press.

Nishiyama, R. T., and A. J. Bedard Jr., 1991: A "quad-disc" static pressure probe for measurement in adverse atmospheres: With a comparative review of static pressure probe designs. *Rev. Sci. Instrum.*, **62**, 2193–2204.

Tennekes, H., and J. L. Lumley, 1972: *A First Course in Turbulence*. Cambridge, MA: MIT Press.

Wilczak, J. M., and A. Bedard, 2004: A new turbulence microbarometer and its evaluation using the budget of horizontal heat flux. *J. Atmos. Ocean. Tech.*, **21**, 1170–1181.

Wyngaard, J. C., O. R. Coté, and Y. Izumi, 1971: Local free convection, similarity, and the budgets of shear stress and heat flux. *J. Atmos. Sci.*, **28**, 1171–1182.

Wyngaard, J. C., A. Siegel, and J. Wilczak, 1994: On the response of a turbulent-pressure probe and the measurement of pressure transport. *Bound.-Layer Meteor.*, **69**, 379–396.

6

Large-eddy dynamics, the energy cascade, and large-eddy simulation

6.1 Introduction

As we have seen, the huge range of spatial scales in large-R_t turbulent flows makes it impossible to solve their equations of motion numerically. In applications we use ensemble- or space-averaged equations that have a drastically reduced range of scales. In Chapter 3 we saw that this averaging produces new terms involving turbulent fluxes, and in Chapter 5 we discussed the evolution equations for the fluxes produced by ensemble averaging.

In this chapter we discuss the space-averaged equations further, present the evolution equations for their fluxes, and discuss the modeling of these fluxes. This modeling is of two broad types depending on the spatial scale Δ of the averaging relative to the size ℓ of the energy-containing eddies. In what we shall call "coarse resolution" applications $\Delta \gg \ell$; in "fine resolution" ones $\Delta \ll \ell$.

The first numerical solutions of the averaged equations in meteorology appear to be those of Charney *et al.* (1950). The domain was a limited area of the earth's surface with 15×18 grid squares each 736 km on a side, with only one level in the vertical. The study of Phillips (1955, 1956) soon followed. Its spatial resolution was similarly coarse, but it had two grid levels in the vertical. By the 1970s general circulation models were being used fairly widely, and "mesoscale" or "limited-area" models were being used in regional domains. Computer size and speed had grown substantially since the 1950s, but these newer models were still in the coarse-resolution category.

The first successful high-resolution application was Deardorff's (1970a) study of turbulent channel flow. He used 6720 uniform grid elements (24 in the streamwise direction, 14 in the lateral, 20 in the vertical), and Lilly's (1967) specification for the effects of the unresolvable (now called subfilter-scale) turbulence. His results generally agreed well with measurements, and also revealed aspects of eddy structure that had not been measured. The calculation attracted strong interest in the engineering fluids community, from which came its present and remarkably precise name "large-eddy simulation" or LES.† It is the principal type of fine-resolution, space-averaged modeling used today.

In this chapter we shall derive and discuss the space-averaged equations of motion and explore their dynamics in the LES limit. We'll also do a "thought problem" in equilibrium homogeneous turbulence that sheds light on the energy cascade in turbulence.

6.2 More on space averaging

6.2.1 A one-dimensional, homogeneous example

Perhaps the simplest space-averaging operator is the local average defined in Eq. (3.29). In its one-dimensional form it averages a function $f(x,t)$ over an x-interval of length Δ to produce a smoothed version:

$$f^{\mathrm{r}}(x,t,\Delta) = \frac{1}{\Delta}\int_{-\Delta/2}^{\Delta/2} f(x+x',t)\,dx'. \tag{6.1}$$

This local average is sometimes called a *running mean.* We use the notation introduced in Chapter 3, denoting the smoothed variable by a superscript r, for *resolvable*.

In Chapter 2 we represented a real, homogeneous function $f(x)$ as a complex, infinite Fourier series. Here we'll use the finite form

$$f(x) = \sum_{n=-N}^{N} \hat{f}(\kappa_n)e^{i\kappa_n x}. \tag{6.2}$$

In the averaging defined by Eq. (6.1) the contributions to the sum in Eq. (6.2) from Fourier components having wavelength small compared to Δ (those with $\kappa_n\Delta \gg 1$) are strongly attenuated, since they have many cycles over the averaging interval. Those of wavelength large compared to Δ (those with $\kappa_n\Delta \ll 1$) are minimally affected, since they are nearly constant over the averaging interval.

We can quantify these smoothing properties of the running-mean operator, Eq. (6.1), by using the Fourier representation (6.2) in the averaging expression (6.1) and integrating:

$$f^{\mathrm{r}}(x) = \frac{1}{\Delta}\int_{x-\Delta/2}^{x+\Delta/2} \sum_{n=-N}^{N} \hat{f}(\kappa_n)e^{i\kappa_n x'}\,dx'$$

† According to Parviz Moin (personal communication), the late Bill Reynolds of Stanford University coined the name *large-eddy simulation*.

$$= \sum_{n=-N}^{N} \hat{f}(\kappa_n) \left(\frac{1}{\Delta} \int_{x-\Delta/2}^{x+\Delta/2} e^{i\kappa_n x'} dx' \right) = \sum_{n=-N}^{N} \frac{\sin(\kappa_n \Delta/2)}{(\kappa_n \Delta/2)} \hat{f}(\kappa_n) e^{i\kappa_n x}. \quad (6.3)$$

If we write this as

$$f^{\mathrm{r}}(x) = \sum_{n=-N}^{N} \hat{f}^{\mathrm{r}}(\kappa_n) e^{i\kappa_n x} = \sum_{n=-N}^{N} \hat{f}(\kappa_n) T(\kappa_n) e^{i\kappa_n x}, \quad (6.4)$$

we see that the *amplitude transfer function* $T(\kappa_n)$ for the running-mean operator is

$$T(\kappa_n) = \frac{\hat{f}^{\mathrm{r}}(\kappa_n)}{\hat{f}(\kappa_n)} = \frac{\sin(\kappa_n \Delta/2)}{(\kappa_n \Delta/2)}, \quad (6.5)$$

which is plotted in Figure 6.1. Since $\sin x/x \to 1$ as $x \to 0$, the averaging minimally affects the Fourier coefficients of wavelength large compared to Δ (those with $\kappa_n \Delta \ll 1$). The attenuation begins at $\kappa_n \Delta \sim 1$ and is severe for $\kappa_n \Delta \gg 1$.

6.2.2 The generalization to spatial filtering

We introduced in Chapter 3 a more general representation of space averaging (or spatial *filtering*, as it is often called in the LES literature). In one dimension this generalizes the local average of Eq. (6.1) to

$$f^{\mathrm{r}}(x) = \int_{-\infty}^{\infty} G(x - x') f(x')\, dx', \quad (6.6)$$

with G called the *filter function*. This integral is called the *convolution* of f and G.

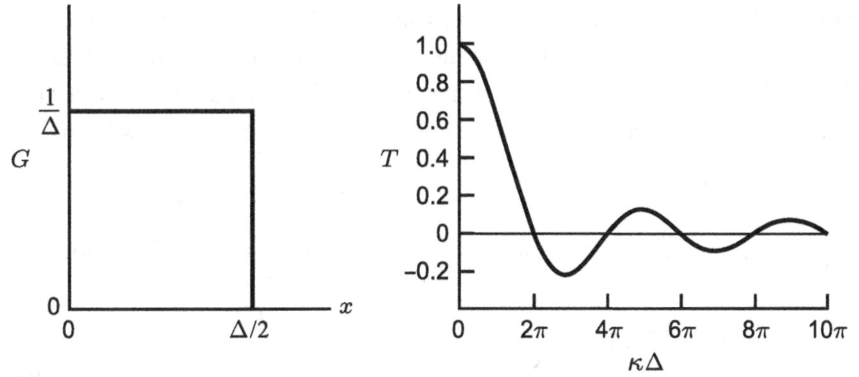

Figure 6.1 The filter function, Eq. (6.9), and transfer function, Eq. (6.5), of a one-dimensional running-mean filter.

Equation (6.6) is a physical-space representation of filtering. With Eqs. (6.2) and (6.4) we can also view it from wavenumber space:

$$f(x) = \sum_{n=-N}^{N} \hat{f}(\kappa_n) e^{i\kappa_n x}, \qquad f^{\mathrm{r}}(x) = \sum_{n=-N}^{N} \hat{f}(\kappa_n) T(\kappa_n) e^{i\kappa_n x}. \tag{6.7}$$

G and T comprise a Fourier transform pair,

$$G(x) = \frac{1}{2\pi} \int_{-\infty}^{\infty} e^{-i\kappa x} T(\kappa)\, d\kappa, \qquad T(\kappa) = \int_{-\infty}^{\infty} e^{i\kappa x} G(x)\, dx. \tag{6.8}$$

For the one-dimensional running-mean operator of Eq. (6.1), G is

$$G(x) = \frac{1}{\Delta}, \quad |x| \leq \frac{\Delta}{2}; \qquad G(x) = 0, \quad |x| > \frac{\Delta}{2}, \tag{6.9}$$

which is plotted in Figure 6.1. This is the Fourier transform of T given in Eq. (6.5) (Problem 6.12).

6.2.3 The wave-cutoff filter

If we filter $f(x)$ a second time it follows from Eq. (6.7) that

$$(f^{\mathrm{r}})^{\mathrm{r}}(x) = \sum_{n=-N}^{N} T^2(\kappa_n) \hat{f}(\kappa_n) e^{i\kappa_n x}, \tag{6.10}$$

so the amplitude transfer function is T^2. If we filter n times the amplitude transfer function is T^n. We saw in Chapter 2 that subsequent applications of the ensemble-averaging operator have no effect, which prompts the question: Is there a spatial filter with this property?

For subsequent applications of a filter to have no effect it is necessary that $T(\kappa) \times T(\kappa) \cdots \times T(\kappa) = T(\kappa)$, which means that T can have only the values 0 and 1. This is the *wave-cutoff* filter. One that passes Fourier components of smaller wavenumbers and rejects those of larger wavenumbers (the *low-pass* form) is

$$T(\kappa) = 1, \quad \kappa \leq \kappa_c; \qquad T(\kappa) = 0, \quad \kappa > \kappa_c, \tag{6.11}$$

with κ_c the *cutoff wavenumber*. From Eq. (6.8) its filter function is (Problem 6.9)

$$\begin{aligned} G(x) &= \frac{1}{2\pi} \int_{-\infty}^{\infty} e^{-i\kappa x} T(\kappa)\, d\kappa = \frac{1}{2\pi} \int_{-\kappa_c}^{\kappa_c} e^{-i\kappa x}\, d\kappa \\ &= \frac{\sin \kappa_c x}{\pi x} = \left(\frac{\kappa_c}{\pi}\right) \frac{\sin \kappa_c x}{\kappa_c x}. \end{aligned} \tag{6.12}$$

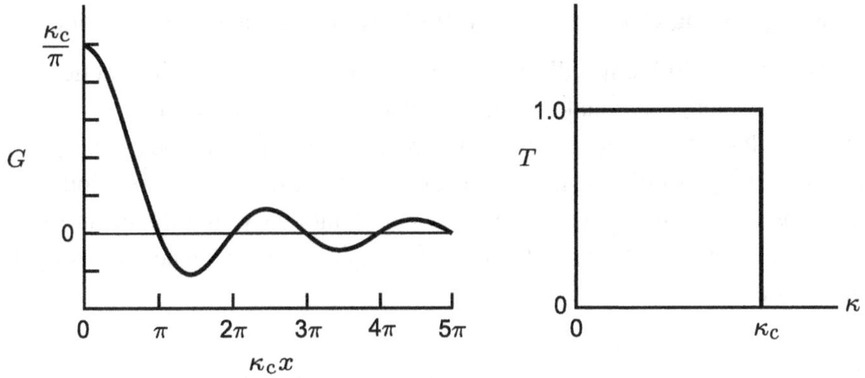

Figure 6.2 The filter function, Eq. (6.12), and transfer function, Eq. (6.11), of a one-dimensional, low-pass, wave-cutoff filter.

T and G for a one-dimensional, low-pass wave-cutoff filter are plotted in Figure 6.2.

From Eq. (6.6) low-pass wave-cutoff filtering gives in physical space

$$f^{\mathrm{r}}(x) = \int_{-\infty}^{\infty} G(x - x') f(x')\, dx' = \int_{-\infty}^{\infty} \frac{\sin[\kappa_c(x - x')]}{\pi(x - x')} f(x')\, dx'. \tag{6.13}$$

Figure 6.2 shows that this filter function G is negative in some regions. This is required in order that the filter have a sharp cutoff in the wavenumber domain.

The *high-pass* wave-cutoff filter rejects Fourier components of smaller wavenumber and passes those of larger wavenumbers:

$$T(\kappa) = 0, \qquad \kappa \leq \kappa_c; \qquad T(\kappa) = 1, \qquad \kappa > \kappa_c. \tag{6.14}$$

6.2.4 The Gaussian filter

Both the running-mean filter, Figure 6.1, and the wave-cutoff filter, Figure 6.2, have negative values in physical or wavenumber space. One that does not is the Gaussian filter,

$$G(x) = \frac{1}{\sqrt{2\pi}\,\sigma} e^{-\frac{x^2}{2\sigma^2}}. \tag{6.15}$$

Its transfer function is also Gaussian (Problem 6.25).

6.3 A "thought problem": equilibrium homogeneous turbulence

A simple thought problem will reveal some important concepts in kinetic energy transfer in turbulence and make clear the principles that underlie LES.

Imagine a large "box" of fluid kept in turbulent motion by a body force per unit mass $\beta_i(\mathbf{x}, t)$ that is stochastic in space and time, is homogeneous and stationary, and has zero mean. The turbulence is homogeneous, stationary, and the mean velocity vanishes everywhere: $\tilde{u}_i(\mathbf{x}, t) \equiv U_i(\mathbf{x}, t) + u_i(\mathbf{x}, t) = u_i(\mathbf{x}, t)$. The governing fluid equations are

$$u_{i,t} + (u_i u_j)_{,j} = -\frac{1}{\rho} p_{,i} + \nu u_{i,jj} + \beta_i, \quad u_{i,i} = 0. \tag{6.16}$$

For compactness we shall use the comma notation for differentiation.

In this flow the TKE equation is the stationary, homogeneous form of Eq. (5.42) with an added forcing term:

$$\frac{1}{2}\frac{\partial}{\partial t}\overline{u_i u_i} = \overline{\beta_i u_i} - \epsilon = 0. \tag{6.17}$$

The mean rate of TKE input through the forcing, $\overline{\beta_i u_i}$, is balanced by the mean rate of viscous dissipation of TKE.

6.3.1 The filtered equations

Now we write the velocity field u_i in Eqs. (6.16) as the sum of r and s parts

$$u_i(\mathbf{x}, t) = u_i^{\mathrm{r}}(\mathbf{x}, t) + u_i^{\mathrm{s}}(\mathbf{x}, t),$$

with the r denoting the part of the field that passes through a spatial filter:

$$u_i^{\mathrm{r}}(\mathbf{x}, t) = \int\int\int_{-\infty}^{\infty} G(\mathbf{x} - \mathbf{x}') u_i(\mathbf{x}', t)\, dx_1'\, dx_2'\, dx_3'. \tag{6.18}$$

The eddies removed by the filter are denoted with a superscipt s, for *subfilter-scale*.

As with ensemble averaging, we require that the filter:

1. commute with temporal and spatial differentiation;
2. remove fluctuations of scale much less than the filter cutoff scale Δ;
3. leave variations of scale much greater than Δ unchanged;
4. be a *wave-cutoff* filter – one that produces r and s components with no Fourier components in common.

To verify that the filtering operation (6.18) meets the first requirement we take the partial derivative of Eq. (6.18) with respect to time:

$$\frac{\partial u_i^{\mathrm{r}}}{\partial t} = \int G(\mathbf{x}-\mathbf{x}')\frac{\partial u_i(\mathbf{x}',t)}{\partial t}d\mathbf{x}' = \left(\frac{\partial u_i}{\partial t}\right)^{\mathrm{r}}. \tag{6.19}$$

In an unbounded domain filtering commutes with spatial differentiation as well (Problem 6.2).

To meet the second requirement we choose the filter scale Δ to be much less than the scale ℓ of the large eddies so that the filter does not affect them significantly. We also choose Δ to be much greater than the dissipative eddy scale η so it removes the eddies that are impossible to resolve numerically at large R_{t}. The filter-scale requirement is then $\ell \gg \Delta \gg \eta$, which requires that $\ell/\eta \sim R_{\mathrm{t}}^{3/4}$ be sufficiently large.

Satisfying the third requirement means, in particular, that filtering a constant C leaves the constant unchanged. This implies that

$$\begin{aligned} C^{\mathrm{r}} = C &= \int\int\int_{-\infty}^{\infty} G(\mathbf{x}-\mathbf{x}')\,C\,dx_1'\,dx_2'\,dx_3' \\ &= C\int\int\int_{-\infty}^{\infty} G(\mathbf{x}-\mathbf{x}')\,dx_1'\,dx_2'\,dx_3', \end{aligned} \tag{6.20}$$

so that

$$\int\int\int_{-\infty}^{\infty} G(\mathbf{x}-\mathbf{x}')\,dx_1'\,dx_2'\,dx_3' = 1, \tag{6.21}$$

which poses a requirement on G.

Using this filter on the equations of motion (6.16) yields

$$u_{i,t}^{\mathrm{r}} + (u_i u_j)_{,j}^{\mathrm{r}} = -\frac{1}{\rho}p_{,i}^{\mathrm{r}} + \nu u_{i,jj}^{\mathrm{r}} + \beta_i^{\mathrm{r}}, \quad u_{i,i}^{\mathrm{r}} = 0. \tag{6.22}$$

We take the stochastic forcing β_i to be of a spatial scale large enough so that it is resolved perfectly by the filter; i.e., $\beta_i^{\mathrm{r}} = \beta_i$. In the energy-containing range of the filtered motion, where the eddy velocity and length scales are u and ℓ, viscous effects are negligible because R_{t} is large. At the smaller resolvable scales we can use the result of Chapter 2, Subsection 2.5.4, that eddies of scale r have a velocity scale $u(r) \sim u(r/\ell)^{1/3}$. This implies that the ratio of the inertial and viscous terms for these eddies is $u(r)r/\nu \sim (r/\ell)^{4/3}R_{\mathrm{t}}$, which is smallest when $r \sim \Delta$, the cutoff scale. If $(\Delta/\ell)^{4/3}$ is 10^{-2} and R_{t} exceeds 10^4, say, then even at the cutoff scale the viscous term in Eq. (6.22) is negligible. We neglect it, taking as our large-eddy equations

$$u_{i,t}^{\mathrm{r}} + (u_i u_j)_{,j}^{\mathrm{r}} = -\frac{1}{\rho}p_{,i}^{\mathrm{r}} + \beta_i, \qquad u_{i,i}^{\mathrm{r}} = 0. \tag{6.23}$$

If we now write the velocity as a Fourier series in three dimensions, as in Chapter 2,

$$u_i(\mathbf{x}, t) = \sum_{\boldsymbol{\kappa}} \hat{u}_i(\boldsymbol{\kappa}, t) e^{i\boldsymbol{\kappa}\cdot\mathbf{x}}, \tag{6.24}$$

then $u_i u_j$ is (suppressing the dependence on t)

$$u_i u_j = \sum_{\boldsymbol{\kappa}} \hat{u}_i(\boldsymbol{\kappa}) e^{i\boldsymbol{\kappa}\cdot\mathbf{x}} \sum_{\boldsymbol{\kappa}'} \hat{u}_j(\boldsymbol{\kappa}') e^{i\boldsymbol{\kappa}'\cdot\mathbf{x}} = \sum_{\boldsymbol{\kappa}} \sum_{\boldsymbol{\kappa}'} \hat{u}_i(\boldsymbol{\kappa}) \hat{u}_j(\boldsymbol{\kappa}') e^{i((\boldsymbol{\kappa}+\boldsymbol{\kappa}')\cdot\mathbf{x})}. \tag{6.25}$$

Equation (6.25) says that the wavenumber vectors of the Fourier components of the product $u_i u_j$ are the sums of the wavenumber vectors of the Fourier components of u_i and u_j. The Fourier components of the filtered product $(u_i u_j)^{\mathrm{r}}$ have wavenumbers of magnitude less than κ_{c},

$$(u_i u_j)^{\mathrm{r}} = \sum_{\boldsymbol{\kappa}} \sum_{\boldsymbol{\kappa}'} \hat{u}_i(\boldsymbol{\kappa}) \hat{u}_j(\boldsymbol{\kappa}') e^{i((\boldsymbol{\kappa}+\boldsymbol{\kappa}')\cdot\mathbf{x})}, \quad |\boldsymbol{\kappa} + \boldsymbol{\kappa}'| < \kappa_{\mathrm{c}}. \tag{6.26}$$

But Fourier components of the velocity field having wavenumbers $\boldsymbol{\kappa}$ and $\boldsymbol{\kappa}'$ of magnitude larger than κ_{c} contribute to $(u_i u_j)^{\mathrm{r}}$ (Figure 6.3). We can see this by writing $(u_i u_j)^{\mathrm{r}}$ as

$$(u_i u_j)^{\mathrm{r}} = [(u_i^{\mathrm{r}} + u_i^{\mathrm{s}})(u_j^{\mathrm{r}} + u_j^{\mathrm{s}})]^{\mathrm{r}} = (u_i^{\mathrm{r}} u_j^{\mathrm{r}})^{\mathrm{r}} + (u_i^{\mathrm{r}} u_j^{\mathrm{s}})^{\mathrm{r}} + (u_i^{\mathrm{s}} u_j^{\mathrm{r}})^{\mathrm{r}} + (u_i^{\mathrm{s}} u_j^{\mathrm{s}})^{\mathrm{r}}. \tag{6.27}$$

The last three terms in (6.27) involve u_i^{s} and, hence, involve Fourier components of the velocity field having wavenumbers beyond the cutoff (Figure 6.3).

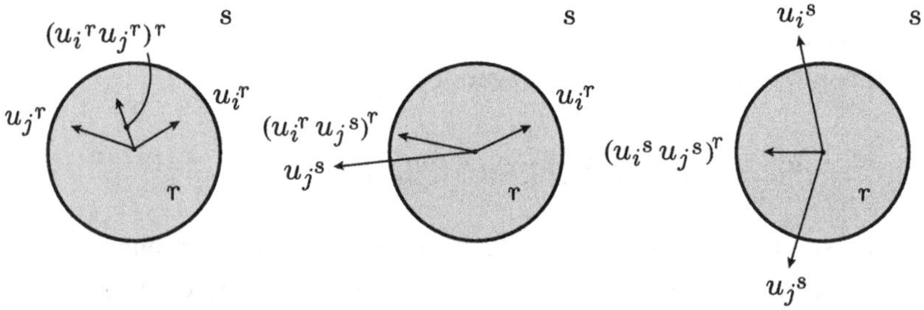

Figure 6.3 An illustration in κ_1, κ_2 space of the three types of interactions of Fourier components of the velocity field that according to Eq. (6.27) contribute to the resolvable-scale product $(u_i u_j)^{\mathrm{r}}$. Resolvable (r) wavenumbers lie within the circles; subfilter-scale (s) wavenumbers lie outside. The left panel shows wavenumber vectors of Fourier components of u_i^{r} and u_j^{r}, respectively. By Eq. (6.26) their sum, which lies within the circle, is the wavenumber of a Fourier component of the filtered product $(u_i^{\mathrm{r}} u_j^{\mathrm{r}})^{\mathrm{r}}$. Here the product of two r modes produces another r mode, as in the first term on the far rhs of Eq. (6.27). The second panel illustrates the second and third terms in (6.27), where the product of an r mode and an s mode produces an r mode. The third panel illustrates the final term, where the product of two s modes produces an r mode.

6.3.2 TKE budgets for the resolvable and subfilter-scale motions

6.3.2.1 The decomposition of TKE

The decomposition $u_i = u_i^{\mathrm{r}} + u_i^{\mathrm{s}}$ implies that TKE has three components,

$$\frac{\overline{u_i u_i}}{2} = \frac{\overline{(u_i^{\mathrm{r}} + u_i^{\mathrm{s}})(u_i^{\mathrm{r}} + u_i^{\mathrm{s}})}}{2} = \frac{\overline{u_i^{\mathrm{r}} u_i^{\mathrm{r}}}}{2} \quad \text{(TKE of resolvable-scale motion)}$$
$$+ \overline{u_i^{\mathrm{r}} u_i^{\mathrm{s}}} \quad \text{(joint TKE)}$$
$$+ \frac{\overline{u_i^{\mathrm{s}} u_i^{\mathrm{s}}}}{2} \quad \text{(TKE of subfilter-scale motion)}. \tag{6.28}$$

The cross term $\overline{u_i^{\mathrm{r}} u_i^{\mathrm{s}}}$ vanishes with a wave-cutoff filter because the resolvable and subfilter-scale fields have no Fourier modes in common and, hence, are uncorrelated. This is a mathematical property of the Fourier representation of homogeneous fields that is directly related to the property (discussed in Chapter 2, Section 2.5.1) that Fourier coefficients of different wavenumbers are uncorrelated. We make this current property plausible in Appendix 6.1. Thus, the wave-cutoff filter yields $\overline{u_i u_i} = \overline{u_i^{\mathrm{r}} u_i^{\mathrm{r}}} + \overline{u_i^{\mathrm{s}} u_i^{\mathrm{s}}}$, neatly decomposing the TKE into what we shall call the *resolvable-scale* TKE and the *subfilter-scale* TKE.

6.3.2.2 The resolvable-scale TKE budget

Multiplying the filtered Navier–Stokes equation (6.23) by u_i^{r} and ensemble averaging yields

$$\frac{1}{2}(\overline{u_i^{\mathrm{r}} u_i^{\mathrm{r}}})_{,t} = -\frac{1}{\rho}\overline{p^{\mathrm{r}}_{,i} u_i^{\mathrm{r}}} + \overline{\beta_i u_i^{\mathrm{r}}} - \overline{u_i^{\mathrm{r}} (u_i u_j)^{\mathrm{r}}_{,j}}. \tag{6.29}$$

Using the property $u^{\mathrm{r}}_{i,i} = 0$ the pressure covariance here can be rewritten as $(\overline{p^{\mathrm{r}} u_i^{\mathrm{r}}})_{,i}$, which vanishes by homogeneity. Thus, the resolvable-scale TKE budget of our equilibrium, homogeneous turbulence is simply

$$\frac{1}{2}(\overline{u_i^{\mathrm{r}} u_i^{\mathrm{r}}})_{,t} = \overline{\beta_i u_i^{\mathrm{r}}} - \overline{u_i^{\mathrm{r}} (u_i u_j)^{\mathrm{r}}_{,j}} = 0. \tag{6.30}$$

Comparing Eq. (6.30) with the TKE equation (6.17) for the full motion shows that the term $\overline{u_i^{\mathrm{r}} (u_i u_j)^{\mathrm{r}}_{,j}}$ plays the same role for resolvable-scale TKE that viscous dissipation plays for TKE.

As the filter scale $\Delta \to 0$, the resolvable-scale velocity approaches its unfiltered counterpart ($u_i^{\mathrm{r}} \to u_i$) and the rate-of-loss term in Eq. (6.30) vanishes by homogeneity:

$$-\overline{u_i^{\mathrm{r}} (u_i u_j)^{\mathrm{r}}_{,j}} \to -\overline{u_i (u_i u_j)_{,j}} = \frac{-(\overline{u_i u_i u_j})_{,j}}{2} = 0. \tag{6.31}$$

In this limit the viscous term in the u_i^{r} equation of motion (6.23) is no longer negligible. The term it produces in the resolvable TKE budget (6.30) converges to $-\epsilon$ and Eqs. (6.17) and (6.30) become identical.

6.3.2.3 The subfilter-scale TKE budget

Subtracting Eq. (6.22) for u_i^{r} from Eq. (6.16) for u_i yields the equation for u_i^{s}:

$$u^{\mathrm{s}}_{i,t} + (u_i u_j)^{\mathrm{s}}_{,j} = -\frac{1}{\rho} p^{\mathrm{s}}_{,i} + \nu u^{\mathrm{s}}_{i,jj}. \tag{6.32}$$

Multiplying by u_i^{s}, ensemble averaging, and rewriting the viscous term yields the subfilter-scale TKE budget:

$$\frac{1}{2}(\overline{u_i^{\mathrm{s}} u_i^{\mathrm{s}}})_{,t} = -\overline{u_i^{\mathrm{s}}(u_i u_j)^{\mathrm{s}}_{,j}} + \frac{\nu}{2}(\overline{u_i^{\mathrm{s}} u_i^{\mathrm{s}}})_{,jj} - \nu \overline{u^{\mathrm{s}}_{i,j} u^{\mathrm{s}}_{i,j}}. \tag{6.33}$$

By using $u_i^{\mathrm{s}} = u_i - u_i^{\mathrm{r}}$ we can write the final term, molecular destruction, as

$$\nu \overline{u^{\mathrm{s}}_{i,j} u^{\mathrm{s}}_{i,j}} = \nu \overline{u_{i,j} u_{i,j}} - 2\nu \overline{u_{i,j} u^{\mathrm{r}}_{i,j}} + \nu \overline{u^{\mathrm{r}}_{i,j} u^{\mathrm{r}}_{i,j}}. \tag{6.34}$$

The first term on the right is ϵ. The second and third terms have had the most dissipative eddies filtered out of one and both of their factors, respectively. As a result these terms are quite small at large R_{t}, so we neglect them and write

$$\nu \overline{u^{\mathrm{s}}_{i,j} u^{\mathrm{s}}_{i,j}} \simeq \nu \overline{u_{i,j} u_{i,j}} = \epsilon. \tag{6.35}$$

This allows us to neglect the molecular diffusion term in Eq. (6.33) (Problem 6.15).

Equation (6.33) then becomes

$$\frac{1}{2}(\overline{u_i^{\mathrm{s}} u_i^{\mathrm{s}}})_{,t} = -\overline{u_i^{\mathrm{s}}(u_i u_j)^{\mathrm{s}}_{,j}} - \epsilon. \tag{6.36}$$

This is the simplified subfilter-scale TKE budget. The first term on the right side must be a source term that balances viscous dissipation.

Let's summarize the TKE budgets for our equilibrium, homogeneous, large-R_{t} turbulence:

$$\text{Resolvable-scale TKE:} \quad \frac{1}{2}(\overline{u_i^{\mathrm{r}} u_i^{\mathrm{r}}})_{,t} = \overline{\beta_i u_i^{\mathrm{r}}} - \overline{u_i^{\mathrm{r}}(u_i u_j)^{\mathrm{r}}_{,j}} = 0, \tag{6.37}$$

$$\text{Subfilter-scale TKE:} \quad \frac{1}{2}(\overline{u_i^{\mathrm{s}} u_i^{\mathrm{s}}})_{,t} = -\overline{u_i^{\mathrm{s}}(u_i u_j)^{\mathrm{s}}_{,j}} - \epsilon = 0, \tag{6.38}$$

$$\text{TKE:} \quad \frac{1}{2}(\overline{u_i u_i})_{,t} = \overline{\beta_i u_i} - \epsilon = 0. \tag{6.17}$$

Since a wave-cutoff filter partitions a second moment into its rr and ss parts, the terms in the resolvable-scale TKE budget (6.30) and the subfilter-scale TKE budget (6.36) sum to those terms in the full TKE budget (6.17). The summed third-moment term vanishes by homogeneity,

$$\begin{aligned}
-\overline{u_i^{\mathrm{r}}(u_iu_j)^{\mathrm{r}}_{,j}} - \overline{u_i^{\mathrm{s}}(u_iu_j)^{\mathrm{s}}_{,j}} &= -\overline{(u_i - u_i^{\mathrm{s}})(u_iu_j)^{\mathrm{r}}_{,j}} - \overline{(u_i - u_i^{\mathrm{r}})(u_iu_j)^{\mathrm{s}}_{,j}} \\
= -\overline{u_i(u_iu_j)^{\mathrm{r}}_{,j}} - \overline{u_i(u_iu_j)^{\mathrm{s}}_{,j}} &= -\overline{u_i(u_iu_j)_{,j}} = -\frac{1}{2}(\overline{u_iu_iu_j})_{,j} = 0,
\end{aligned} \tag{6.39}$$

so it does not appear in the TKE budget (6.17). In Eq. (6.39) we have also used the property that r and s quantities are uncorrelated.

The resolvable-scale TKE budget (6.30) lacks a viscous dissipation term, so its $\overline{u_i^{\mathrm{r}}(u_iu_j)^{\mathrm{r}}_{,j}}$ term must represent an equal rate of energy loss that balances its production term. Similarly, since the forcing term does not appear in the subfilter-scale TKE budget, its $\overline{u_i^{\mathrm{s}}(u_iu_j)^{\mathrm{s}}_{,j}}$ term must represent an equal rate of input that balances the loss to viscous dissipation. No evidence of these mechanisms survives when Eqs. (6.30) and (6.36) are added, for the two terms sum to zero in this homogeneous flow. We conclude that the third-moment terms in the resolvable-scale TKE budget (6.30) and the subfilter-scale TKE budget (6.36) are equal in magnitude but opposite in sign and represent *interscale transfer*, the mean rate of transfer of TKE from resolvable to subfilter scales. We explore it in more detail in Appendix 6.2.

6.3.3 Completing the resolvable and subfilter-scale TKE budgets

We can now rewrite and relabel the TKE budgets (6.30) and (6.36) in our thought problem, using the results of Appendix 6.2:

Resolvable-scale TKE budget:

$$\frac{1}{2}\left(\overline{u_i^{\mathrm{r}}u_i^{\mathrm{r}}}\right)_{,t} = \overline{\beta_i u_i^{\mathrm{r}}} - I = 0. \tag{6.40}$$

Subfilter-scale TKE budget:

$$\frac{1}{2}\left(\overline{u_i^{\mathrm{s}}u_i^{\mathrm{s}}}\right)_{,t} = I - \epsilon = 0. \tag{6.41}$$

Here I is the mean rate of interscale transfer given by (Appendix 6.2)

$$I = \overline{u_i^{\mathrm{r}}u_j^{\mathrm{r}}s_{ij}^{\mathrm{s}}} - \overline{u_i^{\mathrm{s}}u_j^{\mathrm{s}}s_{ij}^{\mathrm{r}}}, \tag{6.42}$$

with s_{ij} the turbulent strain-rate tensor

$$s_{ij} = \frac{u_{i,j} + u_{j,i}}{2}. \tag{6.43}$$

Each of two terms contributing to I is a contraction of a kinematic stress tensor and a strain-rate tensor. This is a familiar form: we saw it in the shear-production term of the TKE budget, Eq. (5.42), where it represents the mean rate of loss of kinetic energy of the mean flow by transfer to the turbulence. We saw it also in the viscous dissipation term of the TKE budget (Problem 1.7), where it represents the mean rate of loss of TKE through conversion into internal energy. Guided by those examples, and seeing its role in Eq. (6.40), we interpret I as the net mean rate of loss of resolvable-scale TKE due to two effects: the working of the subfilter-scale stresses against the resolvable-scale strain rate $\left(-\overline{u_i^{\mathrm{s}} u_j^{\mathrm{s}} s_{ij}^{\mathrm{r}}}\right)$, and the working of the resolvable-scale stresses against the subfilter-scale strain rate $\left(\overline{u_i^{\mathrm{r}} u_j^{\mathrm{r}} s_{ij}^{\mathrm{s}}}\right)$.

6.3.4 The flow of turbulence kinetic energy from larger scales to smaller

By adding them we could summarize our TKE budgets (6.40) and (6.41) in a single equation:

$$\frac{1}{2}\left(\overline{u_i^{\mathrm{r}} u_i^{\mathrm{r}}}\right)_{,t} + \frac{1}{2}\left(\overline{u_i^{\mathrm{s}} u_i^{\mathrm{s}}}\right)_{,t} = \frac{1}{2}(\overline{u_i u_i})_{,t} = \overline{\beta_i u_i^{\mathrm{r}}} - \epsilon. \tag{6.44}$$

But this does not reveal the subtlety of the full set. The TKE of the large-scale turbulence field is kept in balance not by viscous dissipation, as the summary equation (6.44) might suggest, but by its mean rate of TKE transfer to the subfilter scales. In steady turbulence these subfilter scales, in turn, dissipate TKE at the same mean rate per unit mass.

A large-R_{t} turbulent velocity field has a wide range of eddies that are small compared to ℓ but large compared to η. These eddies contain negligible TKE, carry negligible flux, contribute negligibly to TKE production, and do negligible viscous dissipation. But they are an essential link between the energy-containing and dissipative eddy ranges. They receive their energy by transfer from the resolvable eddies and in turn transfer energy to subfilter eddies. This range of intermediate eddies, called the inertial subrange, has a mean rate of transfer of kinetic energy per unit mass through it that is independent of scale and numerically equal to ϵ. We call this *spectral* energy transfer.

6.3.5 The flow of scalar variance from larger scales to smaller

Imagine our forced, equilibrium, homogeneous turbulence advecting a conserved scalar having a stochastic, homogeneous, stationary source term $S(\mathbf{x}, t)$ chosen such that the scalar field is homogeneous and steady (Problem 6.13). If S is confined to the larger scales and passes perfectly through the filter, the resolved part of the scalar follows

$$c^{\mathrm{r}}_{,t} + (u_j c)^{\mathrm{r}}_{,j} = \gamma c^{\mathrm{r}}_{,jj} + S. \tag{6.45}$$

Multiplying by c^{r}, ensemble averaging, scaling, and rearranging as we did with the TKE budgets gives the equation for the evolution of the variance of c^{r}:

$$\frac{\partial \overline{c^{\mathrm{r}} c^{\mathrm{r}}}}{\partial t} = 2\overline{S c^{\mathrm{r}}} - I_c = 0. \tag{6.46}$$

I_c is the mean rate of interscale transfer of squared scalar (Appendix 6.2),

$$I_c = 2\overline{c^{\mathrm{r}} u_j^{\mathrm{r}} c^{\mathrm{s}}_{,j}} - 2\overline{c^{\mathrm{s}} u_j^{\mathrm{s}} c^{\mathrm{r}}_{,j}}. \tag{6.47}$$

The mean-gradient production and molecular destruction terms of the scalar variance budget, Eq. (5.7), are the product of a scalar flux and a scalar gradient. I_c has this form as well.

The same process gives the equation for the evolution of $\overline{c^{\mathrm{s}} c^{\mathrm{s}}}$:

$$\frac{\partial \overline{c^{\mathrm{s}} c^{\mathrm{s}}}}{\partial t} = I_c - \chi_c = 0. \tag{6.48}$$

The interpretation here is analogous to that for TKE: the variance of the large-scale part of the scalar is kept in equilibrium by the balance between the mean rate of production by the forcing term and the mean rate of loss by transfer to smaller scales. The variance in those smaller scales is kept in equilibrium as well by the balance between the mean rate of transfer from the larger scales and the mean rate of molecular destruction. Again the mean rate of spectral transfer of variance – the *cascade rate* – is the magnitude of the rate of molecular destruction of variance, here χ_c.

6.4 Application to flows homogeneous in two dimensions

We saw in Chapter 5 that the fluctuating velocity satisfies

$$u_{i,t} + U_j u_{i,j} + u_j U_{i,j} + \left(u_i u_j - \overline{u_i u_j}\right)_{,j} = -\frac{1}{\rho} p_{,i} + \nu u_{i,jj}. \tag{6.49}$$

We'll use this equation in a boundary layer over a homogeneous surface. We choose x_1 to be the mean-flow direction and assume that the inhomogeneity is confined to the x_3 or z (vertical) direction, so we have

$$\begin{aligned} U_i &= U_1(x_3)\,\delta_{i1}, \\ U_{i,j} &= U_{1,3}\,\delta_{i1}\delta_{j3}, \\ (\overline{u_i u_j})_{,j} &= (\overline{u_i u_3})_{,3}, \\ u,\ \ell &= u(x_3),\ \ell(x_3). \end{aligned} \tag{6.50}$$

Filtering Eq. (6.49) in the horizontal plane and using the constraints (6.50) gives, assuming that the filter passes the mean fields perfectly,

$$u^{\mathrm{r}}_{i,t} + U_1 u^{\mathrm{r}}_{i,1} + u^{\mathrm{r}}_3 U_{1,3}\delta_{i1} + \left(u_i u_j\right)^{\mathrm{r}}_{,j} - (\overline{u_i u_3})_{,3} = -\frac{1}{\rho} p^{\mathrm{r}}_{,i}, \tag{6.51}$$

the filtered viscous term being negligible. Multiplying Eq. (6.51) by u^{r}_i, ensemble averaging, and using $u_{i,i} = 0$ yields

$$\frac{1}{2}\left(\overline{u^{\mathrm{r}}_i u^{\mathrm{r}}_i}\right)_{,t} = -U_1 \left(\frac{\overline{u^{\mathrm{r}}_i u^{\mathrm{r}}_i}}{2}\right)_{,1} - U_{1,3}\, \overline{u^{\mathrm{r}}_1 u^{\mathrm{r}}_3} - \frac{1}{\rho}\left(\overline{p^{\mathrm{r}} u^{\mathrm{r}}_3}\right)_{,3} - \overline{(u_i u_j)^{\mathrm{r}}_{,j} u^{\mathrm{r}}_i}. \tag{6.52}$$

The first term on the right side vanishes through horizontal homogeneity. From Appendix 6.2 we can write the last term on the right as

$$-\overline{(u_i u_j)^{\mathrm{r}}_{,j} u^{\mathrm{r}}_i} = -I - \frac{1}{2}\left(\overline{u^{\mathrm{r}}_i u^{\mathrm{r}}_i u_3}\right)_{,3} - \left(\overline{u^{\mathrm{r}}_i u^{\mathrm{s}}_i u^{\mathrm{s}}_3}\right)_{,3}. \tag{6.53}$$

Combining Eqs. (6.52) and (6.53) then gives the resolvable-scale TKE budget:

$$\frac{1}{2}\left(\overline{u^{\mathrm{r}}_i u^{\mathrm{r}}_i}\right)_{,t} = -U_{1,3}\, \overline{u^{\mathrm{r}}_1 u^{\mathrm{r}}_3} - \frac{1}{\rho}\left(\overline{p^{\mathrm{r}} u^{\mathrm{r}}_3}\right)_{,3} - I - \frac{1}{2}\left(\overline{u^{\mathrm{r}}_i u^{\mathrm{r}}_i u_3}\right)_{,3} - \left(\overline{u^{\mathrm{r}}_i u^{\mathrm{s}}_i u^{\mathrm{s}}_3}\right)_{,3}. \tag{6.54}$$

The first four terms on the right represent the mean rates of shear production, pressure transport, interscale transfer, and turbulent transport. The fifth is a variant of turbulent transport resulting from the r–s decomposition. The interscale transfer term here is that derived in Appendix 6.2.

The corresponding equation for u^{s}_i is

$$u^{\mathrm{s}}_{i,t} + U_1 u^{\mathrm{s}}_{i,1} + u^{\mathrm{s}}_3 U_{1,3}\delta_{i1} + \left(u_i u_j\right)^{\mathrm{s}}_{,j} = -\frac{1}{\rho} p^{\mathrm{s}}_{,i} + \nu u^{\mathrm{s}}_{i,jj}. \tag{6.55}$$

Multiplying by u^{s}_i, ensemble averaging, and substituting from Appendix 6.2 gives the subfilter-scale TKE budget:

$$\frac{1}{2}\left(\overline{u^{\mathrm{s}}_i u^{\mathrm{s}}_i}\right)_{,t} = -U_{1,3}\, \overline{u^{\mathrm{s}}_1 u^{\mathrm{s}}_3} - \frac{1}{\rho}\left(\overline{p^{\mathrm{s}} u^{\mathrm{s}}_3}\right)_{,3} + I - \frac{1}{2}\left(\overline{u^{\mathrm{s}}_i u^{\mathrm{s}}_i u_3}\right)_{,3} - \left(\overline{u^{\mathrm{r}}_i u^{\mathrm{s}}_i u^{\mathrm{r}}_3}\right)_{,3} - \epsilon. \tag{6.56}$$

The interpretation of the first five terms on the right side is the same as in the resolvable-scale budget (6.54); the last term is viscous dissipation.

When the resolvable and subfilter-scale TKE equations (6.54) and (6.56) are added the interscale transfer terms disappear, the two pressure-transport terms sum to one, the four turbulent-transport terms also sum to one, and the familiar TKE budget emerges (Problem 6.17):

$$\frac{1}{2}(\overline{u_i u_i})_{,t} = -U_{1,3}\, \overline{u_1 u_3} - \frac{1}{\rho}(\overline{p u_3})_{,3} - \frac{1}{2}(\overline{u_i u_i u_3})_{,3} - \epsilon. \tag{6.57}$$

6.5 The physical mechanisms of interscale transfer

Interscale transfer plays a key role in the balances (6.40), (6.41) for TKE and (6.46), (6.48) for scalar variance in our "thought problem" involving three-dimensional, homogeneous turbulence. The interscale-transfer terms in these equations are third moments (mean values of triple products) that involve both r and s components. We showed in Chapter 2 that second moments can be represented as a sum of contributions from different wavenumbers; we'll now extend that analysis to third moments.

We'll begin with a one-dimensional example, an ensemble of one-dimensional, homogeneous, random functions $a(x;\alpha)$, $b(x;\alpha)$, and $c(x;\alpha)$ that we represent through complex Fourier series. We write a as

$$a(x;\alpha) = \sum_{k=-N}^{N} \hat{a}(\kappa_k;\alpha)e^{i\kappa_k x}, \tag{6.58}$$

and similarly for b and c. Then $\overline{abc}$ is

$$\overline{abc} = \sum_{k=-N}^{N}\sum_{l=-N}^{N}\sum_{m=-N}^{N} \overline{\hat{a}(\kappa_k)\hat{b}(\kappa_l)\hat{c}(\kappa_m)e^{i(\kappa_k+\kappa_l+\kappa_m)x}}. \tag{6.59}$$

Since by homogeneity $\overline{abc}$ does not depend on x, in the terms that contribute to the sum in Eq. (6.59) the exponential must be 1, which happens if and only if $\boldsymbol{\kappa}_k + \boldsymbol{\kappa}_l + \boldsymbol{\kappa}_m = 0$. It follows that the only nonzero contributions to $\overline{abc}$ come from wavenumbers that satisfy $\kappa_m = -(\kappa_k + \kappa_l)$. Such groups of three Fourier components whose wavenumbers sum to zero are called *triads*. Therefore we can write Eq. (6.59) as

$$\overline{abc} = \sum_{k=-N}^{N}\sum_{l=-N}^{N} \overline{\hat{a}(\kappa_k)\hat{b}(\kappa_l)\hat{c}(-\kappa_k-\kappa_l)}. \tag{6.60}$$

The result (6.60) extends to three dimensions, and thus to the expression (6.42) for the rate of interscale transfer in three-dimensional homogeneous turbulence:

$$\begin{aligned} I &= \overline{u_i^{\mathrm{r}} u_j^{\mathrm{r}} s_{ij}^{\mathrm{s}}} - \overline{u_i^{\mathrm{s}} u_j^{\mathrm{s}} s_{ij}^{\mathrm{r}}} \\ &= \sum_{k=-N}^{N}\sum_{l=-N}^{N} \overline{\hat{u}_i\left(\boldsymbol{\kappa}_k^{\mathrm{r}}\right)\hat{u}_j\left(\boldsymbol{\kappa}_l^{\mathrm{r}}\right)\hat{s}_{i,j}\left(\boldsymbol{\kappa}_m^{\mathrm{s}}\right)} - \sum_{n=-N}^{N}\sum_{p=-N}^{N} \overline{\hat{u}_i\left(\boldsymbol{\kappa}_n^{\mathrm{s}}\right)\hat{u}_j\left(\boldsymbol{\kappa}_p^{\mathrm{s}}\right)\hat{s}_{i,j}\left(\boldsymbol{\kappa}_q^{\mathrm{r}}\right)}, \\ &\quad \boldsymbol{\kappa}_k^{\mathrm{r}} + \boldsymbol{\kappa}_l^{\mathrm{r}} + \boldsymbol{\kappa}_m^{\mathrm{s}} = 0, \quad \boldsymbol{\kappa}_n^{\mathrm{s}} + \boldsymbol{\kappa}_p^{\mathrm{s}} + \boldsymbol{\kappa}_q^{\mathrm{r}} = 0, \end{aligned} \tag{6.61}$$

where $\boldsymbol{\kappa}_k^{\mathrm{r}}$, for example, is a wavenumber vector that lies in the resolved domain.

Equation (6.61) implies that the interscale transfer of energy requires organization in the turbulent velocity field. Casual observation of turbulence – in a smoke plume, for example – might not suggest organization if we see only tangled, swirling eddies. But their swirling motions mark them as rotational, and their stretching and amplification by the larger, straining motions in which they are embedded causes transfer of kinetic energy to smaller scales – what we have called interscale transfer – and it is dynamically essential to turbulence.

6.6 Large-eddy simulation

6.6.1 The concept

The goal of the computational technique called *large-eddy simulation*, or LES, is to compute the energy- and variance-containing parts of the turbulent velocity and conserved scalar fields, i.e., their resolvable parts $\tilde{u}_i^{\mathrm{r}}(\mathbf{x}, t)$ and $\tilde{c}^{\mathrm{r}}(\mathbf{x}, t)$. LES starts from the governing equations for velocity and a conserved scalar, which in our constant-density case are

$$\begin{aligned} &\tilde{u}_{i,t} + (\tilde{u}_i\tilde{u}_j)_{,j} = -\frac{1}{\rho}\tilde{p}_{,i} + \nu\nabla^2\tilde{u}_i, \qquad \tilde{u}_{i,i} = 0, \\ &\tilde{c}_{,t} + (\tilde{c}\tilde{u}_j)_{,j} = \gamma\nabla^2\tilde{c}. \end{aligned} \tag{6.62}$$

Applying a spatial filter with cutoff scale $\eta \ll \Delta \ll \ell$ to these equations yields the governing equations for the resolvable fields,

$$\tilde{u}^{\mathrm{r}}_{i,t} + (\tilde{u}_i\tilde{u}_j)^{\mathrm{r}}_{,j} = -\frac{1}{\rho}\tilde{p}^{\mathrm{r}}_{,i}, \qquad \tilde{c}^{\mathrm{r}}_{,t} + (\tilde{c}\tilde{u}_j)^{\mathrm{r}}_{,j} = 0, \tag{6.63}$$

the filtered molecular-diffusion terms being negligible.

As we have seen, the nonlinear term in each of these filtered equations (6.63) involves Fourier components of wavenumbers beyond the cutoff, and so it is unknown (Problem 6.27). It is conventional to rewrite these filtered nonlinear terms by adding and subtracting the product of the filtered variables, as in Eq. (3.40). One such form of the filtered equation set is

$$\begin{aligned} &\tilde{u}^{\mathrm{r}}_{i,t} + \left(\tilde{u}^{\mathrm{r}}_i\tilde{u}^{\mathrm{r}}_j\right)_{,j} = -\frac{1}{\rho}\tilde{p}^{\mathrm{r}}_{,i} + \frac{\tau_{ij,j}}{\rho}, \qquad \frac{\tau_{ij}}{\rho} = \tilde{u}^{\mathrm{r}}_i\tilde{u}^{\mathrm{r}}_j - (\tilde{u}_i\tilde{u}_j)^{\mathrm{r}}, \\ &\tilde{c}^{\mathrm{r}}_{,t} + \left(\tilde{c}^{\mathrm{r}}\tilde{u}^{\mathrm{r}}_j\right)_{,j} + f_{j,j} = 0, \qquad f_j = (\tilde{c}\tilde{u}_j)^{\mathrm{r}} - \tilde{c}^{\mathrm{r}}\tilde{u}^{\mathrm{r}}_j. \end{aligned} \tag{6.64}$$

This set contains the new unknowns τ_{ij} and f_j, the "Reynolds fluxes" produced by the spatial filtering. Today they are often called *subfilter-scale* fluxes.

We can deduce something about these subfilter-scale fluxes by using the ensemble-mean-plus-fluctuating decomposition in their definitions (6.64) and assuming the ensemble means pass perfectly through the spatial filter (Problem 6.23):

$$\frac{\tau_{ij}}{\rho} = (U_i + u_i)^{\mathrm{r}}(U_j + u_j)^{\mathrm{r}} - \left[(U_i + u_i)(U_j + u_j)\right]^{\mathrm{r}} = u_i^{\mathrm{r}} u_j^{\mathrm{r}} - (u_i u_j)^{\mathrm{r}},$$
$$f_j = \left[(C + c)(U_j + u_j)\right]^{\mathrm{r}} - (C + c)^{\mathrm{r}}(U_j + u_j)^{\mathrm{r}} = (c u_j)^{\mathrm{r}} - c^{\mathrm{r}} u_j^{\mathrm{r}}. \tag{6.65}$$

With the further decomposition $u_i = u_i^{\mathrm{r}} + u_i^{\mathrm{s}}$, $c = c^{\mathrm{r}} + c^{\mathrm{s}}$ these become

$$\frac{\tau_{ij}}{\rho} = \left(u_i^{\mathrm{r}} u_j^{\mathrm{r}}\right)^{\mathrm{s}} - \left(u_i^{\mathrm{r}} u_j^{\mathrm{s}} + u_i^{\mathrm{s}} u_j^{\mathrm{r}} + u_i^{\mathrm{s}} u_j^{\mathrm{s}}\right)^{\mathrm{r}},$$
$$f_j = -\left(c^{\mathrm{r}} u_j^{\mathrm{r}}\right)^{\mathrm{s}} + \left(c^{\mathrm{r}} u_j^{\mathrm{s}} + c^{\mathrm{s}} u_j^{\mathrm{r}} + c^{\mathrm{s}} u_j^{\mathrm{s}}\right)^{\mathrm{r}}. \tag{6.66}$$

Equation (6.66) shows that τ_{ij}/ρ and f_j depend only on the filtered turbulence fields, but beyond that the expressions are not easy to interpret. But two aspects are evident:

1. Because the wavenumbers of the Fourier components of a product involve the sum of the wavenumbers of the Fourier components of each term in the product, multiplication of u_i^{r} by u_j^{r}, and c^{r} by u_j^{r}, generates spectral content at wavenumbers of magnitude up to $2\kappa_{\mathrm{c}}$, with κ_{c} the filter-cutoff wavenumber. Thus the first term of each of (6.66) has Fourier components of wavenumber magnitudes from κ_{c} to $2\kappa_{\mathrm{c}}$.
2. The second term of each of (6.66) involves filtered products of the r and s fields; each of these can be nonzero. Interactions of this type are sketched in the second and third panels of Figure 6.3. The subfilter-scale variables in the last term in this group can involve large, computationally unresolvable wavenumbers.

An alternative form of the filtered equation set (6.64) is

$$\tilde{u}_{i,t}^{\mathrm{r}} + \left(\tilde{u}_i^{\mathrm{r}} \tilde{u}_j^{\mathrm{r}}\right)^{\mathrm{r}}_{,j} - \frac{\tau^*_{ij,j}}{\rho} = -\frac{1}{\rho}\tilde{p}^{\mathrm{r}}_{,i}, \quad \frac{\tau^*_{ij}}{\rho} = \left(\tilde{u}_i^{\mathrm{r}} \tilde{u}_j^{\mathrm{r}}\right)^{\mathrm{r}} - (\tilde{u}_i \tilde{u}_j)^{\mathrm{r}},$$
$$\tilde{c}^{\mathrm{r}}_{,t} + \left(\tilde{c}^{\mathrm{r}} \tilde{u}_j^{\mathrm{r}}\right)^{\mathrm{r}}_{,j} + f^*_{j,j} = 0, \qquad f_j^* = (\tilde{c}\tilde{u}_j)^{\mathrm{r}} - \left(\tilde{c}^{\mathrm{r}} \tilde{u}_j^{\mathrm{r}}\right)^{\mathrm{r}}. \tag{6.67}$$

In this case the expressions for the Reynolds fluxes have only the resolved-scale term in Eq. (6.66):

$$\frac{\tau^*_{ij}}{\rho} = -\left(u_i^{\mathrm{r}} u_j^{\mathrm{s}} + u_i^{\mathrm{s}} u_j^{\mathrm{r}} + u_i^{\mathrm{s}} u_j^{\mathrm{s}}\right)^{\mathrm{r}},$$
$$f_j^* = \left(c^{\mathrm{r}} u_j^{\mathrm{s}} + c^{\mathrm{s}} u_j^{\mathrm{r}} + c^{\mathrm{s}} u_j^{\mathrm{s}}\right)^{\mathrm{r}}. \tag{6.68}$$

Both τ_{ij}, τ^*_{ij} and f_j, f_j^* are called *subfilter-scale* fluxes because they involve subfilter-scale fields. From their definition (6.68) τ^*_{ij} and f_j^* are resolved quantities, however. This illustrates the difficulty of naming such variables.

In LES the quantity $\tilde{u}_i^{\mathrm{r}}\tilde{u}_j^{\mathrm{r}}$ or $(\tilde{u}_i^{\mathrm{r}}\tilde{u}_j^{\mathrm{r}})^{\mathrm{r}}$ is calculated as the solution of the filtered equations (6.64) or (6.67) proceeds. If the filter scale Δ is much less than the turbulence scale ℓ, these can contain virtually all of the turbulent flux. This can give LES a strong advantage over computational approaches that model the turbulent flux.

The subfilter-scale fluxes in LES are important, even in cases where they are much smaller than the resolved fluxes, because they are intimately involved in interscale transfer (Problem 6.16). Thus, before the set (6.64) or (6.67) can be solved its subfilter-scale fluxes must be represented through a *subfilter-scale model.*

6.6.2 Conservation equations for subfilter-scale quantities

One can derive the conservation equations for the subfilter-scale fluxes from their definitions in Eq. (6.64) or (6.67). We'll work with (6.64) and assume the Reynolds number of the motion at the cutoff scale is large enough that the molecular diffusion terms are negligible.

6.6.2.1 Stress

It is convenient to work with the modified equation of motion

$$\tilde{u}_{i,t}^{\mathrm{r}} + (\tilde{u}_i^{\mathrm{r}}\tilde{u}_j^{\mathrm{r}})_{,j} = -\left(\frac{\tilde{p}^{\mathrm{r}}}{\rho} + \frac{2}{3}e\right)_{,i} + \tau_{ij,j}^{\mathrm{d}},$$

$$\text{where } \tau_{ij}^{\mathrm{d}} = \tilde{u}_i^{\mathrm{r}}\tilde{u}_j^{\mathrm{r}} - (\tilde{u}_i\tilde{u}_j)^{\mathrm{r}} + \frac{2}{3}e\delta_{ij}; \quad 2e = (\tilde{u}_i\tilde{u}_i)^{\mathrm{r}} - \tilde{u}_i^{\mathrm{r}}\tilde{u}_i^{\mathrm{r}}. \tag{6.69}$$

We use the superscript d because τ_{ij}^{d} is a *deviatoric* kinematic stress tensor, the difference between the kinematic subfilter-scale stress tensor $\tilde{u}_i^{\mathrm{r}}\tilde{u}_j^{\mathrm{r}} - (\tilde{u}_i\tilde{u}_j)^{\mathrm{r}}$ and its isotropic form $-2e\delta_{ij}/3$; e is the subfilter-scale TKE. This deviatoric representation facilitates the modeling of subfilter-scale stress.

The basis of the derivation is the property

$$\frac{\partial}{\partial t}\left[(\tilde{u}_i\tilde{u}_j)^{\mathrm{r}} - \tilde{u}_i^{\mathrm{r}}\tilde{u}_j^{\mathrm{r}}\right] = \left(\tilde{u}_i\frac{\partial \tilde{u}_j}{\partial t} + \tilde{u}_j\frac{\partial \tilde{u}_i}{\partial t}\right)^{\mathrm{r}} - \tilde{u}_i^{\mathrm{r}}\frac{\partial \tilde{u}_j^{\mathrm{r}}}{\partial t} - \tilde{u}_j^{\mathrm{r}}\frac{\partial \tilde{u}_i^{\mathrm{r}}}{\partial t}, \tag{6.70}$$

in which one then uses the evolution equation for $\tilde{u}_i$. The result for τ_{ij}^{d}, as defined in Eq. (6.69), is (Lilly, 1967; Hatlee and Wyngaard, 2007)

$$\begin{aligned}\frac{\partial \tau_{ij}^{\mathrm{d}}}{\partial t} + \tilde{u}_k^{\mathrm{r}}\frac{\partial \tau_{ij}^{\mathrm{d}}}{\partial x_k} = \frac{\partial}{\partial x_k}\Big[&(\tilde{u}_i\tilde{u}_j\tilde{u}_k)^{\mathrm{r}} - \tilde{u}_i^{\mathrm{r}}(\tilde{u}_j\tilde{u}_k)^{\mathrm{r}} - \tilde{u}_j^{\mathrm{r}}(\tilde{u}_i\tilde{u}_k)^{\mathrm{r}} - \tilde{u}_k^{\mathrm{r}}(\tilde{u}_i\tilde{u}_j)^{\mathrm{r}} + 2\tilde{u}_i^{\mathrm{r}}\tilde{u}_j^{\mathrm{r}}\tilde{u}_k^{\mathrm{r}} \\ &- \frac{\delta_{ij}}{3}\left(\left(\tilde{u}_l^2\tilde{u}_k\right)^{\mathrm{r}} - 2\tilde{u}_l^{\mathrm{r}}(\tilde{u}_l\tilde{u}_k)^{\mathrm{r}} - \tilde{u}_k^{\mathrm{r}}\left(\tilde{u}_l^2\right)^{\mathrm{r}} + 2\left(\tilde{u}_l^{\mathrm{r}}\right)^2\tilde{u}_k^{\mathrm{r}}\right)\Big]\end{aligned}$$

$$+\frac{2e}{3}\tilde{s}^{\mathrm{r}}_{ij}-\left[\tau^{\mathrm{d}}_{ik}\frac{\partial\tilde{u}^{\mathrm{r}}_{j}}{\partial x_k}+\tau^{\mathrm{d}}_{jk}\frac{\partial\tilde{u}^{\mathrm{r}}_{i}}{\partial x_k}-\frac{1}{3}\delta_{ij}\tau^{\mathrm{d}}_{kl}\tilde{s}^{\mathrm{r}}_{kl}\right]-\left[\frac{\tilde{p}}{\rho}\tilde{s}_{ij}\right]^{\mathrm{r}}+\frac{\tilde{p}^{\mathrm{r}}}{\rho}\tilde{s}^{\mathrm{r}}_{ij}$$
$$+\frac{1}{\rho}\frac{\partial}{\partial x_k}\left(\delta_{ik}\left[(\tilde{u}_j\tilde{p})^{\mathrm{r}}-\tilde{u}^{\mathrm{r}}_j\tilde{p}^{\mathrm{r}}\right]+\delta_{jk}\left[(\tilde{u}_i\tilde{p})^{\mathrm{r}}-\tilde{u}^{\mathrm{r}}_i\tilde{p}^{\mathrm{r}}\right]-\frac{2}{3}\delta_{ij}\left[(\tilde{u}_k\tilde{p})^{\mathrm{r}}-\tilde{u}^{\mathrm{r}}_k\tilde{p}^{\mathrm{r}}\right]\right). \tag{6.71}$$

The terms on the left side represent local time change and advection by the resolved velocity. The first term on the right is a divergence, so it integrates to zero over the flow; this is a transport term. The second and third terms on the right represent interactions between τ^{d}_{ij} and the resolved strain rate. The first of this pair, a *gradient* production term, produces τ^{d}_{ij} that is aligned with this strain rate. The second of this pair is a *tilting* production term that reorients τ^{d}_{ij}. The fourth and fifth terms are pressure destruction, and the sixth term is pressure transport.

6.6.2.2 TKE

Lilly (1967) also derived the equation for the evolution of $e=(\tilde{u}_i\tilde{u}_i)^{\mathrm{r}}-\left(\tilde{u}^{\mathrm{r}}_i\tilde{u}^{\mathrm{r}}_i\right)/2$, the TKE of the subfilter-scale motion. It reads

$$\frac{\partial e}{\partial t}+\tilde{u}^{\mathrm{r}}_k\frac{\partial e}{\partial x_k}=\frac{\tau^{\mathrm{d}}_{ij}}{2}\tilde{s}^{\mathrm{r}}_{ij}-\frac{\partial}{\partial x_k}\left(\frac{(\tilde{u}_k\tilde{u}_i\tilde{u}_i)^{\mathrm{r}}}{2}-\frac{\tilde{u}^{\mathrm{r}}_k(\tilde{u}_i\tilde{u}_i)^{\mathrm{r}}}{2}-\tilde{u}^{\mathrm{r}}_i(\tilde{u}_k\tilde{u}_i)^{\mathrm{r}}\right.$$
$$\left.+\tilde{u}^{\mathrm{r}}_i\tilde{u}^{\mathrm{r}}_i\tilde{u}^{\mathrm{r}}_k+\frac{(\tilde{u}_k\tilde{p})^{\mathrm{r}}}{\rho}-\frac{\tilde{u}^{\mathrm{r}}_k\tilde{p}^{\mathrm{r}}}{\rho}\right)-\tilde{\epsilon}. \tag{6.72}$$

On the left are local time change and advection; on the right are shear production, a pair of transport terms, and viscous dissipation.

6.6.2.3 Scalar flux

The conservation equation for the subfilter-scale scalar flux f_i, which appears in Eq. (6.64), is (Wyngaard, 2004; Hatlee and Wyngaard, 2007)

$$\frac{\partial f_i}{\partial t}+\tilde{u}^{\mathrm{r}}_j\frac{\partial f_i}{\partial x_j}$$
$$+\frac{\partial}{\partial x_j}\left((\tilde{c}\tilde{u}_i\tilde{u}_j)^{\mathrm{r}}-(\tilde{c}\tilde{u}_i)^{\mathrm{r}}\tilde{u}^{\mathrm{r}}_j-\tilde{c}^{\mathrm{r}}(\tilde{u}_i\tilde{u}_j)^{\mathrm{r}}-\tilde{u}^{\mathrm{r}}_i(\tilde{c}\tilde{u}_j)^{\mathrm{r}}+2\tilde{c}^{\mathrm{r}}\tilde{u}^{\mathrm{r}}_i\tilde{u}^{\mathrm{r}}_j\right)$$
$$+\frac{1}{\rho}\frac{\partial}{\partial x_i}\left((\tilde{p}\tilde{c})^{\mathrm{r}}-\tilde{p}^{\mathrm{r}}\tilde{c}^{\mathrm{r}}\right)=-f_j\frac{\partial\tilde{u}^{\mathrm{r}}_i}{\partial x_j}-R_{ij}\frac{\partial\tilde{c}^{\mathrm{r}}}{\partial x_j}+\frac{1}{\rho}\left(\left(\tilde{p}\frac{\partial\tilde{c}}{\partial x_i}\right)^{\mathrm{r}}-\tilde{p}^{\mathrm{r}}\frac{\partial\tilde{c}^{\mathrm{r}}}{\partial x_i}\right),$$
$$R_{ij}=(\tilde{u}_i\tilde{u}_j)^{\mathrm{r}}-\tilde{u}^{\mathrm{r}}_i\tilde{u}^{\mathrm{r}}_j. \tag{6.73}$$

The interpretation of the terms here is analogous to that for deviatoric stress: on the left side are local time change, advection by the resolved velocity, turbulent transport, and pressure transport; on the right are tilting production, gradient production, and pressure destruction. The gradient production term is interesting in that it produces a scalar flux that need not be in the direction of the gradient; we can call this a "tensor diffusivity" term (Problem 6.29).

6.6.3 Modeling subfilter-scale fluxes

As with any turbulence moment equations, Eqs. (6.71)–(6.73) involve further unknowns and cannot be solved directly. But they can provide useful insight into modeling the subfilter-scale fluxes.

Lilly's (1967) "first-order theory" for Eq. (6.71) in effect assumes a quasi-steady, locally homogeneous state in which the rates of gradient production and pressure destruction of τ_{ij}^{d} are in balance. If we write this as†

$$\frac{2e}{3}\tilde{s}_{ij}^{\mathrm{r}} = \frac{\tau_{ij}^{\mathrm{d}}}{T_{\mathrm{s}}}, \tag{6.74}$$

with T_{s} a time scale, this yields the eddy-viscosity model

$$\tau_{ij}^{\mathrm{d}} = \frac{2e}{3}T_{\mathrm{s}}\tilde{s}_{ij}^{\mathrm{r}} = K\tilde{s}_{ij}^{\mathrm{r}}. \tag{6.75}$$

Lilly suggested that the Smagorinsky (1963) model for eddy viscosity K be used,

$$K = \frac{(k\Delta)^2 D}{\sqrt{2}}, \quad D^2 = \tilde{s}_{ij}^{\mathrm{r}}\tilde{s}_{ij}^{\mathrm{r}}, \tag{6.76}$$

with Δ the grid-mesh spacing (interpreted today as the filter cutoff scale) and k a constant. By requiring that the spectrum of the resolved motion near the cutoff follow the Kolmogorov inertial subrange form (Chapter 7), Lilly showed that k is related to the inertial subrange velocity spectral constant (now generally known as the Kolmogorov constant $\alpha \simeq 1.5$) by

$$k \simeq 0.23\alpha^{-3/4} \simeq 0.17. \tag{6.77}$$

Lilly (1967) also sketched out a "second-order theory" for Eq. (6.71). It includes time-change and advection but again not the tilting-production terms, so it continues to make the subgrid deviatoric stress τ_{ij}^{d} proportional to the resolved strain-rate tensor $\tilde{s}_{ij}^{\mathrm{r}}$. Whether his neglect of tilting production here or in the first-order theory that gave Eq. (6.75) was intentional is not clear from his paper.

† The "WET" model (Chapter 5) used later in second-order closure has this spirit.

The first application of Lilly's ideas appears to be Deardorff's (1970a) numerical study of turbulent channel flow. He used 6720 grid points ($24 \times 14 \times 20$) and the eddy-viscosity closure of Eq. (6.75), with the eddy viscosity K specified as in Eq. (6.76). He found it necessary to reduce the constant k in Eq. (6.76) from its predicted value of 0.17 to 0.10, due in part to the relatively coarse resolution. He next turned to the convective boundary layer (Deardorff, 1970b), using more grid points (16 000) and the same subgrid model but with some refinements. Here he used $k = 0.21$, double his original value and closer to the prediction of Eq. (6.77). In subsequent papers Deardorff (1974a, 1974b) used modeled versions of the SFS conservation equations (6.71) and (6.73) rather than eddy-diffusivity approximations, but at the expense of a 2.5-fold increase in computation time.

More recent applications of LES have almost exclusively used eddy-diffusivity closures, Eq. (6.75) for the deviatoric SFS stress and its counterpart for SFS scalar flux,

$$f_i = -K_c \frac{\partial c^{\mathrm{r}}}{\partial x_i}. \tag{6.78}$$

The eddy diffusivities K and K_c are typically taken as

$$K = C_u \, e^{1/2} \Delta, \qquad K_c = C_c \, e^{1/2} \Delta, \tag{6.79}$$

with C_u and C_c constants. The subfilter-scale TKE, e, is typically obtained from a modeled version of its conservation equation (6.72). Lilly (1967) showed that C_u can be related to the constant for the inertial subrange of the velocity spectrum (Chapter 7). Schumann *et al.* (1980) and Moeng and Wyngaard (1988) showed that when the filter scale lies in the inertial range the constants are related by

$$\frac{C_u}{C_c} = \frac{\beta}{\alpha}, \tag{6.80}$$

with β and α the inertial range three-dimensional spectral constants for the scalar and for velocity, respectively.

6.6.4 Measuring subfilter-scale fluxes

Subfilter-scale fluxes can be measured through the "array technique" (Tong *et al.*, 1998). As discussed in more detail in Chapter 16, here the signals from a horizontal array of anemometers are filtered in the lateral direction and in time (through Taylor's hypothesis the latter is a surrogate for streamwise filtering) to obtain resolved and subfilter-scale variables. The first such experiment, called HATS (Horizontal Array Turbulence Study), was carried out in 2000 near Kettleman City,

California (Horst *et al.*, 2004). It produced the first substantial exposition of measured statistics of the subfilter-scale motion (Sullivan *et al.*, 2003).

Hatlee and Wyngaard (2007) have discussed the rate-equation SFS models

$$\frac{\partial \tau_{ij}^{\mathrm{d}}}{\partial t} = \frac{2e}{3} s_{ij}^{\mathrm{r}} - \left[\tau_{ik}^{\mathrm{d}} \frac{\partial \tilde{u}_j^{\mathrm{r}}}{\partial x_k} + \tau_{jk}^{\mathrm{d}} \frac{\partial \tilde{u}_i^{\mathrm{r}}}{\partial x_k} - \frac{\delta_{ij}}{3} \tau_{k\ell}^{\mathrm{d}} s_{k\ell}^{\mathrm{r}} \right] - \frac{\tau_{ij}^{\mathrm{d}}}{T}, \tag{6.81}$$

$$\frac{\partial f_i}{\partial t} = -f_j \frac{\partial \tilde{u}_i^{\mathrm{r}}}{\partial x_j} - R_{ij} \frac{\partial \tilde{c}^{\mathrm{r}}}{\partial x_j} - \frac{f_i}{T}. \tag{6.82}$$

Each of these retains the pair of production terms on the rhs of the SFS flux conservation equations (6.71) and (6.73) and models the pressure destruction term.

Their tests of these models with the HATS data are summarized in Figures 6.4 and 6.5. In each case the simple rate-equation model is a clear improvement over

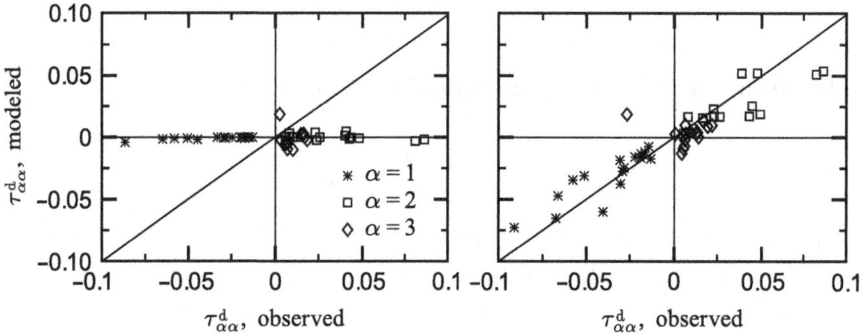

Figure 6.4 Modeled (ordinate) and observed (abscissa) SFS deviatoric kinematic normal stresses (units $\mathrm{m^2\ s^{-2}}$) in the HATS experiment. The left panel uses the Smagorinsky model, Eqs. (6.75)–(6.77); the right panel uses the simple rate-equation model (6.81). From Hatlee and Wyngaard (2007).

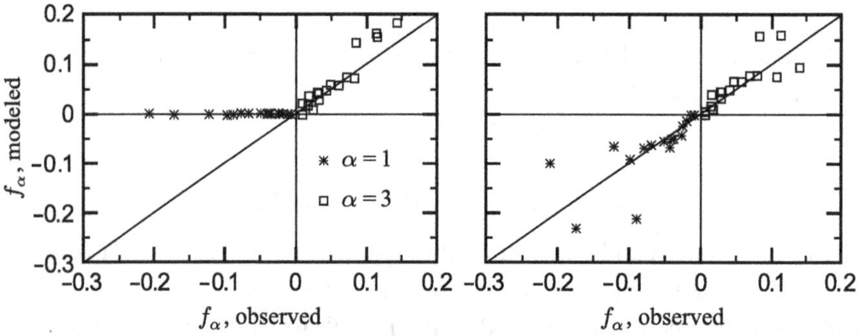

Figure 6.5 Modeled (ordinate) and observed (abscissa) temperature fluxes (units $\mathrm{K\ m\ s^{-1}}$) in HATS. The left panel uses the eddy-diffusivity SFS model (6.78) and (6.79); the right panel uses the simple rate-equation model (6.82). From Hatlee and Wyngaard (2007).

the standard eddy-diffusivity model. The diagonal components of SFS deviatoric stress, Figure 6.4, are maintained largely by the tilting term in Eq. (6.81), which the standard eddy-diffusivity closure neglects. The differences between the predictions of τ_{13} are less dramatic. Similarly, the standard eddy-diffusivity model fails for SFS horizontal scalar flux because it neglects its production mechanism, the tilting and tensor-diffusivity terms in the SFS flux conservation equation (6.82). Chen *et al.* (2009) have presented a formalism for testing SFS models that is based on the pdf evolution equation (Chapter 13).

6.6.5 Sensitivity to flux modeling

As we saw in Chapter 5, one must model all the flux in the ensemble-averaged equations. But one need model only the *unresolved* flux in the space-averaged equations. When the filter scale (grid resolution) lies in the inertial subrange, as is ideal in LES, only a small fraction of the flux is unresolved. We cannot neglect this unresolved flux; it is essential for the transfer of energy and scalar variance out of the resolved scales (Problem 6.16). If the model for this unresolved flux represents this mean rate of transfer well, which Lilly (1967) showed can be accomplished by adjusting its proportionality constant, the parent model can appear to perform well even if the SFS model represents the unresolved flux poorly. As Chen *et al.* (2009) show, statistical analyses based on pdf evolution equations (Chapter 13) can be required to expose SFS model deficiencies.

6.6.6 The effective Reynolds number

The effective Reynolds number of the filtered Navier–Stokes equation (6.64) with an eddy-viscosity closure is of order $UL/\overline{K}$, where $\overline{K}$ is a representative eddy viscosity. If this Reynolds number is large enough the resolved flow will be turbulent. If it is even larger the flow can at least appear to enter the Re-independent range. One can show that this requires $\Delta \ll \ell$ (Problem 6.22). This seems to be achievable in many engineering and atmospheric boundary layer applications, but is difficult in simulations of severe convective storms (Bryan *et al.*, 2003) and has yet to be achieved for a hurricane.

Appendix 6.1

TKE decomposition

Given an ensemble of realizations of a real, homogeneous, one-dimensional, zero-mean random function $u_i(x;\alpha)$ on an interval $0 \le x \le L$, as in Chapter 2, we can approximate any realization of u_i through a Fourier series:

$$u_i(x;\alpha) = \sum_{n=-N}^{N} \hat{u}(\kappa_n;\alpha)e^{i\kappa_n x}, \tag{6.83}$$

with $\kappa_n = 2\pi n/L$. Since u_i is random its Fourier coefficients are different in each realization.

Let us apply a wave-cutoff filter to u_i, with the cutoff wavenumber set at $2\pi N_c/L$, with $N_c \le N$. Fourier coefficients of wavenumber less than or equal to $2\pi N_c/L$ are passed without change; those of wavenumber greater than $2\pi N_c/L$ are made zero. Then we can write the resolvable and subfilter-scale parts of f as

$$
\begin{aligned}
u_i^{\mathrm{r}}(x;\alpha) &= \sum_{n=-N}^{N} \hat{u}_i(\kappa_n;\alpha) T(\kappa_n) e^{i\kappa_n x} = \sum_{n=-N_c}^{N_c} \hat{u}_i(\kappa_n;\alpha) e^{i\kappa_n x},\\
u_i^{\mathrm{s}}(x;\alpha) &= u_i - u_i^{\mathrm{r}} = \sum_{|n|>N_c} \hat{u}_i(\kappa_n;\alpha) e^{i\kappa_n x} \\
&= \sum_{n=-N_c-1}^{-N} \hat{u}_i(\kappa_n;\alpha) e^{i\kappa_n x} + \sum_{n=N_c+1}^{N} \hat{u}_i(\kappa_n;\alpha) e^{i\kappa_n x}.
\end{aligned} \tag{6.84}
$$

Then the covariance $\overline{u_i^{\mathrm{r}} u_i^{\mathrm{s}}}$ is

$$
\overline{u_i^{\mathrm{r}} u_i^{\mathrm{s}}} = \sum_{n=-N_c}^{N_c} \sum_{m=-N_c-1}^{-N} \overline{\hat{u}_i(\kappa_n)\hat{u}_i(\kappa_m)} + \sum_{n=-N_c}^{N_c} \sum_{m=N_c+1}^{N} \overline{\hat{u}_i(\kappa_n)\hat{u}_i(\kappa_m)} = 0, \tag{6.85}
$$

because the wave-cutoff filter ensures that the r and s components have no Fourier modes in common so there are no contributions to the sum. This holds in three spatial dimensions, so the wave-cutoff filter yields $\overline{u_i u_i} = \overline{u_i^{\mathrm{r}} u_i^{\mathrm{r}}} + \overline{u_i^{\mathrm{s}} u_i^{\mathrm{s}}}$, neatly separating the TKE into *resolvable-* and *subfilter-scale* components.

Appendix 6.2

Interscale transfer

In our homogeneous turbulence "thought problem" of Section 6.3 we deduced that the resolvable-TKE budget (6.30) contains a term $-\overline{u_i^{\mathrm{r}}(u_i u_j)^{\mathrm{r}}_{,j}}$ and the subfilter-TKE budget (6.36) contains an analogous term $-\overline{u_i^{\mathrm{s}}(u_i u_j)^{\mathrm{s}}_{,j}}$. We deduced that these represent "interscale transfer" – transfer of energy, without loss, from the r scales to the s scales.

This term appears also in the horizontally homogeneous boundary-layer problem in Section 6.4. In that problem we applied the filter in the horizontal plane, rather than in three dimensions. For that reason we would like to analyze this term without requiring three-dimensional homogeneity.

Since our wave-cutoff filter makes r and s quantities uncorrelated, for the term in question in Eq. (6.30) we can write

$$
\begin{aligned}
-\overline{u_i^{\mathrm{r}}(u_i u_j)^{\mathrm{r}}_{,j}} &= -\overline{u_i^{\mathrm{r}}[(u_i u_j)_{,j} - (u_i u_j)^{\mathrm{s}}_{,j}]} = -\overline{u_i^{\mathrm{r}}(u_i u_j)_{,j}} \\
&= -\overline{u_i^{\mathrm{r}}[(u_i^{\mathrm{r}} + u_i^{\mathrm{s}})(u_j^{\mathrm{r}} + u_j^{\mathrm{s}})]_{,j}} = -\overline{u_i^{\mathrm{r}}[u_i^{\mathrm{r}}u_j^{\mathrm{r}} + u_i^{\mathrm{s}}u_j^{\mathrm{r}} + u_i^{\mathrm{r}}u_j^{\mathrm{s}} + u_i^{\mathrm{s}}u_j^{\mathrm{s}}]_{,j}} \\
&= -\overline{u_i^{\mathrm{r}}u_j^{\mathrm{r}}u_{i,j}^{\mathrm{r}}} - \overline{u_i^{\mathrm{r}}u_j^{\mathrm{r}}u_{i,j}^{\mathrm{s}}} - \overline{u_i^{\mathrm{r}}u_j^{\mathrm{s}}u_{i,j}^{\mathrm{r}}} - \overline{u_i^{\mathrm{r}}u_j^{\mathrm{s}}u_{i,j}^{\mathrm{s}}}.
\end{aligned} \tag{6.86}
$$

The corresponding result for the subfilter-scale TKE budget (6.36) is

$$
-\overline{u_i^{\mathrm{s}}(u_i u_j)^{\mathrm{s}}_{,j}} = -\overline{u_i^{\mathrm{s}}u_j^{\mathrm{r}}u_{i,j}^{\mathrm{r}}} - \overline{u_i^{\mathrm{s}}u_j^{\mathrm{r}}u_{i,j}^{\mathrm{s}}} - \overline{u_i^{\mathrm{s}}u_j^{\mathrm{s}}u_{i,j}^{\mathrm{r}}} - \overline{u_i^{\mathrm{s}}u_j^{\mathrm{s}}u_{i,j}^{\mathrm{s}}}. \tag{6.87}
$$

We expect interscale-transfer terms, like the shear production and viscous dissipation terms in the TKE budget (5.42), to have the form of a stress tensor contracted with a strain-rate tensor. Equations (6.86) and (6.87) contain velocity gradients, not strain rates, but we can express a velocity gradient as a sum of strain-rate and rotation-rate tensors s_{ij} and r_{ij}:

$$
u_{i,j} = \frac{u_{i,j} + u_{j,i}}{2} + \frac{u_{i,j} - u_{j,i}}{2} = s_{ij} + r_{ij}. \tag{6.88}
$$

s_{ij} and r_{ij} are symmetric and antisymmetric, respectively. Since a stress tensor is symmetric, its contraction with r_{ij} vanishes. Thus, on the right side of Eq. (6.87), for example, we can replace $\overline{u_i^{\mathrm{s}}u_j^{\mathrm{s}}u_{i,j}^{\mathrm{r}}}$ with $\overline{u_i^{\mathrm{s}}u_j^{\mathrm{s}}s_{ij}^{\mathrm{r}}}$.

We now rewrite (6.86) and (6.87) as the sum of interscale-transfer and turbulent-transport terms, using some guidelines. The interscale-transfer terms must contain both r and s elements, because they occur in both equations, and they must sum to zero. Keeping the terms in the same order and indicating with brackets terms that we have broken into two parts, we rewrite Eqs. (6.86) and (6.87) as

$$
\begin{aligned}
-\overline{u_i^{\mathrm{r}}(u_i u_j)^{\mathrm{r}}_{,j}} &= -\frac{1}{2}(\overline{u_i^{\mathrm{r}}u_i^{\mathrm{r}}u_j^{\mathrm{r}}})_{,j} - \overline{u_i^{\mathrm{r}}u_j^{\mathrm{r}}s_{ij}^{\mathrm{s}}} - \frac{1}{2}(\overline{u_i^{\mathrm{r}}u_i^{\mathrm{r}}u_j^{\mathrm{s}}})_{,j} \\
&\quad - \left[(\overline{u_i^{\mathrm{r}}u_i^{\mathrm{s}}u_j^{\mathrm{s}}})_{,j} - \overline{u_i^{\mathrm{s}}u_j^{\mathrm{s}}s_{ij}^{\mathrm{r}}}\right],
\end{aligned} \tag{6.89}
$$

$$
\begin{aligned}
-\overline{u_i^{\mathrm{s}}(u_i u_j)^{\mathrm{s}}_{,j}} &= -\left[(\overline{u_i^{\mathrm{r}}u_i^{\mathrm{s}}u_j^{\mathrm{r}}})_{,j} - \overline{u_i^{\mathrm{r}}u_j^{\mathrm{r}}s_{ij}^{\mathrm{s}}}\right] - \frac{1}{2}(\overline{u_i^{\mathrm{s}}u_i^{\mathrm{s}}u_j^{\mathrm{r}}})_{,j} \\
&\quad - [\overline{u_i^{\mathrm{s}}u_j^{\mathrm{s}}s_{ij}^{\mathrm{r}}}] - \frac{1}{2}(\overline{u_i^{\mathrm{s}}u_i^{\mathrm{s}}u_j^{\mathrm{s}}})_{,j}.
\end{aligned} \tag{6.90}
$$

Grouping like parts gives

$$
-\overline{u_i^{\mathrm{r}}(u_i u_j)^{\mathrm{r}}_{,j}} = -\overline{u_i^{\mathrm{r}}u_j^{\mathrm{r}}s_{ij}^{\mathrm{s}}} + \overline{u_i^{\mathrm{s}}u_j^{\mathrm{s}}s_{ij}^{\mathrm{r}}} - \frac{1}{2}(\overline{u_i^{\mathrm{r}}u_i^{\mathrm{r}}u_j})_{,j} - (\overline{u_i^{\mathrm{r}}u_i^{\mathrm{s}}u_j^{\mathrm{s}}})_{,j}, \tag{6.91}
$$

$$
-\overline{u_i^{\mathrm{s}}(u_i u_j)^{\mathrm{s}}_{,j}} = \overline{u_i^{\mathrm{r}}u_j^{\mathrm{r}}s_{ij}^{\mathrm{s}}} - \overline{u_i^{\mathrm{s}}u_j^{\mathrm{s}}s_{ij}^{\mathrm{r}}} - \frac{1}{2}(\overline{u_i^{\mathrm{s}}u_i^{\mathrm{s}}u_j})_{,j} - (\overline{u_i^{\mathrm{r}}u_i^{\mathrm{s}}u_j^{\mathrm{r}}})_{,j}. \tag{6.92}
$$

In each of (6.91) and (6.92) the first pair of terms has the form required for interscale transfer, a contraction of stress and strain-rate tensors. They are equal in magnitude but of opposite signs, also as required. We designate these as the interscale-transfer terms and label them as $-I$ and I, respectively:

$$I = \overline{u_i^{\mathrm{r}} u_j^{\mathrm{r}} s_{ij}^{\mathrm{s}}} - \overline{u_i^{\mathrm{s}} u_j^{\mathrm{s}} s_{ij}^{\mathrm{r}}}. \tag{6.93}$$

As we shall discuss in Chapter 7, interscale transfer of energy exists in both two- and three-dimensional turbulence, but their physical mechanisms differ.

Noting that $\overline{u_i^{\mathrm{r}} u_i^{\mathrm{s}} u_j^{\mathrm{r}}} + \overline{u_i^{\mathrm{r}} u_i^{\mathrm{s}} u_j^{\mathrm{s}}} = \overline{u_i^{\mathrm{r}} u_i^{\mathrm{s}} (u_j^{\mathrm{r}} + u_j^{\mathrm{s}})} = \overline{u_i^{\mathrm{r}} u_i^{\mathrm{s}} u_j}$, the second pairs sum to turbulent transport, as required:

$$\begin{aligned} &-\frac{1}{2}(\overline{u_i^{\mathrm{r}} u_i^{\mathrm{r}} u_j})_{,j} - (\overline{u_i^{\mathrm{r}} u_i^{\mathrm{s}} u_j^{\mathrm{s}}})_{,j} - \frac{1}{2}(\overline{u_i^{\mathrm{s}} u_i^{\mathrm{s}} u_j})_{,j} - (\overline{u_i^{\mathrm{r}} u_i^{\mathrm{s}} u_j^{\mathrm{r}}})_{,j} \\ &\quad = -\frac{1}{2}\left(\overline{u_i^{\mathrm{r}} u_i^{\mathrm{r}} u_j} + 2\overline{u_i^{\mathrm{r}} u_i^{\mathrm{s}} u_j} + \overline{u_i^{\mathrm{s}} u_i^{\mathrm{s}} u_j}\right)_{,j} = -\frac{1}{2}\left(\overline{u_i u_i u_j}\right)_{,j}. \end{aligned} \tag{6.94}$$

Thus, $-\overline{u_i^{\mathrm{r}} (u_i u_j)^{\mathrm{r}}_{,j}}$ and $-\overline{u_i^{\mathrm{s}} (u_i u_j)^{\mathrm{s}}_{,j}}$, Eqs. (6.91) and (6.92), are each the sum of an interscale-transfer term and a turbulent-transport term. In the fully homogeneous problem of Section 6.3 the turbulent-transfer terms vanish. In the horizontally homogeneous boundary layer problem of Section 6.4 there is turbulent transport in the inhomogeneous vertical direction.

The analogous procedure with the conserved scalar conservation equation yields the expression for the rate of interscale transfer of scalar variance:

$$I_c = 2\overline{c^{\mathrm{r}} u_j^{\mathrm{r}} c^{\mathrm{s}}_{,j}} - 2\overline{c^{\mathrm{s}} u_j^{\mathrm{s}} c^{\mathrm{r}}_{,j}}. \tag{6.95}$$

Here the interscale transfer mechanisms are the same in two- and three-dimensional turbulence (Chapter 7).

Questions on key concepts

6.1 Explain why the space average, unlike the ensemble average, does not eliminate randomness.

6.2 Explain why a spatial filter cannot be sharp in both physical and wavenumber space, as is reflected in Figures 6.1 and 6.2.

6.3 The running-mean and wave-cutoff filters, Figures 6.1 and 6.2, are positive definite in one space but not in the other. What filter is positive in both spaces?

6.4 Explain physically why wave-cutoff filtering makes the r and s parts of a filtered variable uncorrelated.

6.5 Explain the relationship between Eqs. (6.71) and (5.39), (5.41) which concern two forms of the Reynolds stress tensor.

6.6 Explain physically why molecular diffusion terms are absent in the filtered Navier–Stokes and scalar conservation equations. What takes over their important role in the variance budgets?

6.7 Explain why ensemble-averaged equations have no terms involving derivatives in a homogeneous direction, but the space-averaged equations do. Why is this important?

6.8 In what limit should the conservation equations for subfilter-scale fluxes, Eqs. (6.71) and (6.73), become their ensemble-averaged counterparts from Chapter 5?

6.9 Outline and explain the concepts underlying LES. Contrast it to second-moment modeling.

6.10 Explain why LES is considered to be an inherently more reliable form of turbulence modeling than that based on the second-moment equations of Chapter 5.

6.11 Compare the computational requirements for LES and second-moment modeling in the same problem – a horizontally homogeneous boundary layer, for example. What can you conclude from this?

6.12 In view of your result in Question 6.11, discuss how LES might be used to improve second-moment modeling. In what applications might this approach be useful?

6.13 Explain the advantages of placing the LES cutoff scale in the inertial subrange.

Problems

6.1 Prove that the running-mean filter in one dimension has the filter function given in Eq. (6.9).

6.2 Show that for the filter (6.1)

$$\left(\frac{\partial u}{\partial x}\right)^{\mathrm{r}} = \frac{\partial u^{\mathrm{r}}}{\partial x};$$

i.e., that it commutes with spatial differentiation.

6.3 Prove that the wave-cutoff filter in one dimension satisfies the constraint (6.21).

6.4 Derive and interpret the equations for the variance of the resolvable and subfilter-scale parts of a conserved scalar. Discuss the variance budget in the inertial subrange of spatial scales.

6.5 Discuss the difference between Δ, the scale of the spatial filter, and Δ_{g}, the mesh size of the spatial grid on which the filtered equations are solved numerically. Interpret the cases $\Delta_g \ll \Delta$ and $\Delta_g \gg \Delta$. How would you choose Δ_g?

6.6 George Crunch found a three-dimensional, time-dependent numerical model on the Web but doesn't understand it fully. George can see that the model uses an eddy diffusivity to close the equations, but there is no indication whether the model deals with an ensemble mean or a spatial average. How should he be able to tell by examining the eddy diffusivity?

6.7 Explain the ideas underlying LES.

6.8 Assume that the subgrid-scale stress tensor in the resolvable-scale velocity equation (6.60) is modeled like the Newtonian viscous stress tensor but with constant "eddy" viscosity. Show that this model does give a qualitatively correct rate of interscale energy transfer. How can it be made quantitatively correct?

6.9 Carry out the integral in Eq. (6.12).

6.10 Derive the conservation equation for the evolution of the mean-squared gradient of a conserved scalar and identify the two leading terms. Which is the principal source? Principal sink?

6.11 Show that ensemble averaging commutes with spatial filtering.

6.12 Show that G and T for a running-mean filter, Eqs. (6.9) and (6.5), are a Fourier transform pair.

6.13 Prove that the scalar field in Subsection 6.3.5 can be homogeneous and steady, as claimed.

6.14 Explain where the scale of the filter in LES would ideally be placed relative to ℓ.

6.15 Show that we can neglect the molecular diffusion term in Eq. (6.33).

6.16 There is no viscous dissipation in LES. Why? Use the momentum equation in Eq. (6.64) and a decomposition of its variables into ensemble-mean and fluctuating parts to derive a TKE equation for LES. What takes the place of the viscous dissipation term? Explain.

6.17 Show that Eq. (6.57) results from adding Eqs. (6.54) and (6.56).

6.18 Osborne Reynolds used spatial averages. Do the "Reynolds averaging rules" apply to them? Do they apply in any limit? Discuss.

6.19 Explain why high effective Reynolds numbers can easily be attained in LES of the atmospheric boundary layer but can be difficult to attain for severe storms.

6.20 If you wanted ensemble-mean statistics from LES of homogeneous boundary-layer flows, how would you produce them? Explain.

6.21 How would you achieve randomness in LES? (Hint: revisit our definition of randomness, Chapter 1.)

6.22 Show that reaching the Re-independent range with LES requires $\Delta \ll \ell$.

6.23 In Eq. (6.65) we assumed that ensemble means pass perfectly through the filter. Is that a reasonable assumption? Explain.

6.24 Explain the intermediate steps in Eq. (6.39).

6.25 Show that the transfer function of the Gaussian filter, Eq. (6.15), is also Gaussian.

6.26 Derive Eq. (6.52).

6.27 Explain why we say that the filtered product $(\tilde{u}_i \tilde{u}_j)^{\mathrm{r}}$ in the set (6.63) is unknown. Show that if we approximate it as $\tilde{u}_i^{\mathrm{r}} \tilde{u}_j^{\mathrm{r}}$, then the set could be solved numerically. Show that its kinetic-energy equation would lose its sink term, however, and so could not give steady solutions.

6.28 Interpret the limit of Eq. (6.56) as the filter scale decreases within the inertial subrange.

6.29 Discuss the nature of the gradient production term in the SFS flux conservation equation (6.73). Why does the SFS scalar flux it produces need not be in the direction of the resolved gradient? Why could we call it a "tensor diffusivity" term?

References

Bryan, G. H., J. C. Wyngaard, and J. M. Fritsch, 2003: On adequate resolution for the simulation of deep moist convection. *Mon. Wea. Rev.*, **131**, 2394–2416.

Charney, J., R. Fjortoft, and J. von Neumann, 1950: Numerical integration of the barotropic vorticity equation. *Tellus*, **2**, 237–254.

Chen, Q., M. J. Otte, P. P. Sullivan, and C. Tong, 2009: *A posteriori* subgrid-scale model tests based on the conditional means of subgrid-scale stress and its production rate. *J. Fluid Mech.*, **626**, 149–181.

Deardorff, J. W., 1970a: A numerical study of three-dimensional turbulent channel flow at large Reynolds numbers. *J. Fluid Mech.*, **41**, 453–480.

Deardorff, J. W., 1970b: Preliminary results from numerical integrations of the unstable planetary boundary layer. *J. Atmos. Sci.*, **27**, 1209–1211.

Deardorff, J. W., 1974a: Three-dimensional numerical study of the height and mean structure of a heated planetary boundary layer. *Bound.-Layer Meteor.*, **7**, 81–106.

Deardorff, J. W., 1974b: Three-dimensional numerical study of turbulence in an entraining mixed layer. *Bound.-Layer Meteor.*, **7**, 199–226.

Hatlee, S. C., and J. C. Wyngaard, 2007: Improved subgrid-scale models for LES from field measurements. *J. Atmos. Sci.*, **64**, 1694–1705.

Horst, T. W., J. Kleissl, D. H. Lenschow, *et al.*, 2004: HATS: Field observations to obtain spatially-filtered turbulence fields from crosswind arrays of sonic anemometers in the atmospheric surface layer. *J. Atmos. Sci.*, **61**, 1566–1581.

Lilly, D. K., 1967: The representation of small-scale turbulence in numerical simulation experiments. *Proceedings of the IBM Scientific Computing Symposium on Environmental Sciences*, IBM Form no. 320-1951, pp. 195–210.

Moeng, C.-H., and J. C. Wyngaard, 1988: Spectral analysis of large-eddy simulations of the convective boundary layer. *J. Atmos. Sci.*, **45**, 3573–3587.

Phillips, N., 1955: The general circulation of the atmosphere: a numerical experiment. Presented at the Conference on Application of Numerical Integration Techniques to the Problem of the General Circulation. *Dynamics of Climate*, R. Pfeffer, Ed., Oxford: Pergamon Press, pp. 18–25.

Phillips, N., 1956: The general circulation of the atmosphere: a numerical experiment. *Quart. J. Roy. Meteor. Soc.*, **82**, 123–164.

Schumann, U., G. Grotzbach, and L. Kleiser, 1980: Direct numerical simulation of turbulence. In *Prediction Methods for Turbulent Flows*, W. Kollman, Ed., Washington: Hemisphere, pp. 123–158.

Smagorinsky, J., 1963: General circulation experiments with the primitive equations: Part 1, The basic experiment. *Mon. Wea. Rev.*, **94**, 99–164.

Sullivan, P. P., T. W. Horst, D. H. Lenschow, C.-H. Moeng, and J. C. Weil, 2003: Structure of subfilter-scale fluxes in the atmospheric surface layer with application to LES modeling. *J. Fluid Mech.*, **482**, 101–139.

Tong, C., J. C. Wyngaard, S. Khanna, and J. G. Brasseur, 1998: Resolvable- and subgrid-scale measurement in the atmospheric surface layer: technique and issues. *J. Atmos. Sci.*, **55**, 3114–3126.

Wyngaard, J. C., 2004: Toward numerical modeling in the Terra Incognita. *J. Atmos. Sci.*, **61**, 1816–1826.

7

Kolmogorov scaling, its extensions, and two-dimensional turbulence

7.1 The inertial subrange

As we saw in Chapter 6, turbulent flows of large R_t have a wide range of eddies smaller than ℓ and larger than η. Being dominated by inertial forces, these eddies lie in the *inertial subrange*.

Eddy sizes in the mid regions of a typical daytime atmospheric boundary layer range from roughly 1000 m to 1 mm, which is six decades. Eddies in the central three decades, from 3 cm to 30 m, say, contribute negligibly to TKE, fluxes, and viscous dissipation and might be taken as the inertial subrange. Turbulent flows in pipes, channels, and jets generally have much smaller R_t than geophysical flows but can have an identifiable, if shorter, inertial subrange.

7.1.1 Energetics

In Chapter 6 we decomposed the velocity fields in a horizontally homogeneous, equilibrium boundary layer into resolvable and subfilter-scale parts and derived the balance equations for their mean kinetic energy per unit mass, $\mathrm{TKE_r}$ and $\mathrm{TKE_s}$, respectively. When the scale Δ that separates r and s eddies is such that $\ell \gg \Delta \gg \eta$, these TKE equations say

sum of rates of mean advection, production, turbulent transport, and pressure transport of $\mathrm{TKE_r}$
= rate of loss of $\mathrm{TKE_r}$ by transfer to subfilter scales, (7.1)

rate of gain of $\mathrm{TKE_s}$ by transfer from resolvable scales
= rate of loss of $\mathrm{TKE_s}$ through viscous dissipation $= \epsilon$. (7.2)

Thus, the TKE balance within the inertial subrange is simply

rate of gain of TKE by transfer from larger scales
= rate of loss of TKE by transfer to smaller scales $= \epsilon$. (7.3)

The scalar-variance budget in the inertial subrange is analogous:

rate of gain of scalar variance by transfer from larger scales
= rate of loss of scalar variance by transfer to smaller scales = χ_c. (7.4)

The balances (7.3) and (7.4) are the foundation of Kolmogorov's scaling arguments for the inertial subrange of the energy spectrum and their extension by Obukhov (1949) and Corrsin (1951) to the scalar spectrum.

7.1.2 The velocity spectrum

In Chapter 2 we introduced the three-dimensional velocity spectrum $E(\kappa)$, which has the property

$$\frac{\overline{u_i u_i}}{2} = \int_0^\infty E(\kappa)\, d\kappa. \tag{2.63}$$

Kolmogorov (1941) hypothesized that the only scaling parameters for its inertial subrange are the energy cascade rate, ϵ, and wavenumber, κ. If so, on dimensional grounds it has the form

$$E(\kappa) = \alpha \epsilon^{2/3} \kappa^{-5/3}, \tag{7.5}$$

with α now known as the Kolmogorov constant. This prediction was first confirmed in the large-R_t flow in a tidal channel (Grant *et al.*, 1962). Kolmogorov further hypothesized that the scales of the smallest, dissipative eddies depend only on ϵ and ν; if so, they must be

$$\text{velocity scale } \upsilon = (\nu\epsilon)^{1/4}, \qquad \text{length scale } \eta = \left(\frac{\nu^3}{\epsilon}\right)^{1/4}. \tag{1.34}$$

In his book on G. I. Taylor, Batchelor (1996) gives some historical background on these developments by Kolmogorov:

when I arrived in Cambridge in April 1945 to work for a PhD on turbulence under his supervision ... I cast about for new ideas, and in the course of searching the literature came across two brief papers published in the literature in the USSR in 1941 in which Kolmogorov put forward the idea of statistical equilibrium of the small-scale components of turbulent motions and which had miraculously made their way safely to a Cambridge library. I was excited by them, and told G.I. what I had found.

A little later, during the summer of 1945, ... German atomic scientists W. Heisenberg and C. F. von Weizsacker described to G.I. some new ideas on ... the transfer of energy from large to small-scale components, and ... it was evident to G.I. and me that they had points in common with Kolmogorov's theory. ... it was later noticed that in 1945 the distinguished physical chemist L. Onsager at Yale University had published an abstract in which essentially the same idea of statistical equilibrium of the small-scale components was put forward independently.†

On looking back at this incident I recall that G.I. did not draw attention ... to the closeness of both Kolmogorov's theory and these ideas of Heisenberg and von Weisacker to some of his own much earlier work. ... The important idea that he [Taylor] appears to have missed was that the statistical quasi-equilibrium of the small-scale motions depends on such a small number of parameters, namely two, the rate of energy dissipation and the fluid viscosity, that dimensional arguments alone yield explicit results.

Kolmogorov's hypotheses might now appear as straightforward or unremarkable, but they were a profound advance in our conceptualization of turbulence dynamics.

7.1.3 The extent of the inertial subrange

We saw in Section 2.5 that $u(r)$, the velocity scale of eddies of size r, is

$$u(r) \sim (\epsilon r)^{1/3}. \tag{7.6}$$

The *eddy Reynolds number* $Re(r) = u(r)r/\nu$ is then

$$Re(r) \sim \frac{(\epsilon r)^{1/3} r}{\nu} \sim \left(\frac{r}{\eta}\right)^{4/3}, \tag{7.7}$$

where we have used the definition (1.34) of the Kolmogorov microscale η.

One expects that the inertial subrange ends at eddy scale $r_{\text{end}} \sim 1/\kappa_{\text{end}}$ where $Re(r)$ decreases to a minimum value Re_{min}. If so, then from (7.7) we have

$$\frac{r_{\text{end}}}{\eta} \sim (Re_{\text{min}})^{3/4}; \qquad \kappa_{\text{end}}\, \eta \sim (Re_{\text{min}})^{-3/4} \quad \text{(velocity spectrum)}. \tag{7.8}$$

If Re_{min} is in the range 10–100, from Eq. (7.8) that would correspond to eddy scales $r_{\text{end}} \simeq 6\eta$–$30\eta$ and $\kappa_{\text{end}}\, \eta \simeq 0.03$–$0.2$, which is consistent with observations.

7.1.4 The scalar spectrum

The spectral physics of a conserved scalar (Chapter 6) is broadly like that of velocity: large-scale turbulence acting on a mean scalar gradient generates scalar fluctuations; distortion of these fluctuations by turbulence "cascades" scalar variance to smaller scales; the smoothing effects of molecular diffusion bring the cascade to an end in the smallest scales.

Obukhov (1949) and Corrsin (1951) extended Kolmogorov's scaling arguments to a conserved scalar, assuming that in the inertial range its spectral density $E_c(\kappa)$

† Eyink and Sreenivasan (2006) have written at length on Onsager's original and deep contributions to turbulence.

depends on the scalar variance cascade rate, which is equal to the molecular destruction rate χ_c; wavenumber κ; and the energy cascade rate, which is equal to the viscous dissipation rate ϵ. On dimensional grounds it follows that

$$E_c = \beta \chi_c \epsilon^{-1/3} \kappa^{-5/3}, \tag{7.9}$$

with β a constant. Equivalently, for eddy scales r in the inertial range the scalar intensity $c(r)$ is hypothesized to depend only on χ_c, ϵ, and r, so that

$$c(r) \sim \chi_c^{1/2} \epsilon^{-1/6} r^{1/3}, \tag{7.10}$$

which is the counterpart of Eq. (2.66) for $u(r)$.

Turbulent advection and diffusion effects on a conserved scalar eddy of size r are in the ratio†

$$\frac{\text{turbulent advection}}{\text{molecular diffusion}} = \frac{c(r)u(r)/r}{\gamma c(r)/r^2} = \frac{u(r)r}{\gamma} \equiv Co(r). \tag{7.11}$$

When γ is the thermal diffusivity the dimensionless group $u(r)r/\gamma$ is an eddy Péclet number. When γ is the diffusivity of any conserved scalar we'll define it as the eddy *Corrsin* number.‡ It is for scalar eddies what the eddy Reynolds number $Re(r)$ is for velocity eddies. In analogy with Re_t, we'll also define $Co_t = u\ell/\gamma$ as the turbulence Corrsin number.

As we did for the eddy Reynolds number in Eq. (7.7), we can write the eddy Corrsin number, Eq. (7.11), as

$$Co(r) \sim \frac{u(r)r}{\gamma} \sim \left(\frac{r}{\eta_{oc}}\right)^{4/3}, \tag{7.12}$$

with $\eta_{oc} = (\gamma^3/\epsilon)^{1/4}$ a microscale made from the scalar diffusivity γ. Introduced independently by Obukhov (1949) and Corrsin (1951), it is called the *Obukhov–Corrsin* scale.

7.1.5 The scalar spectrum beyond the inertial subrange

When γ is a mass diffusivity the ratio ν/γ is called the Schmidt number Sc; when it is the thermal diffusivity it is called the Prandtl number Pr. This ratio can range from very small values to very large. Yueng *et al.* (2002) indicate that in diverse applications Sc can range from 10^{-3} to thousands; likewise, Pr ranges from very

† This is a scaling approach used by Corrsin (1951).

‡ Stanley Corrsin (1920–1986), an engineering professor at Johns Hopkins University, was a well-known turbulence researcher (Lumley and Davis, 2003).

small values in mercury to very large values in organic fluids. We'll consider next the behavior of the scalar spectrum beyond the inertial subrange in the limits of very small and very large values of ν/γ, and then also briefly discuss the velocity spectrum beyond the inertial subrange.

7.1.5.1 The inertial-diffusive subrange: $\nu/\gamma \ll 1$

Here molecular diffusion begins to act on the scalar field at wavenumbers at which the velocity field is still effectively inviscid. The rate of destruction of scalar fluctuations by this diffusion causes the scalar variance cascade rate to decrease with increasing κ, attenuating the scalar spectrum at wavenumbers in the inertial subrange of the velocity field. If we assume that Co_{min}, the eddy Corrsin number at the end of the scalar inertial subrange, equals Re_{min}, then from (7.12) we have

$$\frac{r_{\text{end}}}{\eta_{\text{oc}}} \sim (Re_{\text{min}})^{3/4} \quad \text{(scalar spectrum).} \tag{7.13}$$

Corrsin (1964) assumed that in the initial part of this subrange the scalar spectrum continues to behave as in the inertial range, Eq. (7.9), but with a wavenumber-dependent scalar cascade rate $T(\kappa)$:

$$E_c \sim T(\kappa)\epsilon^{-1/3}\kappa^{-5/3}. \tag{7.14}$$

The rate of decrease of $T(\kappa)$ due to smoothing of the scalar field by the molecular diffusivity is given by

$$\frac{dT}{d\kappa} = -2\gamma\kappa^2 E_c(\kappa). \tag{7.15}$$

Corrsin (1964) found the solution of the system (7.14) and (7.15) to be

$$T(\kappa) = \chi_c \exp\left[-\frac{3}{2}\beta(\kappa\eta_{\text{oc}})^{4/3}\right], \qquad E_c = \beta\chi_c\epsilon^{-1/3}\kappa^{-5/3}\exp\left[-\frac{3}{2}\beta(\kappa\eta_{\text{oc}})^{4/3}\right]. \tag{7.16}$$

E_c behaves like the Obukhov–Corrsin form (7.9) at the beginning of the inertial range, but the diminishing spectral cascade rate causes it to fall off exponentially from that beginning at the Obukhov–Corrsin wavenumber $1/\eta_{\text{oc}}$.

7.1.5.2 The viscous-convective subrange: $\nu/\gamma \gg 1$

Here the large kinematic viscosity ν begins to smooth the turbulent velocity field at scales much larger (wavenumbers much smaller) than those at which the small molecular diffusivity γ impacts the scalar field. In the range between the viscous cutoff of the velocity spectrum but well short of the diffusive cutoff of the scalar spectrum, the scalar field is deformed by $(\epsilon/\nu)^{1/2}$, the mean strain rate of the

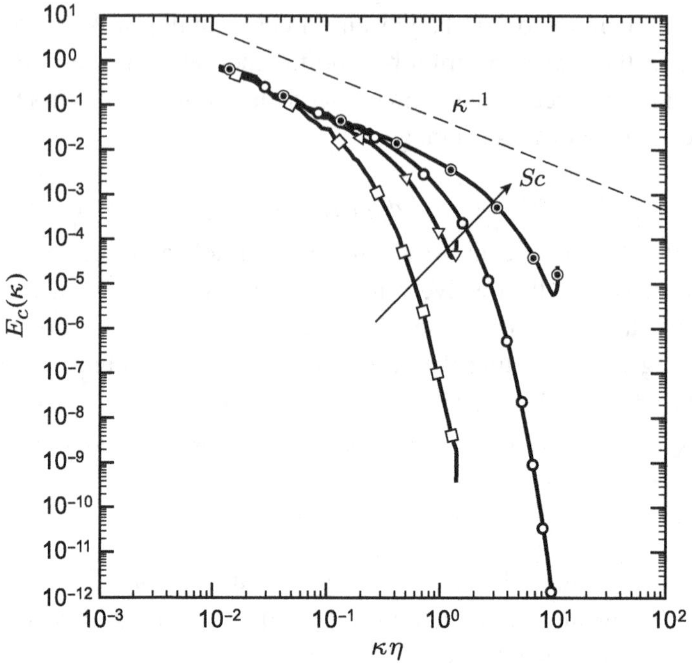

Figure 7.1 The three-dimensional scalar spectrum calculated through direct numerical simulation at four values of the Schmidt number Sc. It is dimensional; the units are scalar variance/wavenumber. □, Sc = 1/8; ▽, Sc =1; ○, Sc = 4; ◉, Sc = 64. The data for curves □ and ▽ were generated with 256^3 grid points; ○ and ◉ used 2048^3. Curve □ shows evidence of the exponential falloff in the inertial-diffusive range, Eq. (7.16), predicted by Corrsin (1964). Curve ◉ shows the κ^{-1} viscous-convective range of Eq. (7.17) predicted by Batchelor (1959). Data courtesy D. Donzis and P. K. Yeung; see Donzis (2007).

much larger, but viscous, eddies, but is not affected by its own molecular diffusivity. Batchelor (1959) hypothesized that here $E_c = E_c(\kappa, \chi_c, (\epsilon/\nu)^{1/2})$. Then on dimensional grounds this yields

$$E_c \sim \chi_c (\epsilon/\nu)^{-1/2} \kappa^{-1}, \tag{7.17}$$

which is observed in experiments (Gibson and Schwartz, 1963), and in direct numerical simulations (Figure 7.1).

Batchelor (1959) hypothesized that in this limit the diffusive cutoff scale in the scalar spectrum, now known as the Batchelor microscale η_{B}, is determined by γ and the strain rate $(\epsilon/\nu)^{1/2}$. This yields

$$\eta_{\mathrm{B}} = f\left(\gamma, (\epsilon/\nu)^{1/2}\right) \sim \left(\frac{\gamma}{(\epsilon/\nu)^{1/2}}\right)^{1/2} \sim \eta \left(\frac{\gamma}{\nu}\right)^{1/2}. \tag{7.18}$$

7.1.5.3 The velocity spectrum

As wavenumber increases and the local eddy Reynolds number $Re(r)$ decreases, the three-dimensional velocity spectrum also falls away from its inertial-range form. This departure is observed to occur in the range $0.01 \leq \kappa\eta \leq 0.1$. There are analytical models for the spectrum in this range, e.g., that of Pao (1965), who used spectral-transfer arguments like those of Corrsin (1964) to find

$$E = \alpha\epsilon^{1/3}\kappa^{-5/3}\exp\left[-\frac{3}{2}\alpha(\kappa\eta)^{4/3}\right]. \tag{7.19}$$

7.1.6 Results from direct numerical simulations

The numerical calculation of turbulent flow directly from the Navier–Stokes equations, which today is called *direct numerical simulation*, or DNS, has advanced steadily since the first trials of Orszag and Patterson (1972) with $32^3 \simeq 3 \times 10^4$ numerical grid points. Today DNS uses as many as $4096^3 \simeq 6 \times 10^{10}$ grid points.

Figure 7.1 shows scalar spectra from DNS runs that produced stationary, homogeneous turbulence of large enough Reynolds number to evidence an inertial subrange. The Schmidt numbers range from 1/8 to 64. The spectrum for $Sc = 1/8$ shows a falloff that is consistent with Corrsin's inertial-diffusive range, Eq. (7.16). That for $Sc = 64$ shows Batchelor's κ^{-1} viscous-convective subrange, Eq. (7.17).

7.2 Applications of inertial-range scaling

The Kolmogorov–Obukhov–Corrsin hypothesis that the governing parameters for scalar turbulence in the scale range $\ell \gg r \gg \eta$ reduce to r and the cascade rates ϵ and χ_c has application to a wide range of problems.

7.2.1 Atmospheric diffusion

Batchelor (1950) was one of the first to apply the Kolmogorov (1941) hypotheses to atmospheric diffusion. He identified two types of problems: diffusion from a fixed, continuous source, and diffusion of a discrete cloud (today called a puff) of effluent. The first is the "Taylor problem" we discussed in Chapter 4; the Kolmogorov (1941) hypotheses are not applicable to it (Problem 7.23).

The second problem can be summarized as follows. An effluent puff released in homogeneous turbulence is distorted as it is moved about. How does the ensemble-mean puff diameter D vary with time in the period when $\ell \gg D \gg \eta$?

Eddies much larger than the puff move it bodily without distorting it; eddies much smaller than the puff only wrinkle it. Eddies of the puff size D distort it, strengthening its concentration gradients, increasing its effective surface area and greatly enhancing molecular diffusion of the effluent; they dominate the ensemble-mean puff growth. Their velocity scale is $u(D)$, so we write

$$\frac{dD}{dt} \sim u(D) \sim (D\epsilon)^{1/3}. \tag{7.20}$$

It follows that at times large enough that the initial puff size is not important, D behaves as (Batchelor, 1950)

$$D \sim \epsilon^{1/2} t^{3/2}, \tag{7.21}$$

which has been observed (Gifford, 1957).†

7.2.2 Structure-function parameters

The *structure functions* of a velocity component and a conserved scalar are

$$S_u = \overline{\left[\tilde{u_\alpha}(\mathbf{x}, t) - \tilde{u_\alpha}(\mathbf{x} + \mathbf{r}, t)\right]^2}, \qquad S_c = \overline{\left[\tilde{c}(\mathbf{x}, t) - \tilde{c}(\mathbf{x} + \mathbf{r}, t)\right]^2}. \tag{7.22}$$

By its nature a structure function involves only the turbulent component of a homogeneous field, the mean being removed by the subtraction. The Russian school has long used structure functions in theoretical developments in turbulence and in electromagnetic and acoustic wave propagation through turbulence. For separations $r = |\mathbf{r}|$ in the inertial range of scales the dominant contributions to S_u and S_c come from turbulent eddies of scale $r = |\mathbf{r}|$, so that

$$\begin{aligned} S_u &= S_u(\epsilon, r) = C_1 \epsilon^{2/3} r^{2/3} \equiv C_{v^2} r^{2/3}, \\ S_c &= S_c(\chi_c, \epsilon, r) = C_2 \chi_c \epsilon^{-1/3} r^{2/3} \equiv C_{c^2} r^{2/3}, \end{aligned} \tag{7.23}$$

with C_1 and C_2 constants that can be determined formally from the difference operator and the spectra (Chapter 15). C_{v^2} and C_{c^2} are called the *structure-function parameters* for velocity and scalar, respectively; in the case of velocity there is some dependence of the structure-function parameter on the relative orientations of the velocity and the separation vector $\mathbf{r}$ (Part III).

† Sawford (2001) says that except for the cited analysis of Gifford, there are no convincing demonstrations of the behavior predicted by Eq. (7.21). He feels this is due to the difficulty of making reliable dispersion measurements under the conditions required by the theory.

If we carry this to third order we have

$$T_u = \overline{\left[\tilde{u_\alpha}(\mathbf{x}, t) - \tilde{u_\alpha}(\mathbf{x} + \mathbf{r}, t)\right]^3} \equiv C_3 \epsilon r, \tag{7.24}$$

with C_3 a constant that again depends on the relative orientations of the velocity and $\mathbf{r}$. For the special case where they are parallel we have "Kolmogorov's four-fifths law" with $C_3 = 4/5$. It is an exact result for isotropic turbulence (Frisch, 1995; Hill, 1997).

7.2.3 Subfilter-scale modeling

Eddy-diffusivity models are often used to represent the subfilter-scale fluxes τ_{ij}^{d} and f_j in LES (Chapter 6). Standard models are

$$\tau_{ij}^{\mathrm{d}} = K s_{ij}^{\mathrm{r}}, \qquad f_j = -K_c \frac{\partial c^{\mathrm{r}}}{\partial x_j}, \tag{7.25}$$

with $K \sim K_c \sim u_{\mathrm{s}} \Delta$, u_{s} being the velocity scale of the subfilter-scale eddies and Δ the filter cutoff scale. If Δ lies in the inertial subrange, $u_{\mathrm{s}} = u(\Delta) = (\Delta\epsilon)^{1/3}$. Hence, these eddy diffusivities are of order $u(\Delta)\Delta \sim \Delta^{4/3}\epsilon^{1/3}$. The effective Reynolds number of LES with eddy-diffusivity closure is hence $(\ell/\Delta)^{4/3}$ (Problem 7.8).

7.3 The dissipative range

An isotropic field is one whose statistics are invariant to translation, reflection, and rotation of the coordinate axes.† Since its production mechanisms are anisotropic (Chapter 5) naturally occurring turbulence cannot be isotropic, but isotropic computational turbulence can be generated (Yeung *et al.*, 2002). In introducing his hypothesis of *local isotropy*, or isotropy confined to the dissipative-range scales, Kolmogorov (1941) wrote:

> ...we think it rather likely that in an arbitrary turbulent flow with a sufficiently large Reynolds number (R_{t}) the hypothesis of local isotropy is realized with good approximation in sufficiently small domains... not lying near the boundary of the flow or its other singularities ...

Kolmogorov (1941) proposed two further similarity hypotheses for turbulence statistics in the dissipative range at large R_{t}. The first is that they are determined only by ϵ and ν. If so, the statistics of the fine-scale velocity field in large-R_{t} turbulence, when made dimensionless with the length and velocity scales η and υ, are the same

† We discuss the tensor implications of isotropy in Chapter 14.

in all turbulent flows. His second hypothesis is that for spatial scales in the inertial subrange, statistics depend only on ϵ and not on ν. Batchelor (1960) named these hypotheses the *universal equilibrium theory.*

7.3.1 Local isotropy

Kolmogorov's hypothesis of local isotropy has in the past been considered to be generally consistent with observations, but the more recent, persistent evidence of anisotropy in the fine structure of turbulent scalar fields has caused a shift of opinion. The most often cited evidence is the nonzero skewness of the streamwise derivative of fluctuating temperature $\theta_{,x}$ in shear flows (Problem 7.20),

$$S = \frac{\overline{\theta_{,x}\theta_{,x}\theta_{,x}}}{\left(\overline{\theta_{,x}\theta_{,x}}\right)^{3/2}}. \tag{7.26}$$

A typical $\theta(t)$ signal measured with a fine-wire temperature sensor is shown in Figure 7.3. Its streamwise derivative is typically found through "Taylor's hypothesis" (Taylor, 1938) in the form

$$\theta_{,t} \simeq -U\theta_{,x}, \tag{7.27}$$

with U the mean streamwise velocity. With adequate bandwidth $\theta_{,t}$ and, with Eq. (7.27), its streamwise derivative $\theta_{,x}$ are easily measured. Since as we saw in Chapter 5 their principal contributions come from eddies beyond the inertial subrange, derivative statistics are candidates for the local-isotropy assumption.

One such is $\overline{(\theta_{,x})^3}$, which changes sign under reversal of the x-direction and, hence, must vanish in a locally isotropic field. But according to Warhaft (2000), measurements in the laboratory and the atmosphere show that flows with mean shear and a mean temperature gradient have temperature-derivative skewness of magnitude $\simeq 1$.

Figure 7.2 shows a turbulent temperature signal measured in a heated jet. Its "ramp-cliff" structure, which has been observed in many other flows, shows dominant, one-signed contributions to the temperature derivative, the sign being dictated by the imposed large-scale temperature gradient. This suggests a direct connection between the statistics of very large and very small structures in the scalar field.

This evidence of local anisotropy startled the turbulence community. In a turbulent flow the eddy size range $\ell/\eta \sim R_t^{3/4}$ can be very large; it was plausible that eddies of such disparate scales are only weakly connected through the long cascades of energy and scalar variance and do not directly interact. Thus the notion of local

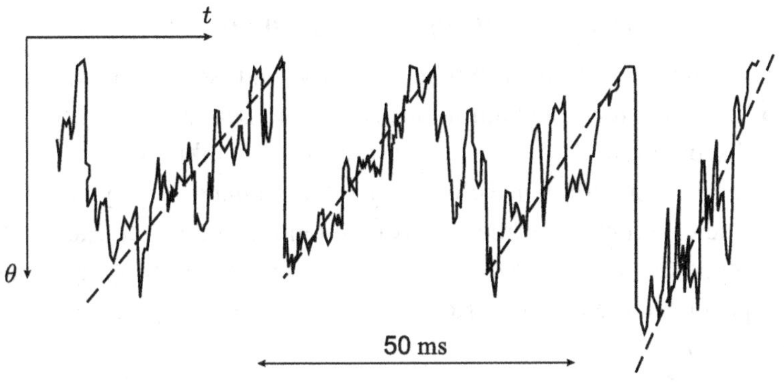

Figure 7.2 A time series of temperature measured in a heated jet showing "ramp-cliff" structure. From Warhaft (2000). Reprinted, with permission, from *Annual Review of Fluid Mechanics*, **32**, ©2000 by Annual Reviews, www.annualreviews.org.

isotropy at very large R_t was also plausible. But as measurements revealed instantaneous details of scalar fine structure – e.g., Figure 7.2 – it became clear that the largest and smallest eddies in a turbulent flow can directly and persistently interact. Kang and Meneveau (2001), for example, have found that in moderate Reynolds number ($R_\lambda = 350$) laboratory turbulence local anisotropy is more pronounced in the scalar field than in the velocity field.

Given the inherent anisotropy of the energy-containing range, Kolmogorov's (1941) local isotropy is an asymptotic, large-R_t hypothesis. At the modest R_t values typical of laboratory turbulence some local anisotropy is to be expected, and the skewness of a temperature derivative, say, is precisely zero there only as a consequence of flow symmetry. Consider, for example, the turbulent wake behind a long cylinder in crossflow, with x the streamwise direction and y along the cylinder axis. Here, by symmetry, statistics are invariant to a reversal of the y-direction, and so the skewness of $\theta_{,y}$ is zero. Local isotropy would imply that the skewness of $\theta_{,x}$ *approaches zero* at large R_t. But Sreenivasan (1991) pointed out that it shows little evidence of this asymptotic behavior, and he concluded:

Experiments suggest that local isotropy is not a natural concept for scalars in shear flows, except perhaps at such extreme Reynolds numbers that are of no practical relevance on Earth.

Warhaft (2000) has written:

We surmise that had we not been imbued with the Kolmogorov phenomenology, ... a fundamentally anisotropic scalar field would seem natural. We would expect that, as the large eddies act on the scalar gradient, large discontinuities might occur. It is the very strong attraction to universality that has diverted our attention.

We'll consider local isotropy in more detail in Chapter 14.

7.3.2 Structure of the dissipative regions

The "spiky" appearance of turbulent velocity and scalar derivative signals, Figure 7.3, indicates that their amplitudes alternate between zero and very large values. As discussed in Part III, one measure of the "spikiness" of a signal is its flatness factor F. For turbulent velocity or scalar signals F is typically near the Gaussian value of 3, but F for their derivative tends to be larger. Early laboratory measurements at low R_t showed that for a given derivative F did appear to approach the R_t-independence predicted by Kolmogorov's universal equilibrium hypothesis (Batchelor, 1960).

By the late 1960s measurements at much larger R_t made it evident that the flatness factors of velocity and temperature derivatives increase continuously with R_t. A physical interpretation is that unlike the energy-containing eddies, dissipative eddies are not uniformly distributed in space at a given time; instead they tend to be concentrated in spatially localized, transient "bursts" in a way that becomes more prominent as R_t increases. This was named *dissipation intermittency.*

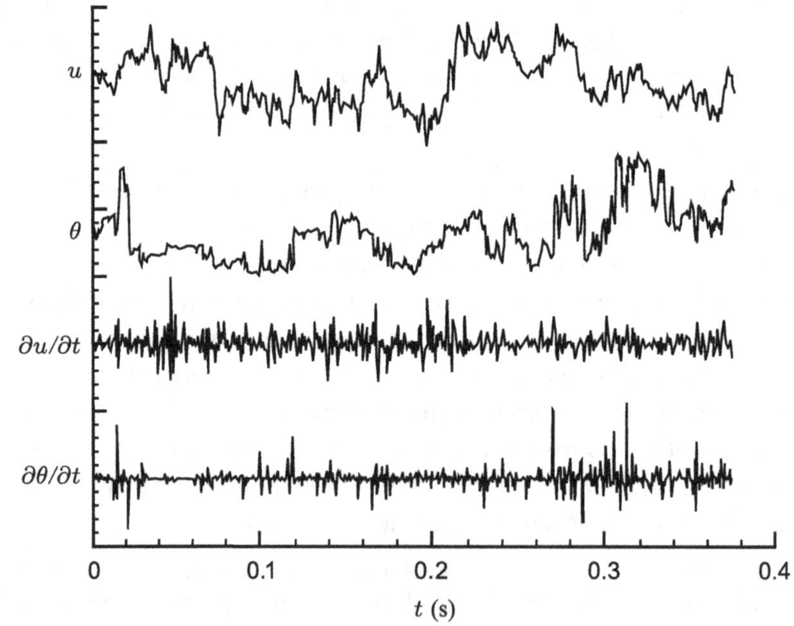

Figure 7.3 Measured fluctuating signals of (from top) streamwise velocity, temperature, and their time derivatives in a laboratory turbulent flow with a transverse mean temperature gradient. Through Taylor's hypothesis the time derivative is interpreted as proportional to the streamwise derivative. From Warhaft (2000). Reprinted, with permission, from *Annual Review of Fluid Mechanics*, **32**, ©2000 by Annual Reviews, www.annualreviews.org.

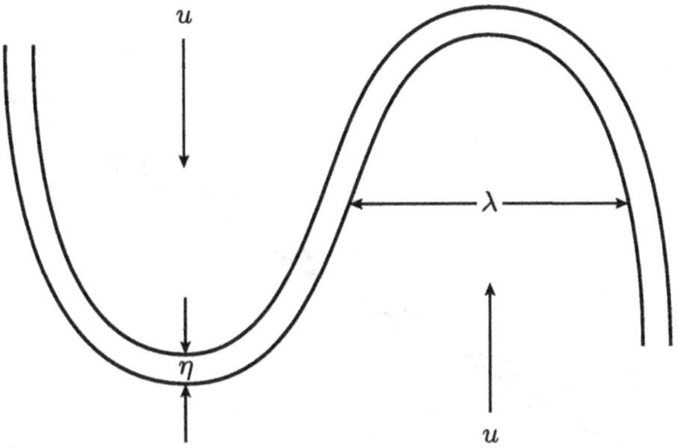

Figure 7.4 The Tennekes (1968) model of the dissipative regions in turbulence.

Based on the Batchelor and Townsend (1949) study, Corrsin (1962) hypothesized that at large enough R_t a turbulence field "has a binary character, with relatively large regions of nearly potential flow with negligible fine structure, and relatively small regions of intense fine structure where the viscous dissipation occurs." Corrsin suggested that these dissipative regions were thin sheets and made a simple model of the resulting intermittency statistics. Tennekes (1968) argued that Corrsin's model was internally inconsistent and suggested instead that the regions of dissipative activity (Figure 7.4) are tubes of diameter η and length λ, the Taylor microscale (Section 1.9); he hypothesized there was of order one of these tubes per volume λ^3.

Current thinking on dissipative structure includes sheets distributed in more complex ways, sheets with wrinkles of many length scales, and a mixture of such sheets and tubes. Here we'll pursue the implications of Tennekes' tube model because it allows some simple but illuminating estimates of fine-structure statistics.

We begin with the relation $\epsilon \sim u^3/\ell \sim \nu u^2/\lambda^2$, which implies that $\ell/\lambda \sim R_t^{1/2} \sim R_\lambda$. The volume fraction occupied by the dissipative regions is therefore

$$\text{volume fraction} \sim \frac{\eta^2\lambda}{\lambda^3} \sim \frac{\eta^2}{\lambda^2} \sim \frac{\eta^2}{\ell^2} R_t \sim R_t^{-1/2} \sim R_\lambda^{-1}, \tag{7.28}$$

using Eq. (1.35). The model predicts that the dissipative regions occupy a decreasing volume fraction of the turbulent fluid as R_t increases.

The model also predicts that the velocity derivative signal grows in spikiness as R_t increases. The highest-amplitude velocity derivatives are predicted to be of order

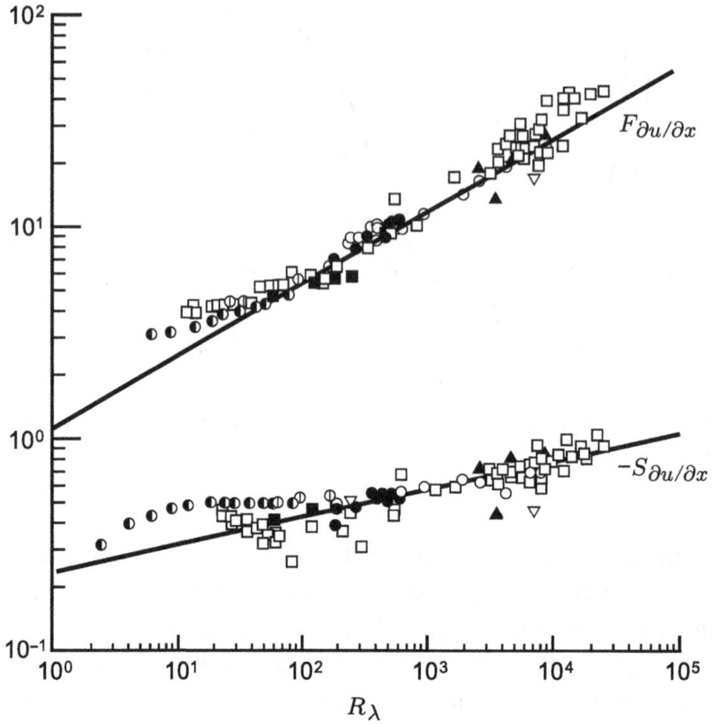

Figure 7.5 Observations of the skewness and flatness factors of the streamwise velocity derivative in a variety of turbulent flows as a function of R_λ. The lines are the prediction of Eq. (7.43), $-S_{\partial u/\partial x} \sim (F_{\partial u/\partial x})^{3/8}$, that is based on the revised Kolmogorov hypothesis. From Sreenivasan and Antonia, 1997. Reprinted, with permission, from *Annual Review of Fluid Mechanics*, **29**, ©1997 by Annual Reviews, www.annualreviews.org.

$\partial u/\partial x \sim u/\eta$, so the predicted mean value of the fourth power of the derivative is of order

$$\overline{\left(\frac{\partial u}{\partial x}\right)^4} \sim \frac{u^4}{\eta^4} \times \text{volume fraction} \sim \frac{u^4}{\eta^4} \times \frac{\eta^2}{\lambda^2} \sim \frac{u^4}{\eta^2\lambda^2}. \tag{7.29}$$

By the definition of the Taylor microscale λ the derivative variance is $\overline{(\partial u/\partial x)^2} \sim u^2/\lambda^2$, so the flatness factor of the velocity derivative goes as

$$F_{\partial u/\partial x} = \frac{\overline{\left(\frac{\partial u}{\partial x}\right)^4}}{\left[\overline{\left(\frac{\partial u}{\partial x}\right)^2}\right]^2} \sim \left(\frac{\lambda}{\eta}\right)^2 \sim R_t^{1/2} \sim R_\lambda. \tag{7.30}$$

Observations in a wide variety of flows (Figure 7.5) indicate that while $F_{\partial u/\partial x}$ does increase with R_t, this simple model considerably overpredicts that rate of increase.

The line in Figure 7.5 is $F_{\partial u/\partial x} \sim R_t^{0.17} \sim R_\lambda^{0.34}$, the exponent being one-third of that predicted by Eq. (7.30).

7.4 Revised Kolmogorov scaling

Moffatt (2002) has described the "sense of frustration" that afflicted G. K. Batchelor and many others working in turbulence in the early 1960s:

These frustrations came to the surface at the now legendary meeting held in Marseille (1961) to mark the opening of the former Institut de Méchanique Statistique de la Turbulence. This meeting, for which Batchelor was a key organizer, turned out to be a most remarkable event. Kolmogorov was there, together with Obukhov, Yaglom, and Millionshchikov…; von Karman and G. I. Taylor were both there – the great father figures of prewar research on turbulence – and the place was humming with all the current stars of the subject – Stan Corrsin, John Lumley, Philip Saffman, Les Kovasznay, Bob Kraichnan, Ian Proudman, and George Batchelor himself, among many others.

One of the highlights … was when Bob Stewart presented … the first convincing measurements to show several decades of a $\kappa^{-5/3}$ spectrum and to provide convincing support for Kolmogorov's theory…. But then, Kolmogorov gave his lecture, which I recall was in the sort of French that was as incomprehensible to the French themselves as to the other participants. However, the gist was clear: He said that … Landau had pointed out to him a defect in the theory …. Kolmogorov showed that the exponent $(-5/3)$ should be changed slightly and that higher-order statistical quantities would be more strongly affected ….

I still see the 1961 Marseille meeting as a watershed for research in turbulence. The very foundations of the subject were shaken by Kolmogorov's presentation; and the new approaches … were of such mathematical complexity that it was really difficult to retain that essential link between mathematical description and physical understanding, which is so essential for real progress.

Given that Batchelor was already frustrated by the mathematical intractability of turbulence, it was perhaps the explicit revelation that all was not well with Kolmogorov's theory that finally led him to abandon turbulence in favor of other fields.†

Kolmogorov (1962) replaced the *ensemble mean* dissipation rate ϵ in his 1941 hypotheses with a *locally averaged dissipation rate* ϵ_r defined as the average of the instantaneous, local dissipation rate $\tilde{\epsilon}$ over a sphere of radius r:

$$\epsilon_r(\mathbf{x}, t) = \frac{3}{4\pi r^3} \int_{|\mathbf{h}| \le r} \tilde{\epsilon}(\mathbf{x} + \mathbf{h}, t)\, d\mathbf{h}. \tag{7.31}$$

The local dissipation rate $\tilde{\epsilon}$ is defined as

$$\tilde{\epsilon} = \frac{\nu}{2} \left(\frac{\partial u_i}{\partial x_j} + \frac{\partial u_j}{\partial x_i} \right) \left(\frac{\partial u_i}{\partial x_j} + \frac{\partial u_j}{\partial x_i} \right), \tag{7.32}$$

† Coincidentally, shortly after the Marseille meeting D. Lilly submitted a paper (Lilly, 1962) that was prescient in advocating and demonstrating the use of numerical simulation in turbulence research (Wyngaard, 2004).

so its ensemble average is ϵ. If the averaging scale r is small compared to ℓ, so that any spatial inhomogeneity has a negligible effect, the ensemble average of the locally averaged dissipation rate ϵ_r is also ϵ.

Obukhov (1962) discussed the application of this newer scaling hypothesis. The velocity scale of eddies of spatial scale r, which was originally taken as $(\epsilon r)^{1/3}$, now depends on ϵ_r and so needs a different symbol:

$$u_{\epsilon_r}(r) = (\epsilon_r r)^{1/3}, \quad \epsilon_r \text{ fixed}. \tag{7.33}$$

A new notion here is *conditional scaling*. That is, $u_{\epsilon_r}(r) = (\epsilon_r r)^{1/3}$ is hypothesized to be the velocity scale for the subset of measurements with the specified ϵ_r value. Similarity expressions made from $u_{\epsilon_r}(r)$ must then undergo a final averaging over all values of ϵ_r. This causes the predictions of the original and revised Kolmogorov hypotheses to differ; for a similarity expression $f(\epsilon_r, r)$, say, $\overline{f(\epsilon_r, r)} \neq f(\overline{\epsilon_r}, r) = f(\epsilon, r)$. Only if f is linear in ϵ_r, which is not often the case, does $\overline{f(\epsilon_r, r)} = f(\overline{\epsilon_r}, r)$.

The original Kolmogorov hypothesis holds that velocity derivative moments scale with ϵ and ν so that, for example,

$$\overline{\left(\frac{\partial u}{\partial x}\right)^2} \sim \frac{\epsilon}{\nu}, \quad \overline{\left(\frac{\partial u}{\partial x}\right)^3} \sim \left(\frac{\epsilon}{\nu}\right)^{3/2}, \quad \overline{\left(\frac{\partial u}{\partial x}\right)^4} \sim \left(\frac{\epsilon}{\nu}\right)^2. \tag{7.34}$$

The skewness and flatness factor of the velocity derivative were thus originally predicted to be constants,

$$S_{\partial u/\partial x} = \frac{\overline{\left(\frac{\partial u}{\partial x}\right)^3}}{\left(\overline{\left(\frac{\partial u}{\partial x}\right)^2}\right)^{3/2}} = \text{constant}, \quad F_{\partial u/\partial x} = \frac{\overline{\left(\frac{\partial u}{\partial x}\right)^4}}{\left(\overline{\left(\frac{\partial u}{\partial x}\right)^2}\right)^{2}} = \text{constant}, \tag{7.35}$$

which is not observed (Figure 7.5). Under the revised hypothesis these moments become

$$\overline{\left(\frac{\partial u}{\partial x}\right)^2} \sim \frac{\overline{\epsilon_r}}{\nu} = \frac{\epsilon}{\nu}, \quad \overline{\left(\frac{\partial u}{\partial x}\right)^3} \sim \frac{\overline{\epsilon_r^{3/2}}}{\nu}, \quad \overline{\left(\frac{\partial u}{\partial x}\right)^4} \sim \frac{\overline{\epsilon_r^2}}{\nu}, \tag{7.36}$$

so this skewness and flatness factor are predicted to be

$$S_{\partial u/\partial x} = \frac{\overline{\epsilon_r^{3/2}}}{\epsilon^{3/2}}, \quad F_{\partial u/\partial x} = \frac{\overline{\epsilon_r^2}}{\epsilon^2}. \tag{7.37}$$

As discussed in Part III, variables that have high intermittency (such as derivatives of turbulent velocity and temperature at large R_t) are observed to have probability densities that are more "peaked" near the origin and have broader "tails" than the Gaussian. Squaring such a variable makes it positive definite while maintaining its intermittent character, and presumably the locally averaged dissipation rate ϵ_r behaves this way. A useful model of such a variable is the "log normal," a random variable whose logarithm has a Gaussian or "normal" probability density. Thus Obukhov (1962) wrote

$$\epsilon_r = \epsilon_0\, e^{v}, \qquad \ln \epsilon_r = \ln \epsilon_0 + v. \tag{7.38}$$

Here ϵ_0 is the *geometrical mean* dissipation rate and v is a Gaussian random variable with zero mean and variance σ^2. For this simple intermittency model the mean value of the locally averaged dissipation rate raised to a power p, say, is

$$\overline{\epsilon_r^p} = \epsilon_0^p\, \overline{e^{pv}} = \epsilon_0^p\, e^{\frac{p^2\sigma^2}{2}}. \tag{7.39}$$

Equation (7.39) then implies

$$\overline{\epsilon_r} = \epsilon_0\, e^{\frac{\sigma^2}{2}}, \qquad \overline{\epsilon_r^{3/2}} = \epsilon_0^{3/2}\, e^{\frac{9\sigma^2}{8}}, \qquad \overline{\epsilon_r^2} = \epsilon_0^2\, e^{2\sigma^2}. \tag{7.40}$$

The corresponding moments of ϵ are

$$\epsilon = \overline{\epsilon_r} = \epsilon_0\, e^{\frac{\sigma^2}{2}}, \qquad \epsilon^{3/2} = (\overline{\epsilon_r})^{3/2} = \epsilon_0^{3/2}\, e^{\frac{3\sigma^2}{4}}, \qquad \epsilon^2 = (\overline{\epsilon_r})^2 = \epsilon_0^2\, e^{\sigma^2}. \tag{7.41}$$

These demonstrate that $\overline{\epsilon_r^\alpha} \neq \epsilon^\alpha$ if $\alpha \neq 1$. Thus, with Obukhov's log-normal model for the locally averaged dissipation rate the predictions (7.37) of the revised hypothesis become

$$S_{\partial u/\partial x} \sim \frac{e^{\frac{9\sigma^2}{8}}}{e^{\frac{3\sigma^2}{4}}} \sim e^{\frac{3\sigma^2}{8}}, \qquad F_{\partial u/\partial x} \sim \frac{e^{2\sigma^2}}{e^{\sigma^2}} \sim e^{\sigma^2}, \tag{7.42}$$

so that

$$S_{\partial u/\partial x} \sim (F_{\partial u/\partial x})^{3/8}, \tag{7.43}$$

which agrees well with the observations (Figure 7.5).

Kolmogorov (1962) suggested that σ^2, the variance of $\ln \epsilon_r$, Eq. (7.38), behaves as

$$\sigma^2 = A(\mathbf{x}, t) + \mu_1 \ln \frac{\ell}{r}, \tag{7.44}$$

where $A(\mathbf{x}, t)$ depends on the large-scale structure of the flow and μ_1 is a universal constant. This predicts that the intensity of the dissipation fluctuations increases as r, the scale of the averaging volume decreases; this has been supported by measurements in a wide variety of flows. From Eq. (7.44) we have

$$e^{\sigma^2} = e^A \, e^{\mu_1 \ln \ell/r} \sim e^{\ln(\ell/r)^{\mu_1}} \sim \left(\frac{\ell}{r}\right)^{\mu_1}. \tag{7.45}$$

If, for example, $r \sim \lambda$, the Taylor microscale, then

$$\frac{\ell}{\lambda} \sim \frac{u\lambda}{\nu} = R_\lambda, \tag{7.46}$$

and Eq. (7.45) becomes

$$e^{\sigma^2} \sim R_\lambda^{\mu_1}. \tag{7.47}$$

Thus, the predictions of Eq. (7.42) become, for $r \sim \lambda$,

$$S_{\partial u/\partial x} \sim R_\lambda^{3\mu_1/8}, \qquad F_{\partial u/\partial x} \sim R_\lambda^{\mu_1}. \tag{7.48}$$

The observations (Figure 7.5) imply that $\mu_1 \sim 1/3$, somewhat greater than the consensus measured value of ~ 0.25 (Sreenivasan and Kailasnath, 1993). This area has attracted much attention and more sophisticated models have been proposed (Sreenivasan and Antonia, 1997).

Tennekes and Woods (1973) argued that since collisions of cloud droplets are fostered by high values of shear, the intermittency of fine structure in large-R_t cloud turbulence should significantly enhance the efficiency of turbulent coalescence. Shaw *et al.* (1998) discussed the resulting preferential concentration of cloud droplets, which causes intermittency in the supersaturation field (Shaw, 2000).

Shaw (2003) sketched a mechanism for this preferential concentration.† The turbulent velocity and number density n of droplets are governed by

$$\frac{dv_i}{dt} = \frac{1}{\tau_d}(u_i - v_i) + g_i, \qquad \frac{\partial n}{\partial t} + v_i \frac{\partial n}{\partial x_i} = -n \frac{\partial v_i}{\partial x_i}, \tag{7.49}$$

with τ_d the droplet inertial time scale resulting from Stokes' solution and v_i the droplet velocity. A negative droplet velocity divergence on the rhs of the second of Eq. (7.49) indicates local congregation of droplets.

For small τ_d we can write the solution of the first of (7.49) as

$$v_i = u_i + \tau_d g_i - \tau_d \left(\frac{\partial u_i}{\partial t} + u_j \frac{\partial u_i}{\partial x_j} \right) + O(\tau_d^2), \tag{7.50}$$

† The author is grateful to Raymond Shaw for this tailored discussion of this mechanism.

so to leading order the droplet velocity divergence is

$$\frac{\partial v_i}{\partial x_i} = -\tau_{\rm d}\frac{\partial u_i}{\partial x_j}\frac{\partial u_j}{\partial x_i} = -\frac{\tau_{\rm d}}{4}\left(s_{ij}s_{ij} - r_{ij}r_{ij}\right), \tag{7.51}$$

where we have used $u_{i,j} = s_{ij} + r_{ij}$, the sum of strain- and rotation-rate tensors, Eq. (2.73). Equation (7.51) says that strong local vorticity causes positive cloud droplet divergence; weak local vorticity causes negative divergence, droplet congregation, and therefore higher droplet collision rates. The variance of this droplet velocity divergence is

$$\overline{\frac{\partial v_i}{\partial x_i}\frac{\partial v_j}{\partial x_j}} = \tau_{\rm d}^2\overline{\frac{\partial u_i}{\partial x_j}\frac{\partial u_j}{\partial x_i}\frac{\partial u_k}{\partial x_m}\frac{\partial u_m}{\partial x_k}} \sim \tau_{\rm d}^2 F_{\partial u/\partial x}\left[\overline{\left(\frac{\partial u}{\partial x}\right)^2}\right]^2 \sim \tau_{\rm d}^2 F_{\partial u/\partial x}\left(\frac{\epsilon}{\nu}\right)^2, \tag{7.52}$$

where we have expressed the contracted derivative tensor in terms of $F_{\partial u/\partial x}$. As Figure 7.5 indicates, that flatness factor increases with the turbulence Reynolds number and so is very much larger in cumulus clouds than in laboratory experiments, say.

7.5 Two-dimensional turbulence

Two-dimensional turbulence, flow confined to a plane, has been used as a model for the largest scales of motion in the atmosphere. As we shall see it shares features with three-dimensional turbulence and is much more accessible computationally.

7.5.1 Vorticity and conserved scalars

We'll consider motion in an x, y plane: $\tilde{u}_i = \left[\tilde{u}(x,y,t), \tilde{v}(x,y,t), 0\right]$. Vorticity can have only a component normal to the plane: $\tilde{\omega}_i = \left[0, 0, \tilde{\omega}_3(x,y,t)\right]$. As a result the vortex-stretching term in the vorticity equation (1.28) is identically zero and the equation reads

$$\frac{D\tilde{\omega}_3}{Dt} = \frac{\partial\tilde{\omega}_3}{\partial t} + \tilde{u}_i\frac{\partial\tilde{\omega}_3}{\partial x_i} = \nu\frac{\partial^2\tilde{\omega}_3}{\partial x_i \partial x_i}, \tag{7.53}$$

which is identical to Eq. (1.31) for a conserved scalar. The absence of a vortex-stretching term in Eq. (7.53) makes the velocity field in two-dimensional turbulence differ in some striking ways from that in three-dimensional turbulence (Tennekes, 1978).

The gradient $g_i = c_{,i}$ of a conserved scalar c satisfies

$$\frac{D\tilde{g}_i}{Dt} = \tilde{g}_{i,t} + \tilde{u}_j\tilde{g}_{i,j} = -\tilde{u}_{j,i}\tilde{g}_j + \nu\tilde{g}_{i,jj}. \tag{7.54}$$

The first term on the right side can amplify and tilt the gradient of a conserved scalar in both two- and three-dimensional flow. Satellite photos reveal large-scale, quasi-two-dimensional turbulent water vapor fields that because of this stretching and tilting process can look very much like the familiar three-dimensional ones. Because in two-dimensional flow the single component of vorticity $\tilde{\omega}_3$ is a conserved scalar, its gradient $\tilde{\omega}_{3,i}$ satisfies Eq. (3.22).

The absence of vortex stretching in two-dimensional turbulence means there is no downscale energy cascade, but there is a cascade of scalar variance. Vertical vorticity, being a scalar, participates in this cascade. The distortion of the vorticity field generates spatial structure in vorticity that extends to the dissipative scales, with an intervening inertial range having a cascade of vorticity variance rather than kinetic energy.

7.5.2 Interscale transfer of enstrophy

If we represent vorticity in two-dimensional turbulence, a scalar, as the sum of ensemble-mean and fluctuating parts, $\tilde{\omega}_3 = \Omega + \omega$, the equation for fluctuating vorticity ω is

$$\frac{\partial \omega}{\partial t} + \Omega_{,j} u_j + \omega_{,j} U_j + \omega_{,j} u_j - \overline{\omega_{,j} u_j} = \nu \omega_{,jj}. \tag{7.55}$$

Multiplying by ω, ensemble averaging, rewriting the molecular term, and neglecting viscous diffusion gives the evolution equation for one-half the mean-square fluctuating vorticity, or *enstrophy*:†

$$\frac{1}{2}\frac{\partial \overline{\omega^2}}{\partial t} + U_j \left(\frac{\overline{\omega^2}}{2}\right)_{,j} + \overline{u_j \omega}\, \Omega_{,j} + \left(\frac{\overline{u_j \omega^2}}{2}\right)_{,j} = -\nu \overline{\omega_{,j} \omega_{,j}}. \tag{7.56}$$

This is identical to Eq. (5.7) for scalar variance. A displacement in the presence of a mean vorticity gradient $\Omega_{,j}$ generates a vorticity fluctuation, so we recognize $\overline{u_j \omega}\, \Omega_{,j}$ as a mean-gradient production term. The mean advection and turbulent transport terms, being divergences, simply move vorticity variance around on the plane. We conclude that the steady, global, physical-space balance of enstrophy implied by Eq. (7.56) is

rate of production of enstrophy through mean vorticity gradient

= rate of destruction of enstrophy through viscosity. (7.57)

† According to Frisch (1995), this term is due to C. Leith.

Let us now assess the spatial scales at which the molecular destruction of enstrophy occurs. We again denote by u and ℓ the scales of the energy-containing velocity fluctuations. Let us scale the fluctuating vorticity as ω, the mean vorticity gradient as ω/ℓ, and the fluctuating vorticity gradient as ω/λ_2, with the magnitude of the length scale λ_2 to be determined. Our scaling estimates for the leading terms in Eq. (7.56) are

$$\Omega_{,j}\overline{u_j\omega} \sim \frac{\omega^2 u}{\ell}, \qquad \nu\,\overline{\omega_{,j}\omega_{,j}} = \chi_\omega \sim \nu\frac{\omega^2}{\lambda_2^2},$$

so that equating the two yields

$$\frac{\omega^2 u}{\ell} \sim \nu\frac{\omega^2}{\lambda_2^2}; \qquad \frac{\lambda_2}{\ell} \sim \left(\frac{\nu}{u\ell}\right)^{1/2} \sim R_t^{-1/2}. \tag{7.58}$$

When the turbulence Reynolds number R_t is large it follows from (7.58) that there is a large scale range between ℓ and λ_2, as in three-dimensional turbulence.

As with the Taylor microscale and viscous dissipation in three-dimensional turbulence, we must be cautious in interpreting λ_2 as the spatial scale of the eddies in which the molecular destruction of enstrophy occurs. Rather, ω/λ_2 is an estimate of the vorticity gradients. Since the rms vorticity in the dissipative eddies is less than ω, their length scale is less than λ_2.

7.5.3 Inertial-range cascades and scaling

Equation (7.58) implies that large-R_t two-dimensional turbulence has a large range of eddy scales r such that $\ell \gg r \gg \lambda_2$. These eddies contain little enstrophy and do little molecular destruction of it. Interpreted for enstrophy in this inertial range in two-dimensional turbulence, the scalar variance budget, Eq. (7.4), indicates that the enstrophy cascade rate is independent of wavenumber and numerically equal to χ_ω, the mean rate of molecular destruction of enstrophy. If $\Psi(\kappa)$ is the vorticity spectrum,

$$\int_0^\infty \Psi(\kappa)d\kappa = \overline{\omega^2}, \tag{7.59}$$

then the Kolmogorov (1941) ideas imply that in this inertial subrange

$$\Psi = \Psi(\kappa, \chi_\omega) \sim \chi_\omega^{2/3}\kappa^{-1}, \tag{7.60}$$

as pointed out by Kraichnan (1967) and Batchelor (1969). Since the vorticity spectrum is κ^2 times the energy spectrum $E(\kappa)$, it follows that the inertial-range energy spectrum is

$$E(\kappa) \sim \chi_\omega^{2/3}\kappa^{-3}. \tag{7.61}$$

If E_c is the spectral density of a conserved scalar (other than vorticity) in two-dimensional turbulence, then the corresponding Kolmogorov hypothesis for its inertial subrange is $E_c = E_c(\kappa, \chi_c, \chi_\omega)$. Then on dimensional grounds it follows that

$$E_c \sim \chi_c\, \chi_\omega^{-1/3} \kappa^{-1}, \tag{7.62}$$

which is a generalization of Eq. (7.60).

For the eddy velocity and scalar intensity scales $u(r)$ and $c(r)$ in the inertial range of two-dimensional turbulence we write

$$u(r) = u(r, \chi_\omega) \sim r\chi_\omega^{1/3}, \qquad c(r) = c(r, \chi_\omega, \chi_c) \sim \chi_c^{1/2} \chi_\omega^{-1/6}. \tag{7.63}$$

The r-dependencies of these two-dimensional scales and the three-dimensional ones, Eqs. (2.66) and (7.10), are strikingly different. In two dimensions the velocity scale $u(r)$ decreases much faster ($\sim r$ as opposed to $r^{1/3}$) with decreasing scale. The difference is manifested in the $\kappa^{-5/3}$ energy spectrum in three-dimensional turbulence but its much steeper κ^{-3} falloff in two dimensions.

7.5.4 Forward and inverse cascades

McWilliams (2006) and Vallis (2006) discuss the properties of two-dimensional turbulence driven by forcing over a narrow range of eddy scales. They show that the energy- and enstrophy-conservation arguments applicable during the early, inviscid stages imply there is a downscale, or *forward* (toward smaller scales) cascade of enstrophy and an upscale, or *inverse* cascade of energy. Numerical experiments confirm this picture, showing forward and inverse cascades emanating from the wavenumber κ_{f} of the forcing. Thus the energy spectrum behaves as

$$E(\kappa) \sim \epsilon^{2/3}\kappa^{-5/3}, \qquad \kappa < \kappa_{\mathrm{f}}; \qquad E(\kappa) \sim \chi_\omega^{2/3}\kappa^{-3}, \qquad \kappa > \kappa_{\mathrm{f}}. \tag{7.64}$$

From Eqs. (6.93) and (6.95) the energy and scalar variance cascade rates ϵ and χ_c can be written

$$\epsilon = I = \overline{u_i^{\mathrm{r}} u_j^{\mathrm{r}} s_{ij}^{\mathrm{s}}} - \overline{u_i^{\mathrm{s}} u_j^{\mathrm{s}} s_{ij}^{\mathrm{r}}}, \qquad \chi_\omega = I_c = 2\overline{c^{\mathrm{r}} u_j^{\mathrm{r}} c^{\mathrm{s}}_{,j}} - 2\overline{c^{\mathrm{s}} u_j^{\mathrm{s}} c^{\mathrm{r}}_{,j}}. \tag{7.65}$$

Since it involves contractions of scalar flux and scalar gradient, I_c has the same nature in two- and three-dimensional turbulence. But as a contraction of a stress and a strain rate I is quite different in the two. In three dimensions I has two types of terms, four involving interactions entirely in the horizontal plane and five that involve vertical velocity; the latter group is presumably related to vortex stretching. In two dimensions only the first four can act, and so presumably the mechanisms involved in interscale transfer of kinetic energy are different in the two cases.

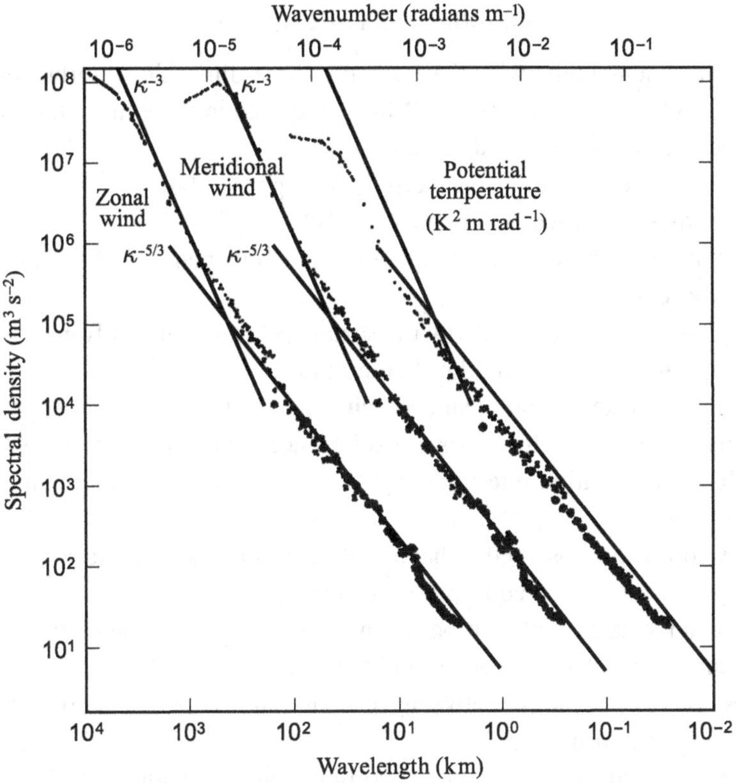

Figure 7.6 Spectra of horizontal velocity components and temperature in the upper atmosphere. From Nastrom and Gage (1985).

7.5.5 Atmospheric observations

Perhaps the most complete observations of velocity spectra above the boundary layer are those reported by Nastrom and Gage (1985). Their results, which are based on measurements made during more than 6000 commercial aircraft flights, are summarized in Figure 7.6. At the smaller wavenumbers the spectra fall as κ^{-3} for perhaps a half-decade, and at larger wavenumbers they show an extended $\kappa^{-5/3}$ range. According to Vallis (2006) the -3 range may be associated with a forward enstrophy cascade, but the origin of the $\kappa^{-5/3}$ range at smaller scales is not clear. But Gage (1979) has hypothesized that two-dimensional turbulence can exhibit a $\kappa^{-5/3}$ range due to an inverse energy cascade, and Gage and Nastrom (1986) state that it seems reasonable to assume that "breaking waves" are an important part of its small-scale energy source. Lilly (1989) shows that turbulence closure models indicate that such forward enstrophy and inverse energy cascades can coexist in two-dimensional turbulence with energy sources at both large and small scales.

Questions on key concepts

(Unless stated otherwise these refer to three-dimensional turbulence.)

7.1 What is the inertial subrange? What is required in order that a turbulent flow have one? Describe its basic physics.

7.2 Discuss the energetics of the inertial subrange in turbulence. What determines the eddy scale at which it begins? Ends?

7.3 Explain why the viscous dissipation rate is independent of the value of the fluid viscosity.

7.4 Explain where, how, and why the scalar spectrum departs from the velocity spectrum when the Prandtl or Schmidt number $\neq 1$.

7.5 Explain how temperature and velocity signals differ qualitatively from their derivative signals. How does this difference evolve as R_t increases?

7.6 What is dissipation intermittency? In what types of problems might it be important?

7.7 How does the observed behavior of turbulent fine structure violate Kolmogorov's universal equilibrium hypothesis?

7.8 Give an example of a process in the lower atmosphere that might be strongly impacted by dissipation intermittency.

7.9 Discuss the analogy between the inertial ranges of two- and three-dimensional turbulence.

7.10 Explain the analogy between the vorticity variance budget in 2-D turbulence and the scalar variance budget in 3-D turbulence.

7.11 Turbulence was studied for many years before Kolmogorov proposed his unifying hypotheses for its small-scale structure. Why do you think it took so long for these hypotheses to emerge? What was the key to his discovery?

7.12 Explain the physical essence of the Kolmogorov–Obukhov revision of the Kolmogorov (1941) ideas. Can you argue physically that it is plausible that local fine structure scales with local rather than global parameters? Can you think of problems in other fields where this would seem to be the case?

7.13 Explain qualitatively the difference between the finest-scale structure of low-R_t laboratory turbulence and that in a cumulus cloud, and how it can impact droplet physics in the two flows.

7.14 Explain how the Kolmogorov ideas of the global dynamics of 3-D turbulence apply to 2-D turbulence. How and why do they differ in detail?

Problems

7.1 Explain why for separations r in the inertial range of scales the dominant contributions to structure functions come from eddies of scale r.

7.2 On what should the Kolmogorov microscales for 2-D turbulence depend? Develop expressions for them. Determine the Reynolds number they define. Use the microscales to evaluate the relative magnitude of the viscous dissipation term in the TKE budget.

7.3 An n-th order velocity structure function is $\overline{[u(\mathbf{x}+\mathbf{r}) - u(\mathbf{x})]^n}$ where u is a velocity component. Assume $r = |\mathbf{r}|$ is in the inertial subrange of scales.
(a) How is it expected to behave under the original Kolmogorov hypothesis?
(b) How is it expected to behave under the revised Kolmogorov hypothesis?
(c) Using the Obukhov model for dissipation fluctuations, compare the two predictions. How does the difference depend on n?

7.4 Assume that Taylor's hypothesis is valid for an eddy whose turnover time is large compared to the time required to advect the eddy past the probe. Justify this assumption. Then use it to develop a criterion for the validity of Taylor's hypothesis for eddies of scale r. Interpret your result.

7.5 The inertial-range scaling result (7.63) has been interpreted for vorticity as follows: vorticity is a conserved scalar in two-dimensional flow, so vorticity fluctuations should move to smaller scales without change in intensity. Comment on this interpretation. What does the argument imply about scalar fluctuations in three-dimensional turbulence? Is that interpretation correct?

7.6 Is the notion of preferred shapes for dissipative regions in turbulence in conflict with local isotropy? Discuss.

7.7 Using an intermittency model as needed, discuss how local, instantaneous values of velocity derviatives in large-R_t turbulence can differ from those predicted by the Kolmogorov (1941) hypothesis.

7.8 Using the results of Section 7.2.3, show that the effective Reynolds number of LES with eddy-diffusivity closure is of order $(\ell/\Delta)^{4/3}$.

7.9 A commercial "ϵ meter" determines ϵ from the mean-squared difference of wind speed measured at two points in space. Explain the concept.

7.10 Explain physically the feedback mechanism (Section 7.1.2) that adjusts the dissipative eddy scales in turbulent flow.

7.11 Explain physically why and how the cutoff scale of the inertial subrange of the scalar spectrum in a given turbulent flow changes as the molecular diffusivity of the scalar changes.

7.12 Interpret the Corrsin spectrum, Eq. (7.14), physically. Can you explain why it is sometimes called the "leaky pipe" model?

7.13 Show that Corrsin's expression (7.16) is a solution of the system (7.14) and (7.15).

7.14 Interpret physically the meaning of the negative exponent of ϵ in Eq. (7.23) for the temperature structure–function parameter.

7.15 Show that $R_t^{1/2} \sim R_\lambda$.

7.16 Let $\tilde{u}$ be a random variable with standard deviation σ and let $u = \tilde{u}/\sigma$. Use the identities $\overline{(u^2 - u)^2} \geq 0$ and $\overline{(u^2 + u)^2} \geq 0$ to show that $|S| \leq (F+1)/2$, with F and S the flatness factor and skewness of u and $\tilde{u}$.

7.17 Show how the inertial-range spectral forms, Eqs. (7.5) and (7.9), follow from their underlying assumptions on dimensional grounds.

7.18 Derive the fluctuating vorticity equation. Use it to derive a vorticity-variance budget. Following the development of the TKE budget in Chapter 5, show that the leading terms in the budget are the rate of production through vortex stretching and the rate of destruction through viscous dissipation.

7.19 Show from Eq. (7.65) that the nature of I_c is the same in two- and three-dimensional turbulence, but that I differs. What is the nature of the differences in I?

7.20 The vector wavenumbers of the Fourier components contributing to a third moment add to zero (Chapter 6, Section 6.5). Demonstrate that this constraint allows $\overline{(\partial\theta/\partial x_1)^3}$ to have contributions from the energy-containing range. Then use the scaling implied by the observation that the skewness of $\partial\theta/\partial x_1$ is $\sim$1 to show that its principal contributions have all three wavenumbers in the dissipative range, however.

7.21 We showed in Eq. (5.35) that $s/s_{\mathrm{d}} \sim R_{\mathrm{t}}^{1/4}$, where s and s_{d} are the scalar intensity scales in the variance-containing and dissipative ranges, respectively. We assumed there that $\gamma \sim \nu$. Show that in the general case this generalizes to $s/s_{\mathrm{d}} \sim Co_{\mathrm{t}}^{1/4}$.

7.22 Explain physically the predicted faster-than-linear growth of a puff in the inertial range of scales, Eq. (7.21).

7.23 Explain why the Kolmogorov (1941) hypotheses are applicable to puff dispersion but not to the "Taylor problem," diffusion from a continuous point source (Chapter 4).

References

Batchelor, G. K., 1950: The application of the similarity theory of turbulence to atmospheric diffusion. *Quart. J. Roy. Meteor. Soc.*, **76**, 133–146.

Batchelor, G. K., 1959: Small-scale variation of convected quantities like temperature in turbulent fluid. Part I. General discussion and the case of small conductivity. *J. Fluid Mech.*, **5**, 113–133.

Batchelor, G. K., 1960: *The Theory of Homogeneous Turbulence.* Cambridge University Press.

Batchelor, G. K., 1969: Computation of the energy spectrum in homogeneous two-dimensional turbulence. *Phys. Fluids Suppl. II*, **12**, 233–239.

Batchelor, G. K., 1996: *The Life and Legacy of G. I. Taylor.* Cambridge University Press.

Batchelor, G. K., and A. A. Townsend, 1949: The nature of turbulent motion at large wave-numbers. *Proc. R. Soc. London A*, **199**, 238–255.

Corrsin, S., 1951: On the spectrum of isotropic temperature fluctuations in an isotropic turbulence. *J. Appl. Phys.*, **22**, 469–473.

Corrsin, S., 1962: Turbulent dissipation fluctuations. *Phys. Fluids*, **5**, 1301–1302.

Corrsin, S., 1964: Further generalizations of Onsager's model for turbulent spectra. *Phys. Fluids*, **7**, 1156–1159.

Donzis, D. A., 2007: Scaling of turbulence and turbulent mixing using terascale numerical simulations. Ph.D. thesis, School of Aerospace Engineering, Georgia Institute of Technology.

Eyink, G. L., and K. R. Sreenivasan, 2006: Onsager and the theory of hydrodynamic turbulence. *Rev. Mod. Phys.*, **78**, 87–135.

Frisch, U., 1995: *Turbulence, The Legacy of A. N. Kolmogorov*. Cambridge University Press.

Gage, K. S., 1979: Evidence for a $\kappa^{-5/3}$ law inertial range in mesoscale two-dimensional turbulence. *J. Atmos. Sci.*, **36**, 1950–1954.

Gage, K. S., and G. D. Nastrom, 1986: Theoretical interpretation of atmospheric wavenumber spectra of wind and temperature observed by commercial aircraft during GASP. *J. Atmos. Sci.*, **43**, 729–740.

Gibson, C. H., and W. H. Schwartz, 1963: The universal equilibrium spectra of turbulent velocity and scalar fields. *J. Fluid Mech.*, **16**, 365–384.

Gifford, F. Jr., 1957: Atmospheric diffusion of smoke puffs. *J. Meteorol.*, **14**, 410–414.

Grant, H. L., R. W. Stewart, and A. Moilliet, 1962: Turbulence spectra from a tidal channel. *J. Fluid Mech.*, **12**, 241–263.

Hill, R. J., 1997: Applicability of Kolmogorov's and Monin's equations of turbulence. *J. Fluid Mech.*, **353**, 67–81.

Kang, H. S., and C. Meneveau, 2001: Passive scalar anisotropy in a heated turbulent wake: new observations and implications for large-eddy simulations. *J. Fluid Mech.*, **442**, 161–170.

Kolmogorov, A. N., 1941: The local structure of turbulence in incompressible viscous fluid for very large Reynolds numbers. *C. R. Acad. Sci., U.R.S.S.*, **30**, 301–305.

Kolmogorov, A. N., 1962: A refinement of previous hypotheses concerning the local structure of turbulence in a viscous incompressible fluid at high Reynolds number. *J. Fluid Mech.*, **13**, 82–85.

Kraichnan, R., 1967: Inertial ranges in two-dimensional turbulence. *Phys. Fluids*, **10**, 1417–1423.

Lilly, D. K., 1962: On the numerical simulation of buoyant convection. *Tellus*, **14**, 148–172.

Lilly, D. K., 1989: Two-dimensional turbulence generated by energy sources at two scales. *J. Atmos. Sci.*, **46**, 2026–2030.

Lumley, J. L., and S. H. Davis, 2003: Stanley Corrsin: 1920–1986. *Ann. Rev. Fluid Mech.*, **35**, 1–10.

Moffatt, H. K., 2002: G. K. Batchelor and the homogenization of turbulence. *Ann. Rev. Fluid Mech.*, **34**, 19–35.

McWilliams, J. C., 2006: *Fundamentals of Geophysical Fluid Dynamics*, Cambridge University Press.

Nastrom, G. D., and K. S. Gage, 1985: A climatology of atmospheric wavenumber spectra of wind and temperature observed by commercial aircraft. *J. Atmos. Sci.*, **42**, 950–960.

Obukhov, A. M., 1949: Structure of the temperature field in turbulent streams. *Izv. Akad. Nauk SSSR, Geogr. Geofiz.*, **13**, 58.
Obukhov, A. M., 1962: Some specific features of atmospheric turbulence. *J. Fluid Mech.*, **13**, 77–81.
Orszag, S. A., and G. S. Patterson Jr., 1972: Numerical simulation of three-dimensional homogeneous isotropic turbulence. *Phys. Rev. Lett.*, **28**, 76–79.
Pao, Y.-H., 1965: Structure of turbulent velocity and scalar fields at large wavenumber. *Phys. Fluids*, **8**, 1063–1075.
Sawford, B., 2001: Turbulent relative dispersion. *Ann. Rev. Fluid Mech.*, **33**, 289–317.
Shaw, R. A., 2000: Supersaturation intermittency in turbulent clouds. *J. Atmos. Sci.*, **57**, 3452–3456.
Shaw, R. A., 2003: Particle-turbulence interactions in atmospheric clouds. *Ann. Rev. Fluid Mech.*, **35**, 183–227.
Shaw, R. A., W. Reade, L. Collins, and J. Verlinde, 1998: Preferential concentration of cloud droplets by turbulence: effects on the early evolution of cumulus cloud droplet spectra. *J. Atmos. Sci.*, **55**, 1965–1976.
Sreenivasan, K. R., 1991: On local isotropy of passive scalars in turbulent shear flows. *Proc. R. Soc. London A*, **434**, 165–182.
Sreenivasan, K. R., and R. A. Antonia, 1997: The phenomenology of small-scale turbulence. *Ann. Rev. Fluid Mech.*, **29**, 435–472.
Sreenivasan, K. R., and P. Kailasnath, 1993: An update on the intermittency exponent in turbulence. *Phys. Fluids A*, **5**, 512–514.
Taylor, G. I., 1938: The spectrum of turbulence. *Proc. R. Soc. London A*, **164**, 476–490.
Tennekes, H., 1968: A simple model for the small-scale structure of turbulence. *Phys. Fluids*, **11**, 669–671.
Tennekes, H., 1978: Turbulent flow in two and three dimensions. *Bull. Am. Meteor. Soc.*, **59**, 22–28.
Tennekes, H., and J. D. Woods, 1973: Coalescence in a weakly turbulent cloud. *Quart. J. Roy. Meteor. Soc.*, **99**, 758–763.
Vallis, G., 2006: *Atmospheric and Oceanic Fluid Dynamics*. Cambridge University Press.
Warhaft, Z., 2000: Passive scalars in turbulent flows. *Ann. Rev. Fluid Mech.*, **32**, 203–240.
Wyngaard, J. C., 2004: Changing the face of small-scale meteorology. In *Atmospheric Turbulence and Mesoscale Meteorology*, E. Fedorovich, R. Rotunno, B. Stevens, Eds., Cambridge University Press.
Yueng, P. K., S. Xu, and K. R. Sreenivasan, 2002: Schmidt number effects on turbulent transport with uniform scalar gradient. *Phys. Fluids*, **14**, 4178–4191.

Part II

Turbulence in the atmospheric boundary layer

8

The equations of atmospheric turbulence

8.1 Introduction

We shall begin by deriving the governing equations for a dry atmosphere. In the shallow-fluid limit they become those traditionally used for laboratory turbulence with heat transfer and buoyancy. We then generalize to moist air (dry air and water vapor) and to cloud air (dry air, water vapor, and water droplets).

Density variations impact turbulence in the lower atmosphere in two ways. First, the decrease of density with height (about 10% in the first kilometer) makes us rethink some of our intuitive notions about turbulent mixing. Second, we'll see that the density fluctuations produced by the surface heat transfer can make daytime and nighttime turbulence as different as night and day.

8.2 The governing equations for a dry atmosphere

8.2.1 An isentropic, hydrostatic base state

Most of the vertical decrease of temperature, pressure, and density in the lower atmosphere can be described through the hydrostatics of an ideal gas. We begin† with a motionless, adiabatic *base state*, denoted with a subscript zero, and write the *ideal gas law* for a dry atmosphere as

$$p_0 = \rho_0 R_{\mathrm{d}} T_0, \tag{8.1}$$

with R_{d} the gas constant for dry air. In Section 8.3 we generalize (8.1) to include water vapor and liquid water. The base-state variables depend only on height (x_3 or z).

In Part I we used a modified pressure whose vertical gradient includes the gravity term in the equation of motion. Because we are now allowing density to vary we shall return to the traditional pressure. The vertical component of the momentum

† This development is an adaptation and extension of that of Lumley and Panofsky (1964).

equation (1.19), written for our motionless base state and with x_3 positive upward so that the gravity vector $g_i = (0, 0, -g)$, is

$$-\frac{dp_0}{dx_3} - \rho_0 g = 0. \tag{8.2}$$

We use the ordinary derivative because x_3 is the only independent variable in the base state.

Since we have three base-state variables, pressure, temperature, and density, we need a third equation. We add the equation for specific entropy (entropy per unit mass) s of air,

$$T\frac{Ds}{Dt} = \frac{Dh}{Dt} - \frac{1}{\rho}\frac{Dp}{Dt}, \tag{8.3}$$

with $h = c_p T$ the specific enthalpy and c_p the specific heat per unit mass at constant pressure. Here we use the substantial derivative D/Dt (Chapter 1) because we shall use frictionless, adiabatic *virtual displacements* of fluid parcels to determine the z-dependence of the base state. For such constant-entropy or *isentropic* displacements we write Eq. (8.3) as

$$T_0\frac{Ds_0}{Dt} = 0, \quad \text{so that} \quad \frac{Dh_0}{Dt} = \frac{1}{\rho_0}\frac{Dp_0}{Dt}. \tag{8.4}$$

This implies that the base-state profiles are related by

$$\frac{dh_0}{dx_3} = \frac{1}{\rho_0}\frac{dp_0}{dx_3}. \tag{8.5}$$

With $h_0 = c_p T_0$ and the vertical equation of motion (8.2), (8.5) yields

$$\frac{dT_0}{dx_3} = -\frac{g}{c_p}, \tag{8.6}$$

the equation for the *adiabatic temperature profile*. The decrease in T_0 is almost exactly 1 K per 100 m height. From the set (8.1), (8.2), and (8.6) the profiles of p_0 and ρ_0 can also be calculated (Problem 8.2).

8.2.2 Flow-induced deviations from the base state: mass conservation

As in Part I we denote properties of the moving, turbulent atmosphere with tildes. We represent them as the sum of the base-state variables plus small deviations, denoted with primes, about the base state:

$$\tilde{p} = p_0(z) + \tilde{p}'(\mathbf{x}, t); \qquad \tilde{T} = T_0(z) + \tilde{T}'(\mathbf{x}, t); \qquad \tilde{\rho} = \rho_0(z) + \tilde{\rho}'(\mathbf{x}, t). \tag{8.7}$$

The deviations also have tildes because they have both a mean and a fluctuating part.

We require that the deviations be small so we can linearize about the base state. Thus, the representation (8.7) is fundamentally different from the decomposition in Part I of turbulence fields into ensemble-mean and fluctuating parts, where the fluctuations need not be small.

We express the density deviation in terms of deviations of temperature and pressure through a linear Taylor series expansion about the base state,

$$\begin{aligned}\tilde{\rho}' = \tilde{\rho}'(\tilde{T}, \tilde{p}) &\simeq \left.\frac{\partial \rho}{\partial T}\right|_0 \tilde{T}' + \left.\frac{\partial \rho}{\partial p}\right|_0 \tilde{p}' \\ &= -\frac{\rho_0}{T_0}\tilde{T}' + \frac{1}{R_\mathrm{d} T_0}\tilde{p}'.\end{aligned} \tag{8.8}$$

Pressure fluctuations in turbulence are observed to be of order ρu^2. If the pressure deviation is of this order as well, then its contribution to $\tilde{\rho}'$ is of order $\gamma \rho_0 u^2/c^2$, where $\gamma = c_p/c_v$ and c is the speed of sound $(\gamma R_\mathrm{d} T)^{1/2}$. Thus, by this argument the contribution of the pressure deviation to Eq. (8.8) is proportional to the square of the Mach number of the turbulence, which is very small. Thus we neglect it and write Eq. (8.8) as

$$\tilde{\rho}' = -\frac{\rho_0}{T_0}\tilde{T}'. \tag{8.9}$$

Equation (1.17) for mass conservation can be written

$$\frac{\partial \tilde{\rho}}{\partial t} + \frac{\partial \tilde{\rho}\tilde{u}_i}{\partial x_i} = \frac{D\tilde{\rho}}{Dt} + \tilde{\rho}\frac{\partial \tilde{u}_i}{\partial x_i} = 0. \tag{8.10}$$

For sufficiently small density deviations this becomes

$$\tilde{u}_3 \frac{d\rho_0}{dx_3} + \rho_0 \frac{\partial \tilde{u}_i}{\partial x_i} \simeq 0, \tag{8.11}$$

and the velocity divergence is

$$\frac{\partial \tilde{u}_i}{\partial x_i} \simeq -\frac{\tilde{u}_3}{\rho_0}\frac{d\rho_0}{dx_3} = \frac{\tilde{u}_3}{H_\rho}; \qquad \frac{1}{H_\rho} \equiv -\frac{1}{\rho_0}\frac{d\rho_0}{dx_3}. \tag{8.12}$$

H_ρ is a *scale height* for density. The ideal gas law then yields (Problem 8.3)

$$H_\rho = \frac{R_\mathrm{d} T_0 \gamma}{g}, \tag{8.13}$$

which is on the order of 10 km.

Presumably we can take the velocity divergence as zero if it is very small compared to u/ℓ. From Eq. (8.12) this requires

$$\frac{\tilde{u}_3}{H_\rho} \ll \frac{u}{\ell}. \tag{8.14}$$

If in a turbulent boundary layer we have $\tilde{u}_3 \sim u$, which is the case in horizontally homogeneous conditions, then the velocity divergence is negligible if $\ell \ll H_\rho$. Since ℓ is of the order of the boundary-layer depth, in order to treat the velocity field as divergence-free the boundary-layer depth must be small compared to the density scale height H_ρ. We'll see that this is often but not always the case.

In neglecting the vertical variation of ρ_0 we are making the *Boussinesq approximation*. It allows the zero velocity divergence assumption, but we shall be on the lookout for its side effects. As discussed by Vallis (2006), the *anelastic approximation* allows ρ_0 to depend on z and is used in deep convection.

8.2.3 Dynamics and thermodynamics

The equation of motion in an atmosphere with variable density is

$$\frac{\partial \tilde{u}_i}{\partial t} + \tilde{u}_j \frac{\partial \tilde{u}_i}{\partial x_j} = -\frac{1}{\tilde{\rho}} \frac{\partial \tilde{p}}{\partial x_i} + g_i - 2\epsilon_{ijk}\Omega_j \tilde{u}_k + \nu \nabla^2 \tilde{u}_i. \tag{8.15}$$

The third term on the right side, which involves the cross product of the earth's rotation vector Ω_j and the fluid velocity vector, is called the *Coriolis* term. It enters because our coordinate system rotates with the earth and is therefore not an *inertial* system. In a flow of velocity scale U and length scale L, the dominant terms in Eq. (8.15) are of order U^2/L. The Coriolis term is therefore important when $\Omega U \sim U^2/L$, where $\Omega = |\Omega_j| \sim 10^{-4}\ \mathrm{s}^{-1}$. This is typically the case in atmospheric flows but not in engineering flows.

In a variable-density flow the kinematic viscosity ν is variable as well. Variations in ν influence the velocity and length scales of the dissipative eddies, but not the dissipation rate itself (Problem 8.4).

We linearize the pressure gradient in the deviations about the base state:

$$\begin{aligned} -\frac{1}{\tilde{\rho}} \frac{\partial \tilde{p}}{\partial x_i} &= -\frac{1}{\rho_0 + \tilde{\rho}'} \left(\frac{\partial p_0}{\partial x_i} + \frac{\partial \tilde{p}'}{\partial x_i} \right) \\ &\simeq -\frac{1}{\rho_0} \frac{\partial p_0}{\partial x_i} - \frac{1}{\rho_0} \frac{\partial \tilde{p}'}{\partial x_i} + \frac{\tilde{\rho}'}{\rho_0^2} \frac{\partial p_0}{\partial x_i}. \end{aligned} \tag{8.16}$$

Using (8.2) and (8.9) then yields

$$-\frac{1}{\tilde{\rho}} \frac{\partial \tilde{p}}{\partial x_i} \simeq g\delta_{3i} - \frac{1}{\rho_0} \frac{\partial \tilde{p}'}{\partial x_i} - g\frac{\tilde{\rho}'}{\rho_0}\delta_{3i} \simeq g\delta_{3i} - \frac{1}{\rho_0} \frac{\partial \tilde{p}'}{\partial x_i} + \frac{g}{T_0}\tilde{T}'\delta_{3i}. \tag{8.17}$$

Using (8.17) the equation of motion (8.15) becomes

$$\frac{\partial \tilde{u}_i}{\partial t} + \tilde{u}_j \frac{\partial \tilde{u}_i}{\partial x_j} = -\frac{1}{\rho_0} \frac{\partial \tilde{p}'}{\partial x_i} - 2\epsilon_{ijk}\Omega_j \tilde{u}_k + \frac{g}{T_0}\tilde{T}'\delta_{3i} + \nu \nabla^2 \tilde{u}_i. \tag{8.18}$$

The equation now has a buoyancy term, the second from the last on the right side.

The earth's surface can be warmer than the air above (as over land on clear days, due to surface absorption of solar radiation) or cooler (as on clear nights, through surface emission). The surface temperature of a body of water also typically differs from that of the air somewhat above. This causes heat transfer between the surface and the air, which in turn causes the air temperature profile near the surface to differ from $T_0(z)$. If an air parcel at height z is warmer than $T_0(z)$, its temperature deviation $\tilde{T}'$ is positive and according to Eq. (8.18) it feels an upward buoyancy force – and so "warm air rises." If a parcel is cooler than its local T_0, $\tilde{T}'$ is negative and it feels a downward buoyancy force. A parcel with $\tilde{T}' = 0$ is *neutrally* buoyant.

It is convenient to use a temperature variable that is constant during isentropic displacements in the atmosphere, like ordinary temperature in laboratory-scale flows. For this purpose the *potential temperature* θ is defined through the specific entropy equation (8.3) rewritten with the ideal gas law,

$$\frac{Ds}{Dt} = \frac{c_p}{T}\frac{DT}{Dt} - \frac{R_\mathrm{d}}{p}\frac{Dp}{Dt} = \frac{c_p}{\theta}\frac{D\theta}{Dt}. \tag{8.19}$$

Thus θ is conserved in an isentropic displacement:

$$\frac{Ds}{Dt} = \frac{c_p}{\theta}\frac{D\theta}{Dt} = 0. \tag{8.20}$$

The solution of Eq. (8.19) is

$$\theta(t) = C[p(t)]^{\frac{-R_\mathrm{d}}{c_p}} T(t), \tag{8.21}$$

with C a constant.

Scaling arguments indicate that the dominant pressure changes along a parcel trajectory in the turbulent lower atmosphere are caused by vertical displacements in the background vertical pressure gradient, rather than turbulent pressure fluctuations (Problem 8.23). Thus we follow the convention of interpreting the independent variable in Eq. (8.21) as z, distance from the surface, rather than time, and we choose the constant C as $[p(0)]^{\frac{R_\mathrm{d}}{c_p}}$. This gives

$$\theta(z) = T(z)\left[\frac{p(0)}{p(z)}\right]^{\frac{R_\mathrm{d}}{c_p}}. \tag{8.22}$$

In this convention the potential temperature at a height z is the temperature that a parcel originating there would have after it traveled isentropically to the surface.

Turbulent motion in the atmosphere involves both viscous dissipation at rate $\tilde{\epsilon}$ and heat transfer at rate $\tilde{Q}$, so the entropy equation for turbulent dry air is

$$\frac{D\tilde{s}}{Dt} = \frac{c_p}{\tilde{\theta}}\frac{D\tilde{\theta}}{Dt} = \frac{\tilde{\epsilon}}{\tilde{T}} - \frac{\tilde{Q}}{\tilde{T}}. \tag{8.23}$$

The local, instantaneous rate of dissipation per unit mass is

$$\tilde{\epsilon} = \frac{\nu}{2}\left(\frac{\partial \tilde{u}_i}{\partial x_j} + \frac{\partial \tilde{u}_j}{\partial x_i}\right)\left(\frac{\partial \tilde{u}_i}{\partial x_j} + \frac{\partial \tilde{u}_j}{\partial x_i}\right). \tag{7.32}$$

While viscous dissipation is always important in the TKE equation, it is important in Eq. (8.23) only at flow speeds much higher than we typically find in the lower atmosphere (Problem 8.5); we neglect it. Thus with the heat transfer expressed as the divergence of the heat fluxes due to conduction and radiation,

$$\tilde{Q} = \frac{1}{\tilde{\rho}}\frac{\partial}{\partial x_i}\left(-k\frac{\partial \tilde{T}}{\partial x_i} + \tilde{R}_i\right), \tag{8.24}$$

with k the thermal conductivity, Eq. (8.23) becomes

$$\frac{D\tilde{\theta}}{Dt} = \frac{\tilde{\theta}}{c_p}\frac{D\tilde{s}}{Dt} = -\frac{\tilde{\theta}}{\tilde{\rho}c_p\tilde{T}}\left(-k\frac{\partial^2 \tilde{T}}{\partial x_i\,\partial x_i} + \frac{\partial \tilde{R}_i}{\partial x_i}\right). \tag{8.25}$$

We can write this as

$$\frac{D\tilde{\theta}}{Dt} = \frac{\tilde{\theta}}{\tilde{T}}\alpha\nabla^2\tilde{T} - \frac{\tilde{\theta}}{\tilde{\rho}c_p\tilde{T}}\frac{\partial \tilde{R}_i}{\partial x_i}, \tag{8.26}$$

with $\alpha = k/\rho c_p$ the thermal diffusivity.

According to the solution (8.21) the factor $\tilde{\theta}/\tilde{T}$ multiplying the conduction term in (8.26) is proportional to $\tilde{p}^{\frac{-R}{c_p}}$. Beyond the energy-containing range the turbulent pressure spectrum falls faster than the velocity and temperature spectra (Chapter 7), so that at the small spatial scales where conduction is important the factor $\tilde{\theta}/\tilde{T}$ varies negligibly in space. Thus, we can take it inside the ∇^2 operator:

$$\frac{\tilde{\theta}}{\tilde{T}}\nabla^2\tilde{T} \simeq \nabla^2\left(\frac{\tilde{\theta}}{\tilde{T}}\right)\tilde{T} = \nabla^2\tilde{\theta}, \tag{8.27}$$

and the potential temperature conservation equation (8.26) becomes

$$\frac{D\tilde{\theta}}{Dt} = \alpha \nabla^2 \tilde{\theta} - \frac{\tilde{\theta}}{\rho c_p \tilde{T}} \frac{\partial \tilde{R}_i}{\partial x_i}. \tag{8.28}$$

If we define a base state and a deviation for potential temperature, $\tilde{\theta} = \theta_0 + \tilde{\theta}'$, then from (8.22) we can write

$$\theta_0 + \tilde{\theta}' = (T_0 + \tilde{T}') \left[\frac{\tilde{p}(0)}{\tilde{p}(z)} \right]^{\frac{R}{c_p}} \simeq (T_0 + \tilde{T}') \left[\frac{p_0(0)}{p_0(z)} \right]^{\frac{R}{c_p}}. \tag{8.29}$$

The base states and the deviations are then related by

$$\theta_0 = T_0 \left[\frac{\tilde{p}(0)}{\tilde{p}(z)} \right]^{\frac{R}{c_p}}, \qquad \tilde{\theta}' = \tilde{T}' \left[\frac{\tilde{p}(0)}{\tilde{p}(z)} \right]^{\frac{R}{c_p}}. \tag{8.30}$$

Equation (8.30) indicates that the deviations of temperature and of potential temperature, and therefore their vertical gradients, differ in general.

8.2.4 Scalar-constituent conservation

The density $\tilde{\rho}_c$ of an advected constituent that has no sources or sinks (i.e., is mass-conserving) and diffuses molecularly within the fluid follows

$$\frac{\partial \tilde{\rho}_c}{\partial t} + \frac{\partial \tilde{\rho}_c \tilde{u}_i}{\partial x_i} = \gamma \frac{\partial^2 \tilde{\rho}_c}{\partial x_i \partial x_i}. \tag{1.29}$$

We can rewrite this as

$$\frac{D\tilde{\rho}_c}{Dt} = -\tilde{\rho}_c \frac{\partial \tilde{u}_i}{\partial x_i} + \gamma \frac{\partial^2 \tilde{\rho}_c}{\partial x_i \partial x_i}, \tag{8.31}$$

and using Eq. (8.12) to express the velocity divergence yields

$$\frac{D\tilde{\rho}_c}{Dt} = -\tilde{\rho}_c \frac{\tilde{u}_3}{H_\rho} + \gamma \frac{\partial^2 \tilde{\rho}_c}{\partial x_i \partial x_i}. \tag{8.32}$$

This says that only in the limit of large density scale height is the density of an advected, mass-conserving constituent conserved. In general it tends to decrease in upward motion and increase in downward motion because of the adiabatic expansion and compression, respectively, accompanying these motions. For that reason turbulent mixing cannot produce vertically uniform mean density of a mass-conserving trace constituent.

However, the *mixing ratio* $\tilde{c} = \tilde{\rho}_c/\tilde{\rho}$ is a conserved variable satisfying (Problem 8.6)

$$\frac{D\tilde{c}}{Dt} = \frac{\partial \tilde{c}}{\partial t} + \tilde{u}_i \frac{\partial \tilde{c}}{\partial x_i} = \frac{\gamma}{\tilde{\rho}} \frac{\partial^2 \tilde{\rho}_c}{\partial x_i \partial x_i} \simeq \gamma \frac{\partial^2 \tilde{c}}{\partial x_i \partial x_i}. \tag{8.33}$$

As a conserved scalar, $\tilde{c}$ can become uniformly mixed in the vertical even if $\tilde{\rho}_c$ cannot be (Problem 8.1).

8.2.5 The dry-air equation set

Our dry-air dynamics, thermodynamics, and conserved-constituent equation set is

$$\frac{\partial \tilde{u}_i}{\partial t} + \tilde{u}_j \frac{\partial \tilde{u}_i}{\partial x_j} = -\frac{1}{\rho_0} \frac{\partial \tilde{p}'}{\partial x_i} - 2\epsilon_{ijk}\Omega_j \tilde{u}_k + \frac{g}{\theta_0}\tilde{\theta}'\delta_{3i} + \nu \nabla^2 \tilde{u}_i, \tag{8.34}$$

$$\frac{\partial \tilde{u}_i}{\partial x_i} = 0, \tag{1.18}$$

$$\frac{\partial \tilde{\theta}}{\partial t} + \tilde{u}_i \frac{\partial \tilde{\theta}}{\partial x_i} = \alpha \nabla^2 \tilde{\theta} - \frac{\tilde{\theta}}{\rho c_p \tilde{T}} \frac{\partial \tilde{R}_i}{\partial x_i}, \tag{8.35}$$

$$\frac{\partial \tilde{c}}{\partial t} + \tilde{u}_i \frac{\partial \tilde{c}}{\partial x_i} = \gamma \frac{\partial^2 \tilde{c}}{\partial x_i \partial x_i}. \tag{8.36}$$

Here $\tilde{\theta}'$ and $\tilde{p}'$ are the deviations of potential temperature and pressure from the adiabatic, motionless, dry background state θ_0 and $p_0(z)$. According to (8.30) $\tilde{T}'/T_0 = \tilde{\theta}'/\theta_0$, so we have used $g\tilde{\theta}'/\theta_0$ as the buoyancy term in (8.34).

8.3 Accounting for water vapor, liquid water, and phase change

The clear atmosphere is a mixture of dry air and water vapor, which we shall call *moist air*. *Cloud air* contains liquid water as well. Water is a key player in atmospheric dynamics and thermodynamics, and we'll now broaden our treatment to include it.

8.3.1 The effects of water vapor

The gas constant R of a mixture having mass M_d of dry air and mass M_v of water vapor is (Problem 8.21)

$$R = \frac{M_d R_d + M_v R_v}{M_d + M_v}, \tag{8.37}$$

with R_d and R_v the gas constants of dry air and water vapor:

$$R_d = R^*/m_d, \qquad R_v = R^*/m_v. \tag{8.38}$$

R^* is the universal gas constant and m_d and m_v are the molecular weights of dry air and water vapor. By introducing the *specific humidity* q, defined as

$$q = \frac{M_v}{M_d + M_v} = \frac{\rho_v}{\rho}, \tag{8.39}$$

we can write Eq. (8.37) as

$$R = (1-q)R_d + qR_v = (1-q)R_d + q\left(\frac{m_d}{m_v}\right)R_d = (1+0.61q)R_d. \tag{8.40}$$

Thus in turbulent air containing water vapor the gas constant R has turbulent fluctuations and the air-density deviation $\tilde{\rho}' = \tilde{\rho}'(\tilde{T})$ of Section 8.2 becomes $\tilde{\rho}' = \tilde{\rho}'(\tilde{T}, \tilde{R})$. If we again linearize about a dry base state, $\tilde{R} = R_d + \tilde{R}'$, from Eq. (8.40) we have $\tilde{R}' = 0.61qR_d$ and we can write

$$\frac{\rho'}{\rho_0} \simeq -\frac{\tilde{T}'}{T_0} - 0.61q, \tag{8.41}$$

which quantifies how both warmer air and moister air are buoyant.

Instead of carrying two contributions to buoyancy in moist air, as in Eq. (8.41), we can write the gas law for moist air in terms of the gas constant for dry air, using Eq. (8.40):

$$p = \rho RT = \rho R_d T(1+0.61q). \tag{8.42}$$

Then if we define a *virtual temperature* T_v for moist air as

$$T_v = T(1+0.61q), \tag{8.43}$$

the gas law (8.42) for moist air is

$$p = \rho R_d T_v. \tag{8.44}$$

Physically, T_v is the temperature of dry air having the same pressure and density as moist air at temperature T.

It is also traditional to generalize the definition of potential temperature, Eq. (8.21), to *virtual* potential temperature:

$$\theta_v(z) = T_v(z)\left[\frac{p(0)}{p(z)}\right]^{\frac{R}{c_p}} = T(1+0.61q)\left[\frac{p(0)}{p(z)}\right]^{\frac{R}{c_p}} = \theta(1+0.61q). \tag{8.45}$$

It is a conserved variable (Problem 8.16). Thus, in moist air we simply use the deviation of *virtual* potential temperature, rather than potential temperature, as the buoyancy variable in the equation of motion (8.34).

8.3.2 The extension to cloud air

The density ρ_{cl} of cloud air is

$$\rho_{cl} = \frac{M_d + M_v + M_l}{\text{volume}} = \rho_d + \rho_v + \rho_l. \tag{8.46}$$

Rewriting in terms of the density of moist air gives

$$\rho_d + \rho_v = \rho_{cl} - \rho_l = \rho_{cl}\left(1 - \frac{\rho_l}{\rho_{cl}}\right) = \rho_{cl}\left(1 - q_l\right), \tag{8.47}$$

where $q_l = \rho_l/\rho_{cl}$ is the *specific liquid water content*. Solving for the cloud air density yields

$$\rho_{cl} = \frac{\rho_d + \rho_v}{(1 - q_l)} = \frac{p}{R_d T_v\,(1 - q_l)} = \frac{p}{R_d T_{vcl}}, \tag{8.48}$$

with T_{vcl} the virtual temperature of cloud air,

$$T_{vcl} = T_v\,(1 - q_l) \simeq T\,(1 + 0.61q - q_l), \tag{8.49}$$

neglecting the higher-order term. Like T_v, T_{vcl} uses the gas constant for dry air in its equation of state:

$$p = \rho_{cl} R_d T_{vcl}. \tag{8.50}$$

The extension to potential temperature is

$$\theta_{vcl}(z) = \theta_v(1 - q_l) \simeq \theta(1 + 0.61q - q_l). \tag{8.51}$$

8.3.3 A conserved temperature in cloud air

We saw that the application of entropy conservation in dry air leads straightforwardly to the concept of potential temperature, which is conserved in isentropic (reversible and adiabatic) processes. But as Bohren and Albrecht (1998) lament, its application to cloud air is much more complicated:

> [Accounting for entropy conservation in cloud air] is tedious and you are almost certain to make errors along the way. We derived the lapse rate for a saturated parcel several times, each time obtaining different results. Indeed, you are likely to find expressions similar but not identical to ours. Nevertheless, we think that what follows is correct.

We shall summarize the Bohren–Albrecht result.

A saturated cloud parcel, of uniform temperature T, contains mass M_d of dry air, whose partial pressure, gas constant, and specific heat at constant pressure are p_d, R_d, c_{pd}. The total water (vapor plus liquid) mixing ratio of the cloud parcel is w_t; the specific heat of its liquid water is c_w; its water vapor mixing ratio is the saturation value, w_{s}; and the latent heat of vaporization is ℓ_{v}. The statement of total entropy conservation, Eq. (6.113) of Bohren and Albrecht (1998), is

$$\frac{1}{M_{\mathrm{d}}}\frac{DS}{Dt} = (w_{\mathrm{t}}c_{\mathrm{w}} + c_{\mathrm{pd}})\frac{1}{T}\frac{DT}{Dt} - \frac{R_{\mathrm{d}}}{p_{\mathrm{d}}}\frac{Dp_{\mathrm{d}}}{Dt} + \frac{D}{Dt}\left(\frac{\ell_{\mathrm{v}}w_{\mathrm{s}}}{T}\right). \tag{8.52}$$

With the weighted specific heat $w_t c_w + c_{pd}$ written simply as c_p, this is

$$\frac{1}{M_{\mathrm{d}}}\frac{DS}{Dt} = \frac{c_p}{T}\frac{DT}{Dt} - \frac{R_{\mathrm{d}}}{p_{\mathrm{d}}}\frac{Dp_{\mathrm{d}}}{Dt} + \frac{D}{Dt}\left(\frac{\ell_{\mathrm{v}}w_{\mathrm{s}}}{T}\right) = \frac{c_p}{\theta_{\mathrm{d}}}\frac{D\theta_{\mathrm{d}}}{Dt} + \frac{D}{Dt}\left(\frac{\ell_{\mathrm{v}}w_{\mathrm{s}}}{T}\right), \tag{8.53}$$

with θ_{d} the "dry-air" potential temperature

$$\theta_{\mathrm{d}}(z) = T_{\mathrm{d}}(z)\left[\frac{p_{\mathrm{d}}(0)}{p_{\mathrm{d}}(z)}\right]^{\frac{R_{\mathrm{d}}}{c_p}}. \tag{8.54}$$

The *equivalent potential temperature* θ_{e} is defined through Eq. (8.53):

$$\frac{1}{M_{\mathrm{d}}}\frac{DS}{Dt} = \frac{c_p}{\theta_{\mathrm{d}}}\frac{D\theta_{\mathrm{d}}}{Dt} + \frac{D}{Dt}\left(\frac{\ell_{\mathrm{v}}w_{\mathrm{s}}}{T}\right) = \frac{c_p}{\theta_{\mathrm{e}}}\frac{D\theta_{\mathrm{e}}}{Dt}. \tag{8.55}$$

It is conserved in isentropic processes in cloud air, including processes with condensation and evaporation. It follows that θ_{e} is (Betts, 1973)

$$\theta_{\mathrm{e}} = \theta_{\mathrm{d}} \exp\left(\frac{\ell_{\mathrm{v}}w_{\mathrm{s}}}{c_pT}\right) = T\left(\frac{p_{\mathrm{d}}(0)}{p_{\mathrm{d}}}\right)^{\frac{R_{\mathrm{d}}}{c_p}} \exp\left(\frac{\ell_{\mathrm{v}}w_{\mathrm{s}}}{c_pT}\right). \tag{8.56}$$

Physically, θ_{e} is approximately the potential temperature a cloud parcel would have if it were moved upwards, reversibly and adiabatically, to a level where the pressure is low enough to allow all its water vapor to condense.

8.4 The averaged equations for moist air

We'll discuss the averaged equations for moist air, considering first the general case where the average can be over space or over an ensemble of realizations. We'll show that the averaging produces some new Reynolds terms involving thermodynamic properties. Then we'll generalize the ensemble-averaged Reynolds-flux equations in Chapter 5 to include buoyancy and Coriolis effects.

8.4.1 The general case

The momentum, continuity, thermodynamics, and vapor-conservation equations are

$$\frac{\partial \tilde{u}_i}{\partial t} + \tilde{u}_j \frac{\partial \tilde{u}_i}{\partial x_j} = -\frac{1}{\rho_0}\frac{\partial \tilde{p}'}{\partial x_i} - 2\epsilon_{ijk}\Omega_j \tilde{u}_k + \frac{g}{\theta_0}\tilde{\theta}'_{\mathrm{v}}\delta_{3i} + \nu \nabla^2 \tilde{u}_i, \tag{8.57}$$

$$\frac{\partial \tilde{u}_i}{\partial x_i} = 0, \tag{1.18}$$

$$\frac{\partial \tilde{\theta}}{\partial t} + \tilde{u}_j \frac{\partial \tilde{\theta}}{\partial x_j} = \alpha \nabla^2 \tilde{\theta} - \frac{\tilde{\theta}}{\rho_0 c_p \tilde{T}}\frac{\partial \tilde{R}_j}{\partial x_j}, \tag{8.35}$$

$$\frac{\partial \tilde{q}}{\partial t} + \tilde{u}_i \frac{\partial \tilde{q}}{\partial x_i} = \gamma \frac{\partial^2 \tilde{q}}{\partial x_i \, \partial x_i}. \tag{8.36}$$

Here the mixing ratio for water vapor, $\tilde{q} \equiv \tilde{\rho}_{\mathrm{v}}/\tilde{\rho}$, also called the *specific humidity*, is the conserved water-vapor variable. ρ_0 and θ_0 are the background density and potential temperature profiles, $\tilde{p}'$ is the deviation in pressure from its background profile p_0, and $\tilde{\theta}'_{\mathrm{v}}$ is the deviation in virtual potential temperature from its background value θ_0:

$$\tilde{\theta}'_{\mathrm{v}} = \tilde{\theta}_{\mathrm{v}} - \theta_0 = \tilde{\theta}(1 + 0.61\tilde{q}) - \theta_0. \tag{8.58}$$

$\tilde{R}_j$ is the radiative heat flux; α is the thermal diffusivity of moist air.

In atmospheric applications these equations have too large a range of scales to be solved numerically. Thus before attempting numerical solutions we reduce the scale range by averaging the equations. Denoting the average, ensemble or space, with an overbar, this gives the set

$$\frac{\partial \overline{\tilde{u}_i}}{\partial t} + \overline{\tilde{u}_j}\frac{\partial \overline{\tilde{u}_i}}{\partial x_j} = -\frac{1}{\rho_0}\frac{\partial \overline{\tilde{p}'}}{\partial x_i} - 2\epsilon_{ijk}\Omega_j \overline{\tilde{u}_k} + \frac{g}{\theta_0}\overline{\tilde{\theta}'_{\mathrm{v}}}\delta_{3i} + \frac{1}{\rho_0}\frac{\partial \tau_{ij}}{\partial x_j}, \tag{8.59}$$

$$\frac{\tau_{ij}}{\rho_0} = \overline{\tilde{u}_i}\,\overline{\tilde{u}_j} - \overline{\tilde{u}_i \tilde{u}_j},$$

$$\frac{\partial \overline{\tilde{u}_j}}{\partial x_j} = 0, \tag{8.60}$$

$$\frac{\partial \overline{\tilde{\theta}}}{\partial t} + \overline{\tilde{u}_j}\frac{\partial \overline{\tilde{\theta}}}{\partial x_j} = -\frac{\partial f_{\theta_j}}{\partial x_j} - \frac{1}{\rho_0 c_p}\left(\overline{\frac{\tilde{\theta}}{\tilde{T}}\frac{\partial \tilde{R}_j}{\partial x_j}}\right), \quad f_{\theta_j} = \overline{\tilde{u}_j \tilde{\theta}} - \overline{\tilde{u}_j}\,\overline{\tilde{\theta}}, \tag{8.61}$$

$$\frac{\partial \overline{\tilde{q}}}{\partial t} + \overline{\tilde{u}_j}\frac{\partial \overline{\tilde{q}}}{\partial x_j} = -\frac{\partial f_{q_j}}{\partial x_j}, \quad f_{q_j} = \overline{\tilde{u}_j \tilde{q}} - \overline{\tilde{u}_j}\,\overline{\tilde{q}}. \tag{8.62}$$

The averaged virtual potential temperature deviation is

$$\overline{\tilde{\theta}'_{\mathrm{v}}} = \overline{\tilde{\theta}} + 0.61\, \overline{\tilde{\theta}\tilde{q}} - \theta_0. \tag{8.63}$$

These equations have the Reynolds terms we met in Part I: a kinematic Reynolds stress τ_{ij} in Eq. (8.59) and Reynolds fluxes of scalars, f_{θ_j} and f_{q_j}, in Eqs. (8.61) and (8.62). The averaging also produces a covariance of radiative flux divergence and $\tilde{\theta}/\tilde{T}$ in Eq. (8.61), but scaling arguments indicate it is quite small and we have neglected it (Problem 8.19). In addition, Eq. (8.63) has a Reynolds term, as does the averaged form of Eq. (8.56) for the conserved temperature in cloud air. Traditionally such "thermodynamic" Reynolds terms have been neglected with little or no discussion, but Larson *et al.* (2001) have shown how this can bias numerical model results.

8.4.2 The ensemble-averaged equations

We now write the dependent variables in our equation set as the sum of ensemble-mean and fluctuating parts, generalizing the water vapor mixing ratio $\tilde{q}$ to a conserved scalar $\tilde{c}$:

$$\begin{aligned} &\tilde{u}_i = U_i + u_i, \qquad \tilde{p}' = P + p, \qquad \tilde{\theta}'_{\mathrm{v}} = \Theta'_{\mathrm{v}} + \theta_{\mathrm{v}}, \\ &\frac{\tilde{\theta}}{\rho_0 c_p \tilde{T}} \frac{\partial \tilde{R}_i}{\partial x_i} = \mathcal{R} + r, \qquad \tilde{c} = C + c, \qquad \tilde{\theta} = \Theta + \theta. \end{aligned} \tag{8.64}$$

The mean equations are then

$$\frac{\partial U_i}{\partial t} + U_j \frac{\partial U_i}{\partial x_j} + \frac{\partial}{\partial x_j}\overline{u_i u_j} = -\frac{1}{\rho_0}\frac{\partial P}{\partial x_i} - 2\epsilon_{ijk}\Omega_j U_k + \frac{g}{\theta_0}\Theta'_{\mathrm{v}}\,\delta_{3i}, \tag{8.65}$$

$$\frac{\partial U_i}{\partial x_i} = 0. \tag{8.66}$$

$$\frac{\partial \Theta}{\partial t} + U_i \frac{\partial \Theta}{\partial x_i} + \frac{\partial \overline{\theta u_i}}{\partial x_i} + \mathcal{R} = 0, \tag{8.67}$$

$$\frac{\partial C}{\partial t} + U_i \frac{\partial C}{\partial x_i} + \frac{\partial \overline{c u_i}}{\partial x_i} = 0. \tag{8.68}$$

We have dropped the molecular diffusion terms because they are very small except at the lower surface.

The fluctuating equations are derived by the process outlined in Part I. That for the fluctuating thermodynamic variable does need special mention, however. It reads

$$\frac{\partial \theta}{\partial t} + U_i \frac{\partial \theta}{\partial x_i} + u_i \frac{\partial \Theta}{\partial x_i} + u_i \frac{\partial \theta}{\partial x_i} - \overline{u_i \frac{\partial \theta}{\partial x_i}} = r + \alpha \nabla^2 \theta. \tag{8.69}$$

The leading terms here are of order s/τ_t, where s is the intensity scale of θ and $\tau_\mathrm{t} \sim \ell/u$ is the turbulence time scale. The fluctuating radiative term is of order r/τ_r, with τ_r a radiative time scale. Following Townsend (1958) we assume that $\tau_\mathrm{r} \gg \tau_\mathrm{t}$, so the fluctuating radiative term in Eq. (8.69) can be neglected.

The equation for the components of the kinematic Reynolds-stress tensor is

$$
\begin{aligned}
\frac{\partial \overline{u_i u_k}}{\partial t} + U_j \frac{\partial \overline{u_i u_k}}{\partial x_j} = \\
& -\overline{u_j u_k}\frac{\partial U_i}{\partial x_j} - \overline{u_j u_i}\frac{\partial U_k}{\partial x_j} \quad \text{(mean-gradient production)} \\
& -\frac{\partial \overline{u_i u_k u_j}}{\partial x_j} \quad \text{(turbulent transport)} \\
& -\frac{1}{\rho_0}\left(\overline{u_k \frac{\partial p}{\partial x_i}} + \overline{u_i \frac{\partial p}{\partial x_k}}\right) \quad \text{(pressure-gradient interaction)} \\
& -2\epsilon_{ijm}\Omega_j \overline{u_m u_k} - 2\epsilon_{kjm}\Omega_j \overline{u_m u_i} \quad \text{(Coriolis)} \\
& +\frac{g}{\theta_0}\left(\overline{\theta_\mathrm{v} u_k}\delta_{3i} + \overline{\theta_\mathrm{v} u_i}\delta_{3k}\right) \quad \text{(buoyant production)} \\
& -\frac{2\epsilon}{3}\delta_{ik} \quad \text{(viscous dissipation).}
\end{aligned} \tag{8.70}
$$

The equation for the flux of a conserved scalar is

$$
\begin{aligned}
\frac{\partial \overline{c u_i}}{\partial t} + U_j \frac{\partial \overline{c u_i}}{\partial x_j} = \\
& -\overline{u_j u_i}\frac{\partial C}{\partial x_j} - \overline{c u_j}\frac{\partial U_i}{\partial x_j} \quad \text{(mean-gradient production)} \\
& -\frac{\partial \overline{c u_i u_j}}{\partial x_j} \quad \text{(turbulent transport)} \\
& -\frac{1}{\rho_0}\left(\overline{c \frac{\partial p}{\partial x_i}}\right) \quad \text{(pressure-gradient interaction)} \\
& -2\epsilon_{ijk}\Omega_j \overline{u_k c} \quad \text{(Coriolis)} \\
& -\frac{g}{\theta_0}\overline{c\theta_\mathrm{v}}\,\delta_{i3} \quad \text{(buoyant production).}
\end{aligned} \tag{8.71}
$$

The equation for conserved-scalar variance is

$$
\frac{\partial \overline{c^2}}{\partial t} + U_j \frac{\partial \overline{c^2}}{\partial x_j} \quad \text{(time derivative following the mean motion)}
$$

$$
\begin{aligned}
&= -2\overline{u_j c}\frac{\partial C}{\partial x_j} \quad \text{(mean-gradient production)} \\
&\quad - \frac{\partial \overline{c^2 u_j}}{\partial x_j} \quad \text{(turbulent transport)} \\
&\quad - 2\gamma \overline{\frac{\partial c}{\partial x_j}\frac{\partial c}{\partial x_j}} \quad \text{(molecular destruction).} \qquad (8.72)
\end{aligned}
$$

In each of these second-moment budgets we have used the large Reynolds number, locally isotropic form of the molecular destruction term (Part III).

Questions on key concepts

8.1 Explain physically why warm air rises.

8.2 Explain how θ is defined through entropy conservation, and how that allows it to be constant on trajectories on which both pressure and temperature vary.

8.3 Explain a *scale height*, such as that for density, physically.

8.4 Explain the meaning of a *conserved* variable.

8.5 Explain why neither the density of air nor the density of an advected constituent is conserved in the atmosphere, but their ratio is conserved.

8.6 Show that T_v is the temperature of dry air having the same pressure and density as moist air at temperature T.

8.7 Explain how the concept of conserved temperature is extended to include phase change.

8.8 Explain what is meant by "thermodynamic" Reynolds terms and why they appear.

Problems

8.1 Explain why it appears on physical grounds that a necessary condition for the uniform mixing of a quantity of constituent added to a turbulent flow (e.g., cream into coffee) is that the constituent be conserved.

8.2 Calculate $p_0(z)$ and $\rho_0(z)$.

8.3 Derive the expression in Eq. (8.13) for the scale height for density.

8.4 Explain why the variation of kinematic viscosity with temperature should have no effect on the dissipation rate in turbulent flow with temperature variations.

8.5 Compare the viscous dissipation term in Eq. (8.23) to the temperature term in the equation. Use the scaling $\tilde{\epsilon} \sim u^3/\ell$, $D\tilde{T}/Dt \sim \theta u/\ell$. Is the dissipation term likely to be important in the boundary layer? In a supercell thunderstorm? In a hurricane?

8.6 Show that mixing ratio is a conserved variable.

8.7 Explain physically why the scalar-flux conservation equation applied to potential temperature indicates a tendency for a negative (downward) temperature flux in an isothermal planetary boundary layer, and zero flux in the presence of the adiabatic temperature gradient.

8.8 Relate a mixing-ratio fluctuation to fluctuations in temperature and species density. When and how can this expression be simplified?

8.9 Derive the conservation equation for the covariance of fluctuations of water vapor mixing ratio and potential temperature. What sign is the covariance over an evaporating surface that is warmer than the air? Over an evaporating surface that is cooler than the air? Near the top of a convective boundary layer that is capped by air of higher potential temperature and lower mixing ratio?

8.10 Discuss the role of the buoyancy term in the TKE equation when the vertical component of the temperature flux is negative. Under what physical situations does this occur? In which component energy equation is its impact felt directly? Do you expect its impact to be felt indirectly by the other components? By what mechanism?

8.11 Assuming that at atmospheric pressure the density of water depends only on temperature, carry out a density expansion like Eq. (8.8) but for water. What is the resulting form of the buoyancy term for water? Does warmer water tend to rise or sink?

8.12 Can eddy-diffusion models apply to both conserved and nonconserved scalar variables? Discuss.

8.13 Show that the Coriolis term in the TKE equation vanishes but that Coriolis effects can transfer energy among velocity components.

8.14 Show that the Coriolis term does not affect the magnitude of the flux of a conserved scalar but can affect its direction.

8.15 Show that θ_e, as defined by Eq. (8.56), does satisfy entropy conservation in cloud air as expressed by Eq. (8.55).

8.16 Show that virtual potential temperature is a conserved variable.

8.17 Dynamically induced pressure fluctuations in turbulence have an rms value of the order of ρu^2. Calculate the vertical distance over which the background pressure in the atmospheric boundary layer changes by ρu^2 if $u = 1$ m s^{-1}.

8.18 Average Eq. (8.56) for equivalent potential temperature. Can one express $\overline{\theta_e}$ in terms of average properties? Using a two-term expansion of the exponential, try to estimate the Reynolds terms it produces when averaged. Could they be important?

8.19 Confirm by scaling arguments that the neglected radiative covariance in Eq. (8.61) is negligible.

8.20 Evaluate the Reynolds term in Eq. (8.63). Could it be important?

8.21 Prove Eq. (8.37) by using the equation of state for each constituent.

8.22 Use the equation of state and the hydrostatic relation to find an expression for $\rho(z)$ in an adiabatic atmosphere.

8.23 Use scaling arguments to show that the dominant pressure changes along a parcel trajectory in the turbulent lower atmosphere are caused by vertical displacements in the background vertical pressure gradient, rather than turbulent pressure fluctuations.

References

Betts, A., 1973: Non-precipitating cumulus convection and its parameterization. *Quart. J. R. Meteor. Soc.*, **99**, 178–196.

Bohren, C. F., and B. A. Albrecht, 1998: *Atmospheric Thermodynamics*. New York: Oxford University Press.

Larson, V. E., R. Wood, P. Field, *et al.*, 2001: Systematic biases in the microphysics and thermodynamics of numerical models that ignore subgrid-scale variability. *J. Atmos. Sci.*, **58**, 1117–1128.

Lumley, J. L., and H. A. Panofsky, 1964: *The Structure of Atmospheric Turbulence*. New York: Interscience.

Townsend, A. A., 1958: The effects of radiative transfer on turbulent flow of a stratified fluid. *J. Fluid Mech.*, **4**, 361–375.

Vallis, G. K., 2006: *Atmospheric and Oceanic Fluid Dynamics*. Cambridge University Press.

9

The atmospheric boundary layer

9.1 Overview

A body in a moving fluid has a *boundary layer* on its surface (Figure 9.1). Near the leading edge this boundary layer is laminar; it becomes turbulent at a downstream distance x_{tran} where the local Reynolds number $U_{\infty}x_{tran}/\nu$ exceeds about 5×10^5. This turbulent boundary layer can be seen with flow-visualization techniques (Van Dyke, 1982).

In the early 1970s the *acoustic sounding* technique pioneered by McAllister *et al.* (1969) and Little (1969) made the atmospheric boundary layer (ABL) visible as well (Figure 9.2). Its instantaneous top (Figure 2.2) can be even thinner than we can easily measure.† Neff *et al.* (2008) discuss the use of the acoustic sounder in the quasi-steady stable ABL at the South Pole.

The ABL has several important features:

- It is a region of continuous turbulence, instantaneously bounded by a thin, sharp top, the convoluted, ever-changing interface between the turbulent motion and the stably stratified, nonturbulent flow above. As a *free-stream boundary* (Figure 2.1) of a turbulent flow, its local, instantaneous thickness could be of the order of the Kolmogorov microscale η (Corrsin and Kistler, 1955; Corrsin, 1972).
- Averaging produces a much thicker *interfacial layer* at the ABL top; in it the mean potential temperature increases smoothly with height, the mean-wind components transition to their free-stream values, and the turbulence level decreases to near zero. Its thickness can be as much as 20–50% of the mean depth of the boundary layer.
- Mean flow in the steady, horizontally homogeneous ABL, the result of a mean balance among turbulent stress divergence, pressure gradient, and Coriolis forces, can vary from nearly parallel to the *isobars* (constant-pressure lines) to nearly perpendicular to them.

† Acoustic sounding provided many meteorologists their first evidence of this sharp instantaneous ABL top, but D. Lenschow (personal communication) believes that many airborne meteorologists had already been made aware of it by the abrupt changes in temperature and turbulence level as they crossed it.

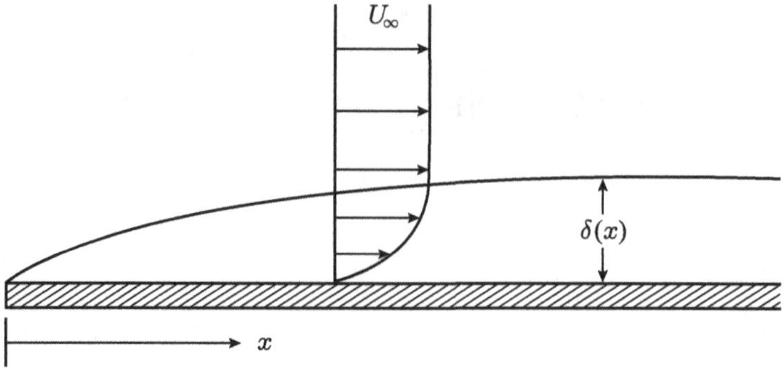

Figure 9.1 A laminar boundary layer on a flat plate in a uniform flow. Its thickness $\delta(x)$ is defined as the distance above the surface where the streamwise velocity reaches a specified fraction of the free-stream velocity U_∞.

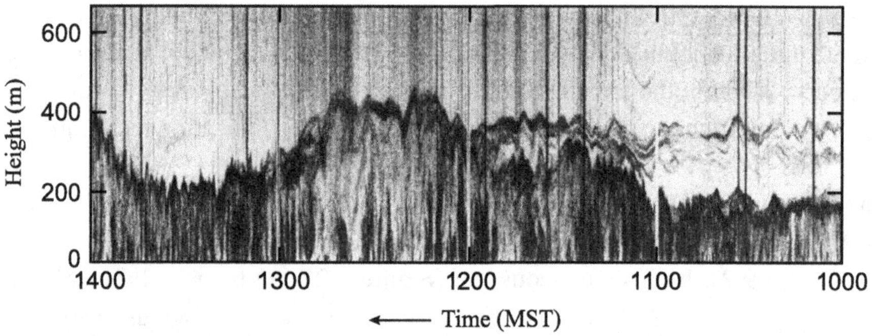

Figure 9.2 An acoustic sounder record from a convective ABL. The ABL is about 200 m deep until 1100 MST, grows to about 400 m at 1230, falls to 200 m at 1330, and grows to 400 m again at 1400. Photo courtesy W. D. Neff. From Wyngaard (1988).

The first could be approached when the stress-divergence term is small (as in a deep convective ABL over a smooth surface, Chapter 11) and the mean horizontal momentum balance (8.65), in vector form, nears the nonturbulent flow limit

$$0 = -\frac{1}{\rho_0}\nabla p - 2\Omega \times \mathbf{u}. \tag{9.1}$$

Since the Coriolis term is perpendicular to $\mathbf{u}$, the dot product of Eq. (9.1) with $\mathbf{u}$ is simply

$$-\frac{1}{\rho_0}\nabla p \cdot \mathbf{u} = 0, \tag{9.2}$$

so that $\mathbf{u}$ is parallel to the isobars, as shown in Figure 9.3. The second bound is approached in turbulent flow when the Coriolis term is small (as with an inversion at height z_i such

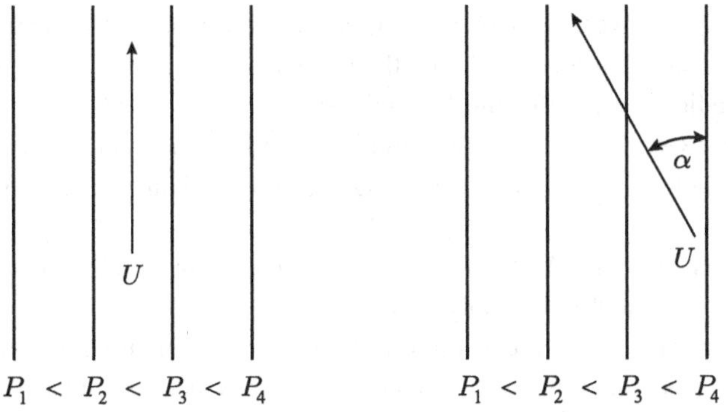

Figure 9.3 Left: Steady, horizontally homogeneous, nonturbulent flow (shown for the northern hemisphere) is parallel to the isobars. Right: With turbulence the mean flow has a cross-isobaric angle α.

that $f z_i / u_* \ll 1$ (Csanady, 1974)). This limit (Problem 9.14) is "channel flow" – i.e., flow down the mean pressure gradient.

The cross-isobaric angle enters ABL mean-flow energetics as follows. In steady, horizontally homogeneous conditions the mean horizontal momentum equation (8.65) in the ABL is

$$0 = -\frac{\partial \overline{u_i u_j}}{\partial x_j} - \frac{1}{\rho_0}\frac{\partial P}{\partial x_i} - 2\epsilon_{ijk}\Omega_j U_k, \qquad i = 1, 2. \tag{9.3}$$

This expresses a mean balance among turbulent stress divergence, pressure-gradient, and Coriolis forces in the horizontal plane. Multiplying this equation by U_i and rewriting the first term yields the steady budget of $U_i U_i/2$, the mean-horizontal-flow kinetic energy per unit mass (MKE):

$$0 = -\frac{\partial}{\partial x_j} U_i \overline{u_i u_j} + \overline{u_i u_j}\frac{\partial U_i}{\partial x_j} - \frac{U_i}{\rho_0}\frac{\partial P}{\partial x_i}. \tag{9.4}$$

The first term on the right, a transport term, integrates to zero over the ABL (Problem 9.13). The second term appears with the opposite sign in the TKE budget (Chapter 5); it represents the mean rate of exchange of kinetic energy between the mean horizontal flow and the turbulence. The final term is the mean rate of production of MKE by mean horizontal flow down the mean pressure gradient. Maintaining the mean ABL flow against its rate of loss of kinetic energy to the turbulence requires this final term to be nonzero. This in turn requires a nonzero cross-isobaric angle α (Figure 9.3).

In the stable ABL (i.e., one with $\partial\Theta/\partial z$ positive, Chapter 12) both buoyant production and viscous dissipation are loss rates of TKE, Eq. (8.70), so the turbulence must extract kinetic energy from the mean flow in order to survive. As a result its cross-isobaric angle tends to be larger, typically on the order of 45 degrees near the surface; the adjustment to geostrophic flow occurs over the depth of the stable ABL. In contrast, most of the TKE production in the convective ABL ($\partial\Theta/\partial z$ negative, Chapter 11) is from buoyancy; its cross-isobaric angle thus tends to be smaller and its much more diffusive turbulence shifts much of this adjustment to the interfacial layer.

- In the ABL the Coriolis terms in the turbulence second-moment budgets (8.70), (8.71) tend to be small compared to the leading terms (Problem 9.17); thus we'll neglect them.
- The mean depth h of the ABL can range from tens of meters to a few kilometers. In clear weather over land h typically has a diurnal cycle driven by the surface energy budget. After sunrise h increases with time as surface heating drives buoyant convection; h typically reaches a maximum in mid to late afternoon. On a clear night a much shallower, stably stratified boundary layer develops at the surface in response to the surface cooling through emitted radiation; the turbulence in what was the convective boundary aloft decays (Chapter 12). The local, instantaneous, stable ABL top is the interface between turbulent and nonturbulent air. This nocturnal ABL can be complicated by breaking gravity waves that generate their own turbulence.
- Buoyancy effects make the convective and stable ABLs strikingly different. The neutral (zero virtual temperature flux, mean virtual potential temperature constant with height) ABL is rare because small virtual temperature differences in the ABL can cause large buoyancy effects. Numerical simulations suggest that (given enough time, measured in units of $1/f$) an equilibrium neutral ABL can establish its own depth that scales with u_*/f, where $u_* = \sqrt{|\text{surface stress}|/\rho}$. However, below this height in the atmosphere there is often an inversion that establishes the ABL depth by extinguishing the turbulence.
- In clear weather over land the mean wind speed in the surface layer can have a diurnal cycle of substantial amplitude, with higher speeds in unstable, daytime conditions and lower speeds in the stable conditions at night.
- The Reynolds number R_t of ABL turbulence is far larger than that of laboratory or computational turbulence. By Reynolds-number similarity (Chapter 2) the energy-containing structure of the ABL is believed not to be substantially different from that of a lower-R_t version, so laboratory simulation, also called *fluid modeling*, can be an effective tool for studying the large-eddy structure of the ABL. But the difference in R_t causes the fine structure of ABL turbulence to differ strongly from that of laboratory flows (Chapter 7).
- The large time scales and high fluctuation levels of ABL turbulence can cause the required averaging times (Chapter 2) for ABL statistics to exceed the times over which the ABL can be considered quasi-steady. This tends to make the scatter in ABL turbulence measurements larger than in engineering flows.
- The near-surface portion of the ABL, the *surface layer*, is the best understood, in part because measurements are easiest to make there. But also its structure and dynamics are dominated by local interactions with the surface, which minimizes the influence of horizontal inhomogeneity and nonstationarity.

We shall elaborate on these points in detail in this and the following chapters.

9.2 The surface energy balance

Both the convective and stably stratified ABLs owe their distinctive structural properties to stability effects linked to the surface energy balance. That energy balance, a first-law statement for a thin slab of the earth's surface (Figure 9.4), reads:

$$\frac{\partial I}{\partial t} = C_t + R_n + C_b, \tag{9.5}$$

$I =$ internal energy of the slab,

$C_t =$ net rate of energy input by conduction through air,

$R_n =$ net rate of energy input by radiation,

$C_b =$ net rate of energy input by conduction through soil.

For a dry surface I is proportional to the depth-averaged temperature of the slab, so for a very thin slab Eq. (9.5) in effect determines the surface temperature. With the near-surface air temperature and wind speed this determines the rate of surface heat transfer, C_t. As in turbulent pipe flow, Chapter 1, this is carried by conduction caused by the temperature gradient at the surface. The turbulent air above minimizes the thickness of the surface sublayer and thereby minimizes the temperature difference across it that is required to support C_t (Problem 9.18).

If the surface is warmer than the air above, C_t is directed upward, making the near-surface air unstably stratified. If the surface is cooler than the air it is downward and the stratification is stable. Figure 9.5 shows the diurnal behavior of the surface heat flux measured over a dry prairie in the 1968 Kansas experiments.

In general the contributions of water substance need to be included in Eq. (9.5) and the vertical flux of *virtual* temperature (Chapter 10) determines the stability state of the overlying air. An evaporating surface has a positive (upward) flux of water vapor mixing ratio; this can cause the flux of virtual temperature to be positive even if the surface is cooler than the air.

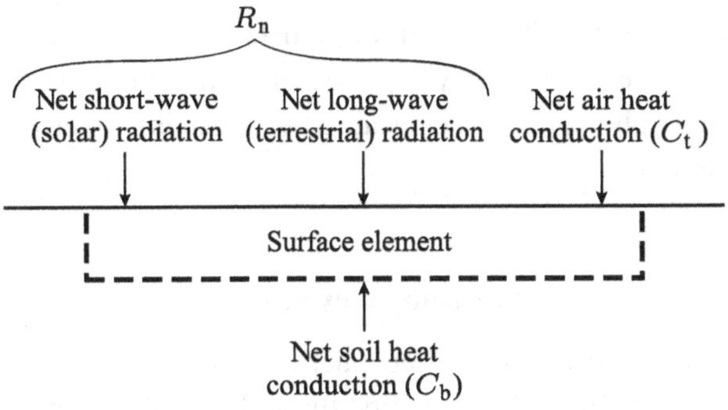

Figure 9.4 A schematic of the surface energy balance, Eq. (9.5).

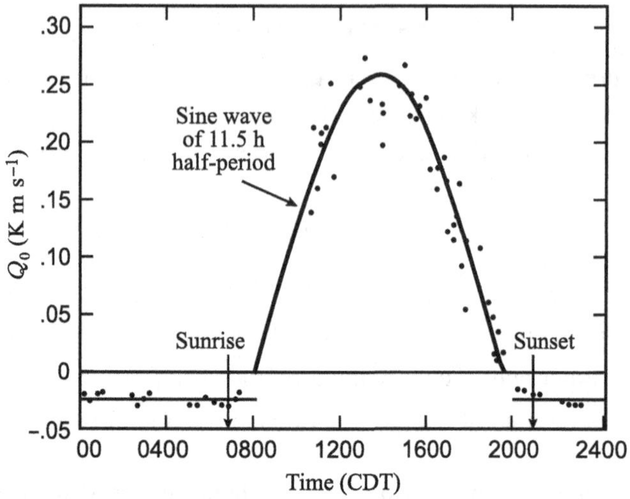

Figure 9.5 The time variation of the surface temperature flux observed in the Kansas experiment. The scatter in these one-hour averages is due to the day-to-day variations during the three weeks of midsummer observations. From Wyngaard (1973).

Because of its key role in determining the surface flux of virtual temperature, which in turn determines the stability state, the surface energy balance is one of the most important influences on the atmospheric boundary layer. Numerical models for weather forecasting include a detailed treatment of it.

The *neutral* state of zero heat flux throughout the ABL is relatively rare (it is sometimes said it exists only in the literature). Under overcast conditions the surface and air temperatures can be very nearly equal so that the surface heat flux is negligible and $z_i/L \simeq 0$. But if there is an overlying inversion within the first kilometer or two of the atmosphere, entrainment from below can diffuse this capping inversion down into the boundary layer. Thus the vertical temperature flux vanishes at the surface but is negative aloft, causing most of the ABL to be stably stratified.

The most likely place to find a neutral planetary boundary layer might be the upper ocean. The state of boundary-layer observations in the ocean substantially lags that in the atmosphere, however, because the ocean is a much more challenging research environment.

9.3 Buoyancy effects

One of the important attributes of atmospheric turbulence is its strong sensitivity to buoyancy forces. We'll explain this first through simple scaling arguments, and then discuss the two limiting stability states.

9.3.1 The sensitivity of turbulence to buoyancy

The vertical component of our Navier–Stokes equation, Eq. (8.57), has a buoyancy term $g\overline{\tilde{\theta}'_{\rm v}}/\theta_0$, with $\overline{\tilde{\theta}'_{\rm v}}$ the local deviation in virtual potential temperature from the background value. If we define a temperature scale $\theta \sim \overline{\tilde{\theta}'_{\rm v}}$, and take the turbulent inertia terms in that equation as of order u^2/ℓ, their ratio is a turbulence Richardson number $Ri_{\rm t}$:

$$Ri_{\rm t} = \frac{\text{buoyancy force/mass}}{\text{inertia force/mass}} \sim \frac{g\theta/\theta_0}{u^2/\ell} \sim \frac{g\theta\ell}{\theta_0 u^2}. \tag{9.6}$$

An indoor air flow with $\theta = 1$ K, $\ell = 1$ m, $u = 1$ m s^{-1} has $Ri_{\rm t} = 1/30$, and turbulent buoyancy is unimportant. An ABL flow with the same θ, u but with $\ell = 1000$ m has $Ri_{\rm t} = 30$; here turbulent buoyancy is quite important.

Atmospheric turbulence tends to have larger ℓ and smaller u than engineering turbulence, both of which tend to give larger $Ri_{\rm t}$. As a result buoyancy effects are typically much more important in atmospheric turbulence. Hereafter, with temperature we will not generally use the modifier *virtual*, or its subscript v, but in general *temperature* is to be interpreted as *virtual temperature*.

9.3.2 Stable and unstable stratification

Equation (8.69) says that when $\partial\Theta/\partial z$[†] is negative, upward motion of a parcel generates a positive potential temperature deviation $\tilde{\theta}'$ and downward motion generates a negative one. In each case the resulting buoyancy force tends to amplify the motion, and so we call this *unstable* stratification. When $\partial\Theta/\partial z$ is positive, parcels undergoing vertical displacements feel opposing buoyant accelerations; this is *stable* stratification.

Through its strong buoyancy effects an elevated inversion can act as an effective "lid" on the ABL. Consider, for example, an upward-moving air parcel of vertical velocity w_i entering the inversion, a region of greater potential temperature. There θ', the difference between the temperatures of the parcel and the environment, becomes negative; this causes a downward buoyancy force $g\theta'/\theta_0$ on the parcel. If that force brings the parcel to rest in a vertical distance d, then equating the work done on the parcel, $gd\theta'/\theta_0$, and the change in parcel kinetic energy, $w_i^2/2$, gives

$$\frac{gd\theta'}{\theta_0} \simeq \frac{w_i^2}{2}, \qquad d \simeq \frac{w_i^2\theta_0}{2g\theta'}. \tag{9.7}$$

For $w_i = 1$ m s^{-1} and $\theta' = 1$ K this gives $d \sim 15$ m, a relatively small excursion if the mean height of the capping inversion is 1000 m. Thus a surprisingly weak

† Hereafter θ and T are, in general, virtual potential temperature and virtual temperature, respectively.

inversion can damp upward motions quite effectively. Passengers in an airplane taking off from a large metropolitan area can often see the dark smudge of pollutants trapped in the boundary layer below, a sign of such an inversion. In Denver, Colorado, this is known as the "brown cloud."

In unstable stratification, $\partial\Theta/\partial z$ negative, we expect $\overline{w\theta}$, the vertical flux of temperature, to be positive, so the buoyancy term represents a rate of gain in the TKE budget, Eq. (8.70). In stable stratification we expect a negative flux and correspondingly a rate of TKE loss to buoyancy. Labeling this TKE budget term as *buoyant production* might be confusing, since that might be taken to imply an energy source term, so where possible we will refer to it under stable conditions as *buoyant destruction.* It represents energy conversion (Problem 9.12).

The TKE equation (6.54) for a horizontally homogeneous resolvable-scale field with buoyancy effects becomes

$$\frac{1}{2}\left(\overline{u_i^{\mathrm{r}}u_i^{\mathrm{r}}}\right)_{,t} = -U_{1,3}\,\overline{u_1^{\mathrm{r}}u_3^{\mathrm{r}}} - \frac{1}{\rho_0}\left(\overline{p^{\mathrm{r}}u_3^{\mathrm{r}}}\right)_{,3} - I - \frac{1}{2}\left(\overline{u_i^{\mathrm{r}}u_i^{\mathrm{r}}u_3}\right)_{,3} - \left(\overline{u_i^{\mathrm{r}}u_i^{\mathrm{s}}u_3^{\mathrm{s}}}\right)_{,3} + \frac{g}{\theta_0}\overline{\theta^{\mathrm{r}}w^{\mathrm{r}}}. \tag{9.8}$$

If the spatial resolution is sufficient to give $\overline{\theta^{\mathrm{r}}w^{\mathrm{r}}} \simeq \overline{\theta w}$, then the buoyancy term here is well resolved. This is the case if $\Delta \ll \ell$.

The corresponding TKE equation for the SFS field is (6.56) generalized to include buoyancy effects:

$$\begin{aligned}\frac{1}{2}\left(\overline{u_i^{\mathrm{s}}u_i^{\mathrm{s}}}\right)_{,t} &= -U_{1,3}\,\overline{u_1^{\mathrm{s}}u_3^{\mathrm{s}}} - \frac{1}{\rho_0}\left(\overline{p^{\mathrm{s}}u_3^{\mathrm{s}}}\right)_{,3} + I - \frac{1}{2}\left(\overline{u_i^{\mathrm{s}}u_i^{\mathrm{s}}u_3}\right)_{,3} \\ &\quad - \left(\overline{u_i^{\mathrm{r}}u_i^{\mathrm{s}}u_3^{\mathrm{r}}}\right)_{,3} + \frac{g}{\theta_0}\overline{\theta^{\mathrm{s}}w^{\mathrm{s}}} - \epsilon.\end{aligned} \tag{9.9}$$

If the filter cutoff separating the resolvable and subfilter-scale fields lies well beyond the energy-containing range, i.e., if $\Delta \ll \ell$, the subfilter-scale field contains negligible variance and flux. Then as we saw in Chapter 6 this TKE equation reduces to

$$\frac{1}{2}\left(\overline{u_i^{\mathrm{s}}u_i^{\mathrm{s}}}\right)_{,t} \simeq I - \epsilon, \tag{9.10}$$

a statement of the energy balance of all eddies beyond the energy-containing range. This was central to Kolmogorov's thinking.

9.3.3 The convective ABL

When a parcel of buoyant air rises from the warmer or moister surface its buoyancy lasts as long as its potential temperature excess θ' does. Distortion of the parcel by turbulent velocity gradients strengthens its temperature gradients, accelerating the removal of this temperature excess by molecular diffusion (Chapter 3, Section 3.3). The deformation rate of the eddies most effective in distorting a parcel of spatial scale r is $u(r)/r$, which according to Kolmogorov scaling is of order $r^{-2/3}\epsilon^{1/3}$; the time scale for parcel distortion is the inverse, $r^{2/3}\epsilon^{-1/3}$. Thus we expect that the largest buoyant parcels retain their temperature excess the longest. The largest eddies in the convective boundary layer (CBL) extend from the surface to the inversion base.

Another prominent feature of the convective ABL is the skewness of its vertical velocity fluctuations, defined by

$$S_w = \frac{\overline{w^3}}{(\overline{w^2})^{3/2}}. \tag{9.11}$$

Skewness is a statistical measure of the differences between positive and negative fluctuations – here, updrafts vs. downdrafts. $S_w \simeq 0.4$–1.0 in the CBL driven by surface heating. The updrafts tend to be stronger than downdrafts, but since w averages to zero they are also rarer. The most likely value of w is slightly negative – a very weak downdraft. S_w is negative in a cloud-capped CBL driven by radiative cooling of cloud top (Moeng and Rotunno, 1990). We shall see in Chapter 11 that its w-skewness gives the CBL some unusual diffusion properties.

Since buoyancy forces tend to make intense, flow-filling turbulent eddies, we expect the convective ABL to be particularly diffusive. We see this in the differing behavior of an instantaneous smokestack plume in the ABL on a sunny day and an overcast day. On an overcast day with primarily *mechanical* turbulence (turbulence generated by shear production) the plume can remain intact for some distance downwind, but on a sunny day it loops and twists wildly under the influence of the large, convectively driven eddies (Figure 9.6). The ensemble-average plume grows most rapidly with downstream distance under convective conditions.

9.3.4 The stable ABL

We can recognize at least three types of stably stratified ABLs. One occurs when ABL air encounters a cooler surface downwind (as in Figure 9.7, upper). A second is a boundary layer that is entraining higher potential temperature air from the capping inversion and has zero surface heat flux (Figure 9.7, lower). This might better be called the "inversion-capped neutral" case, since *neutral* traditionally means zero surface heat flux. The third and perhaps most common type is that over cooler land at night in clear weather. Each of these stable ABLs is typically shallow (as much as

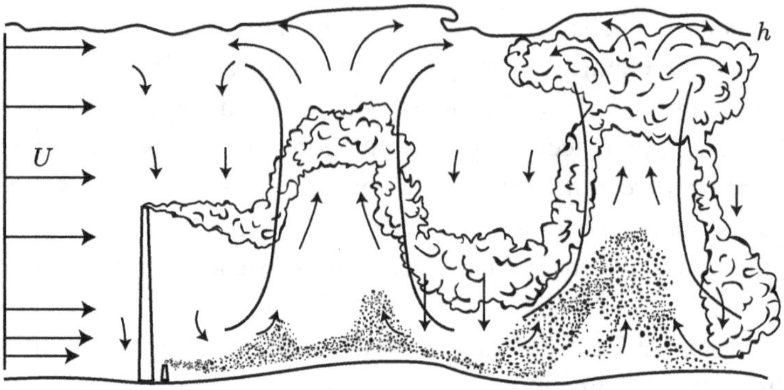

Figure 9.6 An artist's sketch of instantaneous plumes of effluent in a CBL. From Briggs (1988).

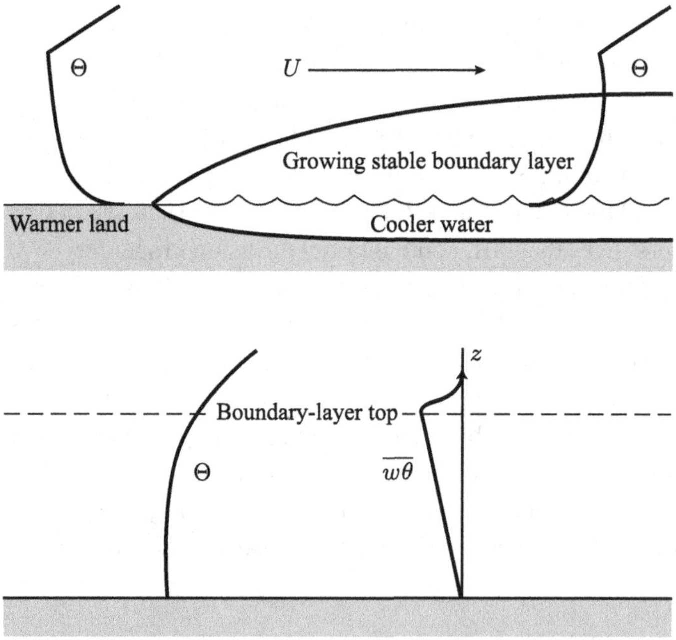

Figure 9.7 Upper: A boundary layer made stably stratified by flowing over a cooler surface. Lower: A boundary layer made stably stratified by entrainment of warmer air aloft.

an order of magnitude thinner than the daytime convective ABL) with large eddies much smaller and weaker than in the convective case.

Under stable conditions the buoyancy term in the TKE equation (8.70) represents a rate of loss through buoyant destruction (BD), the rate of working of the vertical turbulent motion against restoring buoyancy forces; BD supplements the rate of loss through viscous dissipation (ϵ). The principal source term, shear production (SP), represents the rate of TKE input through the interaction of Reynolds stresses and the mean velocity gradient. These can be the dominant terms, in which case $SP \simeq BD + \epsilon$. Turbulence can survive a sufficiently small rate of buoyant destruction, i.e., $BD \ll SP$. But $BD \gg SP$ is a possible steady state only when there is another source term. We shall see that this can be the case in the interfacial layer that caps the convective ABL; there the rate of import of TKE from below can allow the turbulence to exist without shear production.

This opens the possibility of a *critical state* of stably stratified turbulence in which the rates of buoyant destruction and viscous dissipation approximately balance shear production and the turbulence is in equilibrium, but an increase in the rate of buoyant destruction extinguishes the turbulence. To gain some insight, let us examine the implications of equating, in order of magnitude, the rates of shear production and buoyant destruction of TKE. In terms of the scales u, ℓ, and a characteristic potential temperature fluctuation $\theta \sim \ell \partial \Theta / \partial z$ we have

$$\overline{uw}\frac{\partial U}{\partial z} \sim \frac{u^3}{\ell} \sim \frac{g}{\theta_0}\overline{w\theta} \sim \frac{g}{\theta_0}u\left(\ell\frac{\partial \Theta}{\partial z}\right). \tag{9.12}$$

This yields an order-of-magnitude estimate of the size ℓ of the largest eddy that can survive its rate of energy loss to stable stratification:

$$\ell \sim \left(\frac{u^2}{\frac{g}{\theta_0}\frac{\partial \Theta}{\partial z}}\right)^{1/2}. \tag{9.13}$$

Eddies much larger than ℓ are extinguished by the stable stratification; much smaller ones can survive. This length scale is sometimes defined with the vertical velocity variance and given the subscript B,

$$\ell_B = \left(\frac{\overline{w^2}}{\frac{g}{\theta_0}\frac{\partial \Theta}{\partial z}}\right)^{1/2} = \frac{\sigma_w}{N}, \tag{9.14}$$

with $N^2 = g/\theta_0 \partial \Theta / \partial z$ the square of the Brunt–Vaisala frequency. ℓ_B is used in models of stably stratified turbulence (Brost and Wyngaard, 1978; Nieuwstadt, 1984; Hunt, 1985). Mason and Derbyshire (1990) found evidence for this scaling in their LES studies of the stable boundary layer (SBL).

By selectively damping the largest eddies, stable stratification reduces the turbulent diffusivity $K \sim u\ell$ by decreasing both ℓ and u. Measures of its strength include the *flux Richardson number* R_f and the *gradient Richardson number* Ri:

$$R_{\rm f} = \frac{\text{rate of buoyant destruction of TKE}}{\text{rate of shear production of TKE}} = \frac{\frac{g}{\theta_0}\overline{\theta w}}{\overline{u_i u_j}\frac{\partial U_i}{\partial x_j}}, \qquad Ri = \frac{\frac{g}{\theta_0}\frac{\partial \Theta}{\partial z}}{\left(\frac{\partial U}{\partial z}\right)^2}. \quad (9.15)$$

As we'll discuss in Chapter 12, there is evidence that when these Richardson numbers exceed about 0.2 to 0.3 all turbulence can be extinguished.

9.4 Average vs. instantaneous structure

Perhaps you remember lying on your back outdoors on a summer day and watching cumulus clouds drifting overhead. Their three-dimensional, cauliflower-like turbulent structure seems "frozen" because the large-eddy time scale ℓ/u is so large. You need to watch even a modest cloud with $\ell \sim 1000$ m and $u \sim 1$ m s^{-1} on the order of 20 minutes – longer than most of us can – to see it change. As a result we are probably more aware of instantaneous cloud properties than average ones.

Most numerical models of turbulent flows predict average properties, either the ensemble average or a spatial average over a numerical grid cell (Chapter 1). The Gaussian-plume models routinely used for effluent concentrations downwind of sources, for example, are based on G. I. Taylor's ensemble-mean solution (Chapter 4) for dispersion in stationary, homogeneous turbulence. But in practice we must test their predictions against short-term time averages, and so the model evaluations tend to be plagued by scatter. It can be difficult to apportion the scatter between inadequate averaging time and model physics. This is illustrated in Figure 9.8.

9.5 Quasi-steadiness and local homogeneity

The ABL over land is nonstationary due to the diurnal cycle; it also tends to be inhomogeneous in the horizontal due to spatial variations in synoptic conditions and surface properties. But if its time and length scales are small compared to those of the external variations, we can often use a steady, horizontally homogeneous model of ABL turbulence.

We'll illustrate with time changes. Figure 9.5 shows the daily cycle of surface temperature flux Q_0 measured in the Kansas experiment. We expect that the turbulence can be quasi-steady if the time scale of these changes, $Q_0(\partial Q_0/\partial t)^{-1}$, is much larger than the large-eddy turnover time h/u. This is the case if

$$\frac{1}{Q_0}\frac{\partial Q_0}{\partial t}\frac{h}{u} \ll 1 \ \ \text{(quasi-steadiness)}. \qquad (9.16)$$

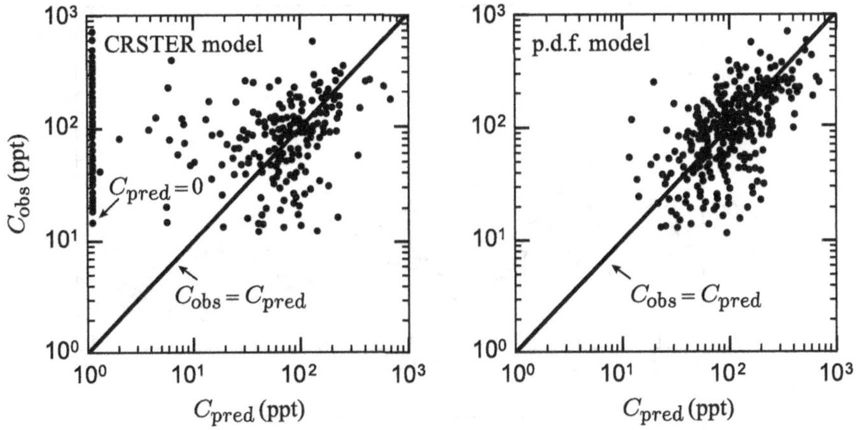

Figure 9.8 Left: Observed versus predicted ground-level SF_6 concentrations for CRSTER, a Gaussian-plume model, at the Kincaid power plant. Observations are one-hour averages; predictions are ensemble means. Right: Same but for an improved model. From Weil (1988).

A parallel analysis shows that time changes in ABL depth h can be unimportant for the turbulence if $\partial h/\partial t \ll u$. The dynamics of the entrainment of a capping inversion typically does limit $\partial h/\partial t$ to a small fraction of u (Chapter 11). Such conditions are typically satisfied in the daytime ABL away from the early-morning and late-afternoon transitions.

Inhomogeneity can be treated in the same way. If L_x is the spatial scale of variability in surface conditions, we expect that when $L_x \gg \ell \sim h$ the ABL turbulence does not "feel" the inhomogeneity and so can be *locally homogeneous*. Requiring $L_x \gg h$ can be more demanding than the quasi-steady criterion, but it can be satisfied in some applications.

Turbulence in the stable ABL tends to have smaller spatial scales and be less diffusive than that in the convective ABL, so we expect that the stable ABL can be much more sensitive to terrain-related inhomogeneities such as the downslope gravity forces on even slightly uneven terrain (Chapter 12). Figure 9.9 illustrates these differences through potential temperature contours in an early morning, stably stratified ABL and in the CBL existing two hours later.

9.6 The mean-momentum equations

Under horizontal homogeneity, mean properties (except mean pressure) do not depend on horizontal position. If so, the mean continuity equation (8.66) reduces to $\partial W/\partial z = 0$. Since W vanishes at the surface, this says it must vanish everywhere

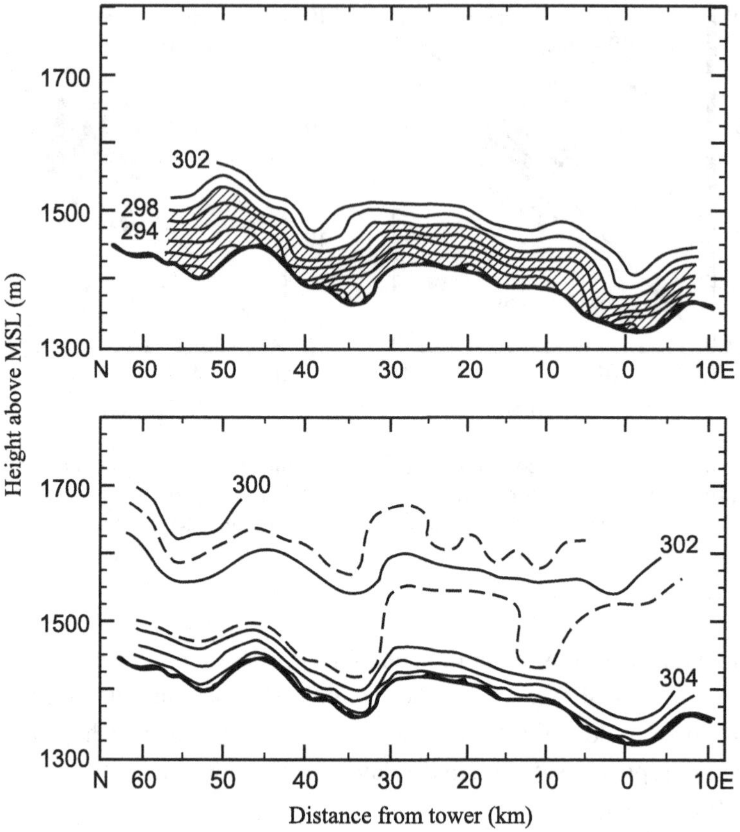

Figure 9.9 Contours of potential temperature measured at two time periods in an ABL over slightly irregular terrain. Upper: before the morning surface heating began. The contours tend to follow the terrain, but with distortion. The lowest surface temperatures are slightly upstream of the valley centers, but the upstream slopes are generally warmer than the downstream slopes. Lower: Two hours later, when the mixing process driven by the surface heating is essentially complete. Turbulent mixing has nearly obliterated the horizontal and vertical temperature gradients that had been established during the night and the surface is now about 5 K warmer than the mixed layer. From Lenschow *et al.* (1979).

and $U_i = [U(z), V(z), 0]$. The horizontal components of the mean momentum equation (8.65) thus reduce to

$$\begin{aligned}\frac{\partial U}{\partial t} + \frac{\partial \overline{uw}}{\partial z} &= -\frac{1}{\rho_0}\frac{\partial P}{\partial x} + 2\Omega_3 V,\\ \frac{\partial V}{\partial t} + \frac{\partial \overline{vw}}{\partial z} &= -\frac{1}{\rho_0}\frac{\partial P}{\partial y} - 2\Omega_3 U.\end{aligned} \tag{9.17}$$

We denote $2\Omega_3 = 2\Omega\sin\theta$, where θ is latitude, as the *Coriolis parameter* f. In steady, laminar *geostrophic* flow these equations reduce to

$$0 = -\frac{1}{\rho_0}\frac{\partial P}{\partial x} + fV, \quad 0 = -\frac{1}{\rho_0}\frac{\partial P}{\partial y} - fU, \tag{9.18}$$

so that

$$U = U_g = -\frac{1}{f\rho_0}\frac{\partial P}{\partial y}, \quad V = V_g = \frac{1}{f\rho_0}\frac{\partial P}{\partial x}. \tag{9.19}$$

This *geostrophic balance* between Coriolis and pressure-gradient forces makes the wind vector perpendicular to the mean pressure gradient:

$$U_i\frac{\partial P}{\partial x_i} = \frac{1}{f\rho_0}\left(-\frac{\partial P}{\partial y}\frac{\partial P}{\partial x} + \frac{\partial P}{\partial x}\frac{\partial P}{\partial y}\right) = 0. \tag{9.20}$$

In the northern hemisphere (where f is positive) geostrophic flow is clockwise around a high-pressure center and counterclockwise around a low; the rotation is reversed in the southern hemisphere.

We often use the solution (9.19) for geostrophic flow to express the components of the horizontal pressure gradient as

$$U_g = -\frac{1}{f\rho_0}\frac{\partial P}{\partial y}, \qquad V_g = \frac{1}{f\rho_0}\frac{\partial P}{\partial x}. \tag{9.21}$$

and to write the mean momentum equations (9.17) as

$$\begin{aligned} \frac{\partial U}{\partial t} + \frac{\partial \overline{uw}}{\partial z} &= f(V - V_g), \\ \frac{\partial V}{\partial t} + \frac{\partial \overline{vw}}{\partial z} &= f(U_g - U). \end{aligned} \tag{9.22}$$

U_g and V_g are not to be interpreted as components of a mean wind in the ABL; rather, fU_g and fV_g are mean kinematic horizontal pressure gradients, as indicated in Eq. (9.19).

The stress-divergence and geostrophic-departure terms in the mean momentum equations (9.22) typically range from 10^{-4} to 10^{-3} m s^{-2} in magnitude. The time-change and advection terms must be much smaller (10^{-5} m s^{-2}, say) for the mean momentum balance to be considered steady and horizontally homogeneous. This would mean the local wind acceleration cannot exceed about 4 cm s^{-1} per hour in order to neglect time changes. If the mean wind speed is 5 m s^{-1}, its horizontal variation cannot exceed about 2 cm s^{-1} per ten kilometers in order to neglect mean advection. These severe requirements suggest that the steady, horizontally homogeneous model of the mean momentum balance does not often apply to real ABL flows.

9.6.1 The Ekman solution

Ekman (1905) assumed there is a constant eddy viscosity K such that $(\overline{uw}, \overline{vw}) = -K\left(\frac{\partial U}{\partial z}, \frac{\partial V}{\partial z}\right)$. If we align the x-axis with the geostrophic wind (which is assumed not to depend on z), then $V_g = 0$. With the boundary conditions that the mean wind vanishes at $z = 0$ and approaches the geostrophic value U_g at very large z, his steady solution to Eq. (9.22) is the *Ekman spiral*

$$U = U_g(1 - e^{-\gamma z}\cos\gamma z), \quad V = U_g e^{-\gamma z}\sin\gamma z, \quad \gamma = (f/2K)^{1/2}. \tag{9.23}$$

For small γz we have $U = V \to \gamma z U_g$, whereas for large γz (i.e., above the boundary layer) $U \to U_g$, $V \to V_g = 0$, so the surface-layer wind is at 45 degrees to the wind above the boundary layer.

This Ekman solution (9.23) is not realistic for geophysical boundary layers for several reasons. First, numerical experiments show that several $1/f$ times – say on the order of one day – can be required for this neutral boundary layer to reach a steady state from arbitrary initial conditions. Steady, horizontally homogeneous conditions persisting that long are rare. Second, observations show that an elevated inversion, rather than the Ekman dynamics, typically establishes the ABL depth. Third, the mean horizontal pressure gradient (the geostrophic wind) often varies significantly with height. Finally, the surface and the effects of stability typically cause both u and ℓ to vary significantly with height, giving $K \sim u\ell$ a pronounced vertical structure. The solution of the mean-momentum equation with K = constant can be quite unphysical, as can be demonstrated in turbulent Couette flow (Problem 9.19).

Only under stably stratified or baroclinic conditions do we tend to see the Ekman wind-turning angle of 45 degrees in the ABL. The Ekman solution is a particularly poor representation of the stress and mean wind profiles under typical convective conditions.

9.6.2 The baroclinic case

Horizontal density gradients can cause the horizontal mean pressure gradient to depend on height z, giving *baroclinity* that can profoundly affect ABL structure. From the definition (9.21) of the geostrophic wind, we see that

$$\begin{aligned} f\frac{\partial U_g}{\partial z} &= -\frac{\partial}{\partial z}\left(\frac{1}{\rho_0}\frac{\partial P}{\partial y}\right) = -\frac{1}{\rho_0}\frac{\partial}{\partial y}\frac{\partial P}{\partial z} + \frac{1}{\rho_0}\frac{\partial \rho_0}{\partial z}\frac{1}{\rho_0}\frac{\partial P}{\partial y} \\ &= \frac{\partial}{\partial y}\left(-\frac{1}{\rho_0}\frac{\partial P}{\partial z}\right) + \frac{fU_g}{H_\rho}, \end{aligned} \tag{9.24}$$

where H_ρ is the density scale height of the atmosphere, about 12 km. For boundary-layer depths small compared to H_ρ the second term on the right side of Eq. (9.24) is not important. To evaluate the first term on the right we turn to the vertical equation of mean motion:

$$\frac{\partial U_3}{\partial t} + U_j \frac{\partial U_3}{\partial x_j} + \frac{\partial \overline{u_3 u_j}}{\partial x_j} = -\frac{1}{\rho_0}\frac{\partial P}{\partial x_3} - 2\epsilon_{3jk}\Omega_j U_k + \frac{g}{\theta_0}\overline{\tilde{\theta}'}. \tag{9.25}$$

Let us first scale the terms in Eq. (9.25). Let horizontal mean velocities scale with U_0, vertical mean velocity with W_0, turbulent velocities with u, horizontal gradients with L_x, local time changes with L_x/U, and vertical gradients with H. The mean continuity equation implies

$$\frac{U_0}{L_x} \sim \frac{W_0}{H}, \tag{9.26}$$

so we have $W_0 \sim HU_0/L_x$. Then the order of the terms in Eq. (9.25) is

$$\begin{aligned} \text{time change, mean advection: } & \frac{U_0^2 H}{L_x^2}, \\ \text{turbulent flux divergence: } & \frac{u^2}{H}, \\ \text{rotation: } & fU_0, \\ \text{buoyancy: } & \frac{g}{\theta_0}\overline{\tilde{\theta}'}. \end{aligned} \tag{9.27}$$

If $\overline{\tilde{\theta}'} \sim 1\,\text{K}$, then the buoyancy term is of order $3 \times 10^{-2}\,\text{m}\,\text{s}^{-2}$. If $U_0 \sim 10\,\text{m}\,\text{s}^{-1}$ and $H \sim 10^3$ m, the rotation and turbulence terms are of order $10^{-3}\,\text{m}\,\text{s}^{-2}$, considerably smaller than the buoyancy term.

The size of the time-change and advection terms depends additionally on the characteristic horizontal length scale L_x. If L_x is of the order of 10^4 m – which implies a horizontal divergence of 10 m s^{-1} per 10 km, a fairly large value – then these inertial terms are of order 10^{-3} m s^{-2}, again much smaller than the buoyancy term.

We conclude that if L_x is sufficiently large we can use the *hydrostatic approximation* to the U_3-equation, writing it as

$$\frac{1}{\rho_0}\frac{\partial P}{\partial z} = \frac{g}{\theta_0}\overline{\tilde{\theta}'}. \tag{9.28}$$

Differentiating (9.28) with respect to y and using the result in (9.24) provides an expression for the z-variation of U_g:

$$f\frac{\partial U_g}{\partial z} = -\frac{g}{\theta_0}\frac{\partial \overline{\tilde{\theta}'}}{\partial y} + \frac{fU_g}{H_\rho} = -\frac{g}{T_0}\frac{\partial T}{\partial y} + \frac{fU_g}{H_\rho}. \tag{9.29}$$

Similarly, we have

$$f\frac{\partial V_{\mathrm{g}}}{\partial z} = \frac{g}{T_0}\frac{\partial T}{\partial x} + \frac{f V_{\mathrm{g}}}{H_\rho}. \tag{9.30}$$

In general the first term in these expressions is by far the more important.

According to Eqs. (9.29) and (9.30), a horizontal gradient of mean virtual temperature of 3 K per 100 km in the lower atmosphere, which is not uncommon, generates a geostrophic wind shear, or baroclinity, of 10 m s^{-1} per 1000 m. Thus, the mean horizontal pressure gradient can change radically across the depth of the ABL. Horizontal density gradients near fronts can cause comparable effects in the upper ocean.

9.6.3 Ekman pumping

In the absence of significant advection of mean momentum and horizontal variability of the geostrophic wind, the mean vertical vorticity equation in the quasi-steady ABL reduces to a balance between turbulent friction and Coriolis effects:

$$\frac{\partial^2 \overline{uw}}{\partial y\,\partial z} - \frac{\partial^2 \overline{vw}}{\partial x\,\partial z} = f\left(\frac{\partial U}{\partial x} + \frac{\partial V}{\partial y}\right) = -f\left(\frac{\partial W}{\partial z}\right). \tag{9.31}$$

Integrating (9.31) from the surface to the top of the ABL at height h, where the turbulent stress vanishes, gives

$$\frac{\partial \overline{vw}(0)}{\partial x} - \frac{\partial \overline{uw}(0)}{\partial y} = \frac{1}{\rho_0}\left(\frac{\partial \tau_{23}^{f}(0)}{\partial x} - \frac{\partial \tau_{13}^{f}(0)}{\partial y}\right) = -fW(h). \tag{9.32}$$

$(\tau_{13}^{f}(0), \tau_{23}^{f}(0))$ is the vector shearing stress *on the fluid* in the surface plane. Its negative, the shearing stress *on the surface*, is conventionally taken as proportional to the product of mean wind speed and the mean wind vector $(U_{\mathrm{ref}}, V_{\mathrm{ref}})$ at some reference height (e.g., at 10 m):

$$(\tau_{13}^{\mathrm{s}}(0), \tau_{23}^{\mathrm{s}}(0)) = -(\tau_{13}^{\mathrm{f}}(0), \tau_{23}^{\mathrm{f}}(0)) = C_{\mathrm{d}}\,\rho_0\, S_{\mathrm{ref}}\,(U_{\mathrm{ref}}, V_{\mathrm{ref}}), \tag{9.33}$$

with C_{d} the *drag coefficient*. For circularly symmetric flow around a pressure minimum or maximum (a *low* or a *high*) with S_{ref} and ρ_0 taken as constant we can then use Eq. (9.33) to write (9.32) as

$$C_{\mathrm{d}} S_{\mathrm{ref}} \overline{\omega_3} = fW(h), \qquad \overline{\omega_3} = \frac{\partial V_{\mathrm{ref}}}{\partial x} - \frac{\partial U_{\mathrm{ref}}}{\partial y}, \tag{9.34}$$

with $\overline{\omega_3}$ the mean vertical vorticity at the reference height. In the northern hemisphere f is positive and $\overline{\omega_3}$ is negative around a high and positive around a low; in the southern hemisphere the signs of f and $\overline{\omega_3}$ are reversed. Thus in both hemispheres (9.34) indicates that $W(h)$ is negative around a high and positive around a low. Its magnitude is typically on the order of 1 cm s^{-1}, which is often significant compared to the rate of deepening of the boundary layer due to entrainment. Other things being equal, ABLs under a high are apt to be shallower and less cloudy as a result.

Questions on key concepts

9.1 Contrast the nature of the instantaneous tops of laminar and turbulent boundary layers.

9.2 Contrast the nature of the instantaneous and average tops of a turbulent boundary layer.

9.3 Explain why the mean flow in the ABL cannot be geostrophic.

9.4 Explain how acoustic sounding is able to detect the instantaneous top of the ABL.

9.5 Explain why the surface energy balance is one of the most important influences on the ABL.

9.6 Explain the interpretation of the terms in Eq. (9.4).

9.7 Explain why characterization of the stability state of an ABL through near-surface measurements alone is apt not to be reliable. How could a stable ABL be misidentified in this way? An unstable ABL?

9.8 What is subgrid TKE? Why need it not be directly affected by buoyancy?

9.9 Discuss three mechanisms by which an ABL can be made stably stratified.

9.10 Explain why stable stratification affects larger eddies more strongly than smaller ones.

9.11 Explain why and how the difference between average and instantaneous ABL structure impacts dispersion modeling.

9.12 Discuss why the Ekman solution is generally not applicable to the ABL.

9.13 Why is the mean horizontal pressure gradient in the ABL apt to depend on height? Is the dependence typically significant? Why does this not occur in engineering flows?

9.14 Explain the physical mechanism of Ekman pumping.

9.15 The flatness factor of the vertical velocity in the CBL does not differ greatly from the Gaussian value of 3. Explain why, by the criterion of Problem 7.16, the $\sim$1 skewness of vertical velocity there is large.

Problems

9.1 Calculate the temperature gradient at the surface at midday in the Kansas experiment, given that the surface temperature flux (Figure 9.5) is due to molecular diffusion. Explain physically why this gradient is so large. (Recall our discussion of the wall fluxes in pipe flow, Chapter 1.)

9.2 The convective ABL is sometimes modeled as a mixture of updrafts and downdrafts of area fractions and vertical velocities $f^{\mathrm{u}}, w^{\mathrm{u}}$ and $f^{\mathrm{d}}, w^{\mathrm{d}}$ respectively. Interpret physically the statements

$$f^{\mathrm{u}} + f^{\mathrm{d}} = 1, \quad w^{\mathrm{u}} f^{\mathrm{u}} + w^{\mathrm{d}} f^{\mathrm{d}} = 0.$$

Two additional equations are

$$(w^{\mathrm{u}})^2 f^{\mathrm{u}} + (w^{\mathrm{d}})^2 f^{\mathrm{d}} = \overline{w^2}, \quad (w^{\mathrm{u}})^3 f^{\mathrm{u}} + (w^{\mathrm{d}})^3 f^{\mathrm{d}} = \overline{w^3}.$$

Interpret these as well. Solve the set and discuss the solution.

9.3 Use estimates of temperature fluctuations, mean temperature gradients, and other quantities as needed to show why the ABL is seldom apt to be in a neutral state (i.e., be unaffected by buoyancy).

9.4 Reconcile the notions of quasi-steadiness and local homogeneity (Section 9.5) and the sensitivity of the mean momentum balance to time changes and horizontal inhomogeneity (Section 9.6).

9.5 Discuss how variation of the horizontal pressure gradient with z changes the nature of the mean momentum balance in a quasi-steady, horizontally homogeneous ABL. Which term would you expect to balance most of this height variation under very convective conditions? Explain. Sketch the vertical profiles of the three terms in the momentum balance in this limiting case. Contrast the situation with the case where the pressure gradient is independent of height.

9.6 Sketch the profile of vertical temperature flux in a quasi-steady convective ABL capped by an inversion.

9.7 Confirm that the Ekman solution (9.23) satisfies the steady, horizontally homogeneous mean-momentum equations with a constant-K closure.

9.8 A radiosonde is a standard instrument for measuring vertical profiles in the lower atmosphere. Sensors mounted below a lighter-than-air balloon measure temperature and water vapor along its rising path and transmit signals back to earth. Discuss the utility of such data in numerical modeling.

9.9 Generalize the Ekman solution (9.23) to the case with geostrophic wind varying linearly with height. Is the solution physical?

9.10 Discuss the determination of eddy diffusivity for momentum in the ABL in the context of required averaging times.

9.11 Explain the time lag between sunrise and the appearance of positive surface temperature flux (Figure 9.5) and between sunset and negative flux.

9.12 In view of conservation of energy, the rate of loss of TKE to buoyant destruction must reappear as a source term somewhere. Rewrite the buoyancy term in the equation of vertical motion in terms of density fluctuations and use the vertical integral of the continuity equation times height to show where it emerges. Interpret this physically.

9.13 What is the physical meaning of the first term on the right side of Eq. (9.4)? Show that it integrates to zero over the ABL.

9.14 Consider flow in a horizontally homogeneous, quasi-steady ABL capped by a strong inversion at height z_i such that the parameter $fz_i/u_* \ll 1$. Scale the two mean horizontal momentum equations and show that they reduce to that for turbulent channel flow. Where does the adjustment to geostrophic flow occur in this limit? What is the change in mean flow direction in this adjustment?

9.15 Write an expression for a turbulence Rossby number, the ratio of typical inertial and Coriolis forces on energy-containing eddies. Estimate its magnitude in the ABL.

9.16 Derive Eqs. (9.29) and (9.30) from Eq. (9.28).

9.17 Show that in the ABL the Coriolis terms in the turbulence second-moment budgets (8.70), (8.71) tend to be small compared to the leading terms.

9.18 Assume the flow over the earth's surface were laminar, not turbulent. Use the surface energy balance to explain why at mid latitudes during clear summer weather, shallow water on the surface could boil during the day and freeze at night.

9.19 Turbulent channel flow (Figure 3.1) with the upper and lower walls moving in opposite directions at speed U_w is called *turbulent Couette flow*. It is simulated in the laboratory with counter-rotating concentric cylinders of diameter R and $R + \delta R$, with $\delta R \ll R$; this allows the use of cartesian coordinates.

(a) Using the streamwise periodicity of this flow, show that the mean streamwise pressure gradient is zero.
(b) Solve the streamwise mean momentum equation for the total stress profile.
(c) Using an eddy-viscosity (K) closure, show that the mean velocity profile depends sensitively on the K profile.
(d) Sketch a K profile that you feel gives a realistic mean velocity profile.
(e) Discuss the implications for K closure.

References

Briggs, G., 1988: Analysis of diffusion field experiments. In *Lectures on Air Pollution Modeling*, A. Venkatram and J. Wyngaard, Eds., Boston: American Meteorological Society.

Brost, R. A., and J. C. Wyngaard, 1978: A model study of the stably stratified planetary boundary layer. *J. Atmos. Sci.*, **35**, 1427–1440.

Corrsin, S., 1972: Random geometric problems suggested by turbulence. *Statistical Models and Turbulence*, Lecture Notes in Physics, **12**, Springer-Verlag, pp. 300–316.

Corrsin, S., and A. L. Kistler, 1955: *Free-stream Boundaries of Turbulent Flows*. NACA TR 1244, NACA, Washington, D.C.

Csanady, G. T., 1974: Equilibrium theory of the planetary boundary layer with an inversion lid. *Bound.-Layer Meteor.*, **6**, 63–79.

Ekman, V. W., 1905: On the influence of the Earth's rotation on ocean currents. *Arch. Math. Astron. Phys.*, **2**, 1–52.

Hunt, J. C. R., 1985: Diffusion in the stably stratified atmospheric boundary layer. *J. Climate App. Meteor.*, **24**, 1187–1195.

Lenschow, D. H., B. B. Stankov, and L. Mahrt, 1979: The rapid morning boundary-layer transition. *J. Atmos. Sci.*, **36**, 2108–2124.

Little, C. G., 1969: Acoustic methods for the remote probing of the lower atmosphere. *Proc. IEEE*, **57**, 571–578.

Mason, P. J., and S. H. Derbyshire, 1990: Large-eddy simulation of the stably stratified atmospheric boundary layer. *Bound.-Layer Meteor.*, **53**, 117–162.

McAllister, L. G., J. R. Pollard, A. R. Mahoney, and P. J. R. Shaw, 1969: Acoustic sounding: a new approach to the study of atmospheric structure. *Proc. IEEE*, **57**, 579–587.

Moeng, C.-H., and R. Rotunno, 1990: Vertical-velocity skewness in the buoyancy-driven boundary layer. *J. Atmos. Sci.*, **47**, 1149–1162.

Neff, W., D. Helmig, A. Grachev, and D. Davis, 2008: A study of boundary layer behavior associated with high NO concentrations at the South Pole using a minisodar, tethered balloon, and sonic anemometer. *Atmos. Env.*, **42**, 2762–2779.

Nieuwstadt, F. T. M., 1984: The turbulent structure of the stable, nocturnal boundary layer. *J. Atmos. Sci.*, **41**, 2202–2216.

Van Dyke, M., 1982: *An Album of Fluid Motion*. Stanford: Parabolic Press.

Weil, J. C., 1988: Dispersion in the convective boundary layer. In *Lectures on Air Pollution Modeling*, A. Venkatram and J. Wyngaard, Eds., Boston: American Meteorological Society.

Wyngaard, J. C., 1973: On surface-layer turbulence. *Workshop on Micrometeorology*, D. A. Haugen, Ed., Boston: American Meteorological Society, pp. 101–149.

Wyngaard, J. C., 1988: Structure of the PBL. In *Lectures on Air Pollution Modeling*, A. Venkatram and J. Wyngaard, Eds., Boston: American Meteorological Society.

10

The atmospheric surface layer

10.1 The "constant-flux" layer

Near a flat, homogeneous land surface the turbulent shear stress vector in the horizontal plane, $-\rho_0 \overline{u_i u_3}$, is aligned with the mean wind vector U_i.† As is traditional in micrometeorology, we take this as the x-direction. Then integrating the steady forms of the horizontally homogeneous mean momentum equations (9.22),

$$\frac{\partial \overline{uw}}{\partial z} = f(V - V_{\rm g}), \qquad \frac{\partial \overline{vw}}{\partial z} = f\left(U_{\rm g} - U\right), \tag{10.1}$$

from just above the surface to the mean top of the ABL at height h, where the turbulence vanishes, gives

$$-\overline{uw}(0^+) = \frac{\tau_0}{\rho_0} = f\int_0^h (V - V_{\rm g})\,dz, \qquad \overline{vw}(0^+) = 0 = f\int_0^h (U - U_{\rm g})\,dz, \tag{10.2}$$

with τ_0 the magnitude of the surface shear stress. Equations (10.1) and the integral constraints (10.2) imply that in the northern hemisphere (where f is positive) and with $(U_{\rm g}, V_{\rm g})$ independent of z, the (U, V) and $(\overline{uw}, \overline{vw})$ profiles in the near-neutral case behave qualitatively as sketched in Figure 10.1 (Problem 10.1).

Equations (10.1) and Figure 10.1 indicate that $U_{\rm g}$ and $V_{\rm g}$ are independent of height and $\partial \overline{uw}/\partial z$ and $\partial \overline{vw}/\partial z$ are largest just above the surface. So why is the surface layer also called the "constant-flux layer" if momentum-flux gradients can be largest there?

The answer lies in the behavior of turbulence near a surface, Figure 10.2. As shown in the Appendix, the horizontal wavenumber spectrum of w (and, hence, of turbulent shear stress) peaks at wavenumbers of order $1/z$. Thus distance from the

† This need not be the case over a moving, wavy sea surface (Grachev *et al.*, 2003).

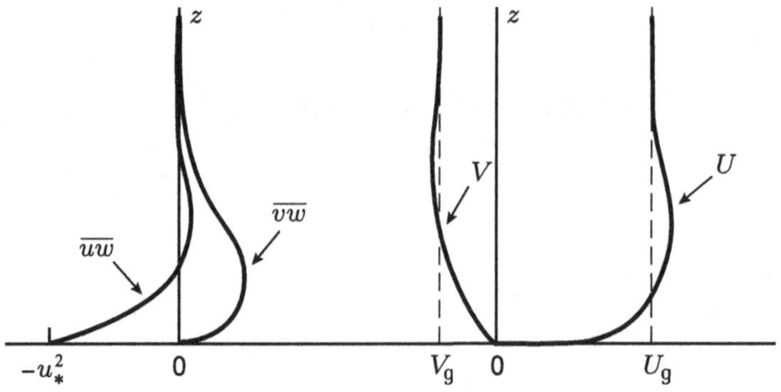

Figure 10.1 A sketch of profiles of kinematic shear stress (left) and mean wind (right) in the near-neutral ABL for z-independent U_g and V_g.

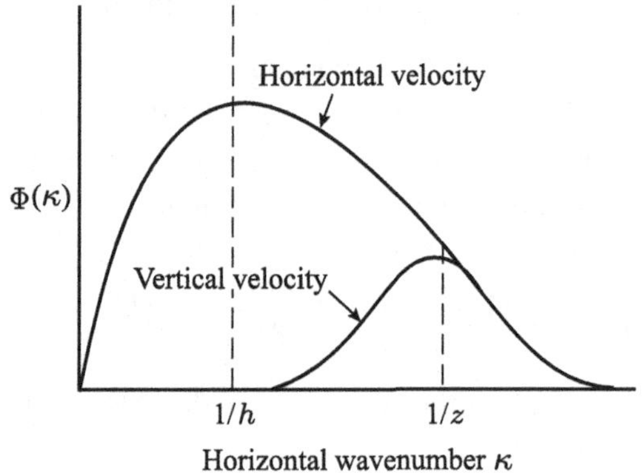

Figure 10.2 A sketch of velocity spectra in the convective surface layer.

surface, z, has traditionally been used as the turbulence length scale in the surface layer. But particularly in the unstable case the largest, ABL-filling eddies cause the horizontal wavenumber spectra of u and v near the surface to peak at wavenumbers of the order of the inverse of the ABL depth h. These large-scale horizontal motions near the surface have been termed *inactive* (Bradshaw, 1967).

The streamwise mean momentum equation (10.1) near the surface scales as

$$\frac{u_*^2}{h} \sim \frac{\partial \overline{uw}}{\partial z} = -f V_g. \tag{10.3}$$

This is small when nondimensionalized with surface-layer scales:

$$\frac{z}{u_*^2}\frac{\partial \overline{uw}}{\partial z} \sim \frac{z}{h} \ll 1. \tag{10.4}$$

Similarly, $\partial \overline{w\theta}/\partial z$ is of order Q_0/h, so when nondimensionalized with surface-layer scales it is also very small:

$$\frac{z}{Q_0}\frac{\partial \overline{w\theta}}{\partial z} \sim \frac{z}{h} \ll 1. \tag{10.5}$$

Because vertical gradients of these turbulent fluxes are negligible when expressed in surface-layer scales, the surface layer has been called the *constant-flux layer*. Horst (1999) has suggested the more appropriate name *surface-flux layer*.

10.2 Monin–Obukhov similarity

10.2.1 Dimensional analysis

The Monin–Obukhov (M-O) similarity hypothesis has provided much of the foundation for our understanding of the atmospheric surface layer. It rests on dimensional analysis and the *Buckingham Pi Theorem*, which prescribes the optimal approach to determining a dependent variable in a physical problem. That theorem says that if one can identify the $m-1$ parameters governing that dependent variable, and if n is the number of dimensions represented by it and the governing parameters, then:

- $m-n$ independent dimensionless quantities can be formed, the nondimensionalization being done with the governing parameters. An *independent* quantity is one that cannot be made from the others. The choice of the dimensionless quantities is arbitrary.
- These $m-n$ independent dimensionless quantities are functionally related, so the dependent variable can be taken to be a function of the governing parameters.

Identifying the governing parameters requires an overall grasp of the physics of the problem and good intuition. We'll illustrate this process with examples from Chapter 1.

Kolmogorov (1941) hypothesized that the length and velocity scales η and υ of dissipative-range turbulence depend only on the viscous dissipation rate per unit mass, ϵ, and the fluid kinematic viscosity, ν:

$$\eta = \eta(\epsilon, \nu), \qquad \upsilon = \upsilon(\epsilon, \nu). \tag{10.6}$$

The dimensional analyses for the unknowns η and υ are similar, so we'll do η.

By hypothesis the dependent variable η is governed by ϵ and ν, so $m-1=2$, $m=3$. The units of η are L; of ϵ, $\mathrm{L}^2\,\mathrm{T}^{-3}$; of ν, $\mathrm{L}^2\,\mathrm{T}^{-1}$. There are $n=2$ dimensions so $m-n=1$ independent dimensionless quantity can be formed from

η, ϵ, and ν; we take it to be $\eta\epsilon^{1/4}/\nu^{3/4}$. Since there is only one quantity, it is a function of nothing – i.e., it is a constant:

$$\frac{\eta\epsilon^{1/4}}{\nu^{3/4}} = \text{constant}, \qquad \eta = \text{constant} \times \left(\frac{\nu^3}{\epsilon}\right)^{1/4}. \tag{10.7}$$

With the constant chosen as 1 this is Kolmogorov's result, Eq. (1.34).

This is as simple a dimensional analysis problem as you'll meet, so simple that it doesn't require a formalism to work out. You can write down the answer: If the length scale η depends only on ϵ and ν, then the result (10.7) follows because it is the only length scale that can be made from ϵ and ν.

The pipe-friction problem in Chapter 1 demonstrates better the power of the Buckingham Pi Theorem. On what parameters does the mean wall stress in turbulent flow in a circular pipe depend? The obvious ones are pipe diameter D, mean velocity averaged over the cross section, $\overline{u}_{\text{ave}}$, fluid density ρ, and kinematic viscosity ν. Another is the characteristic height h_{r} of the roughness elements on the pipe wall, which protrude into the diffusive sublayer and add drag. Therefore (Problem 10.19)

$$\overline{\tau}_{\text{wall}} = \overline{\tau}_{\text{wall}}(D, \overline{u}_{\text{ave}}, \rho, \nu, h_{\text{r}}). \tag{10.8}$$

Here we have $m - 1 = 5$ governing parameters and $n = 3$ dimensions, so there are $m - n = 3$ independent dimensionless quantities that are functionally related. It is conventional to choose these as

$$f = \frac{2\overline{\tau}_{\text{wall}}}{\rho\overline{u}^2_{\text{ave}}}, \quad Re = \frac{\overline{u}_{\text{ave}}D}{\nu}, \quad \frac{h_{\text{r}}}{D}. \tag{10.9}$$

Then we can write the friction factor f as

$$f = f(Re, h_{\text{r}}/D). \tag{10.10}$$

It is conventional to plot f against Re with h_{r}/D as a parameter, as in Figure 1.2.

10.2.2 M-O governing parameters

What we now call the *Monin–Obukhov (M-O) similarity hypothesis*, or simply M-O similarity, rests on the Obukhov (1946) paper and a later one by Monin and Obukhov (1954). Foken (2006) has discussed its history and evolution in detail. In the M-O hypothesis five parameters govern the quasi-steady turbulence structure immediately above a flat, horizontally homogeneous land surface: the length scale

ℓ of the turbulence, taken as distance z from the surface; the velocity scale u of the turbulence, taken as the friction velocity u_*, the square root of the magnitude of the mean kinematic surface stress; the mean temperature flux at the surface, Q_0; the mean surface flux of a conserved scalar constituent, C_0; and the buoyancy parameter g/θ_0. By *surface-layer structure* we denote the mean vertical gradients of wind, potential temperature, and the mixing ratio of conserved scalars, plus the turbulence statistics.

An underlying assumption is that the outer part of the boundary layer does not exert an important influence on the surface layer. Thus, taking $\ell \sim z$ is consistent with the near-surface kinematics of vertical velocity discussed in the Appendix. The *friction velocity* u_* is a natural choice for the turbulent velocity scale because beginning just above the surface the mean shear stress is carried essentially entirely by the turbulence. Likewise, the mean surface flux of a conserved scalar divided by u_* is an appropriate intensity scale for the fluctuations of the scalar. The mean surface temperature flux Q_0 is included because it is proportional to the rate of buoyant production of TKE in the surface layer. Finally, the buoyancy parameter g/θ_0 is included because it appears in the equation of motion (8.56).

In Chapter 8 we defined virtual temperature as $\tilde{T}_{\rm v} = \tilde{T}(1 + 0.61\tilde{q})$, where $\tilde{q}$ is specific humidity. Decomposing into mean and fluctuating parts (and reverting briefly to our previous notation), $\tilde{T} = T + \theta$, $\tilde{q} = Q + q$, shows that the fluctuations are related by $\theta_{\rm v} \simeq \theta + 0.61qT$ and the mean vertical turbulent flux of virtual temperature is (Problem 10.18)

$$\overline{\theta_{\rm v} w} \simeq \overline{\theta w} + 0.61T\,\overline{qw}. \tag{10.11}$$

The contribution from water-vapor flux can be very important, so in general Q_0 is taken to be the mean surface flux of virtual temperature.

Let's discuss why other plausible candidates are not included in the M-O governing group:

- The boundary-layer depth h was not included on the erroneous assumption that it does not directly influence surface-layer turbulence. We'll see shortly, however, that the largest eddies in the CBL do contribute substantially to the horizontal velocity fluctuations near the surface.
- The mean velocity is excluded because a description of turbulence must be invariant to a *Galilean transformation* that moves the coordinate system at a constant velocity, and the mean velocity is not *Galilean invariant.*
- The turbulent Coriolis force/mass (of order fu) is very small compared to the turbulent inertia force/mass (of order u^2/ℓ) when $\ell \sim z$; i.e., the inverse turbulent Rossby number $fz/u \ll 1$. This can also hold when $\ell \sim h$.

- The molecular diffusivities are ignored because the turbulence Reynolds number R_t in the surface layer is so large. Thus M-O similarity is not applicable to the dissipative-range structure, which does depend on the molecular diffusivities.
- The characteristic length of the roughness elements on the surface, called the *roughness length* z_0, is excluded. Thus we restrict M-O similarity to $z \gg z_0$.

10.2.3 The application of M-O similarity

An unknown dependent variable plus the five M-O governing parameters comprise $m = 6$ parameters with $n = 4$ dimensions: length, time, temperature, and c. Thus, there are $m - n = 2$ independent dimensionless quantities that are functionally related. M-O similarity takes one as the dependent variable nondimensionalized with z, u_*, $T_* = -Q_0/u_*$ and $c_* = -C_0/u_*$; the other is taken as z/L, where $L = -u_*^3\theta_0/kgQ_0$ is the *Monin–Obukhov length.*† $k \simeq 0.4$ is the von Kármán constant.‡ The Buckingham Pi Theorem then says that this nondimensionalized dependent variable is a function only of z/L.

L is negative in unstable conditions ($Q_0 > 0$), positive in stable conditions ($Q_0 < 0$), and infinite at neutral ($Q_0 = 0$); thus, the range of the M-O independent variable is $-\infty < z/L < \infty$. Conditions are *near-neutral* when $|z/L|$ is sufficiently small, which occurs near the surface when $z \ll |L|$.

M-O similarity implies that the surface-layer gradients of mean wind speed, mean virtual potential temperature, and mean water vapor mixing ratio behave as

$$\frac{kz}{u_*}\frac{\partial U}{\partial z} = \phi_m\left(\frac{z}{L}\right), \quad -\frac{kzu_*}{Q_0}\frac{\partial \Theta}{\partial z} = \frac{kz}{T_*}\frac{\partial \Theta}{\partial z} = \phi_h\left(\frac{z}{L}\right),$$
$$-\frac{kzu_*}{C_0}\frac{\partial C}{\partial z} = \frac{kz}{c_*}\frac{\partial C}{\partial z} = \phi_c\left(\frac{z}{L}\right), \tag{10.12}$$

with ϕ_m, ϕ_h, and ϕ_c being functions of z/L that are *universal* – the same in all locally homogeneous, quasi-steady surface layers. The minus sign in the definitions of T_* and c_* makes ϕ_h and ϕ_c positive, as ϕ_m is.

It is commonly assumed that $\phi_h = \phi_c$, but Warhaft (1976) has questioned that because the buoyancy term in the scalar flux conservation equation (8.70) differs for potential temperature and for another conserved scalar – water vapor, for example. We'll discuss that in more detail in Chapter 11.

† Businger and Yaglom (1971) and Foken (2006) point out that the Monin–Obukhov length was introduced by Obukhov (1946), so *Obukhov length* is more appropriate.

‡ The value of k was traditionally thought to be $\simeq 0.4$, was measured as 0.35 in the 1968 Kansas experiments, and extensive measurements by Andreas *et al.* (2006) give a value of 0.39.

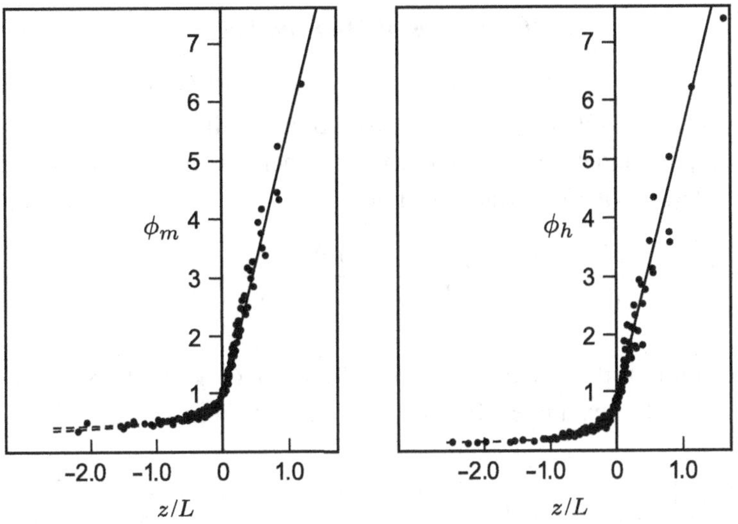

Figure 10.3 The M-O functions for mean wind shear (left) and mean potential temperature gradient (right), Eq. (10.12), from the 1968 Kansas experiment. From Businger *et al.* (1971).

The von Kármán constant k was introduced in the early twentieth century to scale the friction velocity in the fully turbulent region near a wall:

$$k \equiv \frac{(|\tau_0|/\rho)^{1/2}}{z\partial U/\partial z} = \frac{u_*}{z\partial U/\partial z}. \tag{10.13}$$

The M-O function for mean wind shear, Eq. (10.12), is defined with k to make its value in neutral conditions 1.0: $\phi_m(0) = 1.0$. For parallelism ϕ_h and ϕ_c, Eq. (10.12), are defined with a k as well.

The Kansas experiments were among the first in which the mean wind and temperature profiles and their turbulent fluxes were measured in the surface layer. Figure 10.3 shows the resulting M-O plots of ϕ_m and ϕ_h. Each shows a remarkable collapse of data taken over several weeks and a wide range of stability conditions.

Hogstrom (1988) proposed that the underlying cause of the substantial variations in the M-O functions ϕ_m and ϕ_h measured in numerous surface-layer experiments was probe-induced flow distortion (Chapter 16). He presented results from a surface-layer experiment in which its effects were removed and other measurement errors minimized. He suggested the following forms:

$$\begin{aligned} \text{stable:}\quad \phi_m &= 1.0 + 4.8\tfrac{z}{L}, \qquad \phi_h = 1.0 + 7.8\tfrac{z}{L}; \\ \text{unstable:}\quad \phi_m &= \left(1 - 19.3\frac{z}{L}\right)^{-1/4}, \quad \phi_h = \left(1 - 12\frac{z}{L}\right)^{-1/2}. \end{aligned} \tag{10.14}$$

10.2.4 Uses of M-O similarity

10.2.4.1 Wind and temperature profiles

As indicated in Eq. (10.12), M-O similarity predicts that the vertical gradients of mean wind speed and mean virtual potential temperature in a locally homogeneous, quasi-steady atmospheric surface layer are

$$\frac{\partial U}{\partial z} = \frac{u_*}{kz}\phi_m\left(\frac{z}{L}\right), \qquad \frac{\partial \Theta}{\partial z} = \frac{T_*}{kz}\phi_h\left(\frac{z}{L}\right). \tag{10.15}$$

In stable conditions the functions proposed for ϕ_m and ϕ_h tend to be linear and thus integrate easily. For example, Hogstrom's ϕ_m and ϕ_h forms, Eqs. (10.14), give

$$U(z) = \frac{u_*}{k}\left[\ln\frac{z}{z_0} + 4.8\frac{z}{L}\right], \qquad \Theta(z) = \Theta(z_r) + \frac{T_*}{k}\left[\ln\frac{z}{z_r} + 7.8\frac{z}{L}\right], \tag{10.16}$$

with z_r the reference height for the "surface" temperature.

Many different ϕ_m and ϕ_h forms have been proposed for unstable conditions, and they tend to be more difficult to integrate analytically. Based on fits to Hogstrom's (1988) extensive data, Wilson (2001) proposed

$$\phi_{m,h} = (1 + \gamma|z/L|^{2/3})^{-1/2} \tag{10.17}$$

for both ϕ_m and ϕ_h. He shows these give the mean profiles

$$U(z) = \frac{u_*}{k}\left[\ln\frac{z}{z_0} - 3\ln\left(\frac{1+\sqrt{1+\gamma_m|z/L|^{2/3}}}{1+\sqrt{1+\gamma_m|z_0/L|^{2/3}}}\right)\right], \tag{10.18}$$

$$\Theta(z) = \Theta(z_r) + \frac{P_t T_*}{k}\left[\ln\frac{z}{z_r} - 3\ln\left(\frac{1+\sqrt{1+\gamma_h|z/L|^{2/3}}}{1+\sqrt{1+\gamma_h|z_r/L|^{2/3}}}\right)\right], \tag{10.19}$$

and suggests the constants be taken as $k = 0.4$, $P_t = 0.95$, $\gamma_m = 3.6$, $\gamma_h = 7.9$.

10.2.4.2 Inferring a surface-layer property from measurements of others

In a neutral surface layer

$$\frac{\sigma_w}{u_*} = \left.\frac{(\overline{w^2})^{1/2}}{u_*}\right|_{z/L=0} \simeq 1.2. \tag{10.20}$$

Thus, under near-neutral conditions (i.e., at $|z/L| \ll 1$) one can use $\sigma_w/1.2$ as a surrogate for u_*. It is easier to measure σ_w than to measure $u_* \simeq \sqrt{-\overline{uw}}$ and it requires much shorter averaging times (Chapter 2).

10.2.4.3 Diagnosing surface fluxes

Weather and climate models use the M-O mean profiles, Section 10.2.4.1, to diagnose surface fluxes. For surface momentum flux $\tau_0 \equiv \rho_0 u_*^2$, for example, Eqs. (10.16) and (10.18) indicate that $U_{\text{ref}} \equiv U(z_{\text{ref}})$ depends on $\tau_0, \rho_0, z_{\text{ref}}, z_0, L$. Thus if we write $\tau_0 = f(U_{\text{ref}}, z_{\text{ref}}, z_0, L, \rho_0)$, then in the dimensional analysis there are $m - 1 = 5$ governing parameters and $n = 3$ dimensions, M, L, and T. The Pi Theorem tells us there are $m - n = 3$ independent dimensionless groups that are functionally related. If we take these as $\tau_0/\rho_0 U_{\text{ref}}^2$, z_{ref}/z_0, z_{ref}/L, we can write

$$\frac{\tau_0}{\rho_0 U_{\text{ref}}^2} \equiv C_{\text{d}} = C_{\text{d}}(z_{\text{ref}}/z_0, z_{\text{ref}}/L). \tag{10.21}$$

This gives a *drag law* for surface stress: $\tau_0 = C_{\text{d}} \rho_0 U_{\text{ref}}^2$, with C_{d} the *drag coefficient*.

In this way one can also derive a *surface-exchange coefficient* C_h for the surface temperature flux (Problem 10.3). Thus, given U, Θ, and mean water vapor mixing ratio Q at some height z_{ref} above the surface; the surface roughness length z_0 and the corresponding lengths for temperature and water vapor; and Θ and Q at these inner heights, M-O similarity yields the surface fluxes of momentum, heat, and water vapor. Virtually all large-scale meteorological models determine the surface fluxes in this way.

10.2.4.4 Physical interpretation of L

In surface-layer coordinates the rates of shear and buoyant production of TKE are

$$\text{shear production rate} = -\overline{uw}\frac{\partial U}{\partial z},$$
$$\text{buoyant production rate} = \frac{g}{\theta_0}\overline{\theta w} \simeq \frac{g}{\theta_0} Q_0. \tag{10.22}$$

Since the buoyant production rate is independent of z, very near the surface the shear-production rate can greatly exceed it. If we define a height z_{e} at which the buoyant production rate equals the value of the shear production rate under neutral conditions,

$$u_*^2 \left(\frac{u_*}{k z_{\text{e}}}\right) = \frac{g}{\theta_0} Q_0, \tag{10.23}$$

then z_{e} is

$$z_{\text{e}} = \frac{u_*^3 \theta_0}{k g Q_0} = |L|. \tag{10.24}$$

Thus $|L|$ is a rough estimate of the height at which buoyancy effects become dynamically important.

10.2.4.5 Quasi-steadiness, local homogeneity

We'll now apply to the surface layer the general criteria of Chapter 9 for the negligibility of the effects of unsteadiness and inhomogeneity. We can write for the Reynolds stress equation (8.69), for example,

$$\text{time change} \sim \frac{u^2}{\tau_{\mathrm{u}}} \sim \frac{u_*^2}{\tau_{\mathrm{u}}}, \quad \text{rate of production} \sim \frac{u^3}{\ell} \sim \frac{u_*^3}{z}, \tag{10.25}$$

with τ_{u} the time scale of the unsteadiness. The time-change term will be negligible if

$$\frac{u_*^2}{\tau_{\mathrm{u}}} \ll \frac{u_*^3}{z}, \quad \text{or} \quad \frac{z}{\mathrm{u}_* \tau_{\mathrm{u}}} \ll 1. \tag{10.26}$$

If $z = 10$ m, $u_* = 0.3$ m s^{-1}, and $\ll$ means a factor of 10 less than, this requires that τ_{u} be greater than about 5 minutes in order that the surface layer be quasi-steady. This is not difficult to meet in practice.

With horizontal inhomogeneity the Reynolds stress equation has a mean advection term:

$$\text{mean advection} \sim U\frac{\partial \overline{uw}}{\partial x} \sim U\frac{u_*^2}{L_x}, \tag{10.27}$$

where L_x is a horizontal scale of the inhomogeneity. For this to be negligible we need

$$U\frac{u_*^2}{L_x} \ll \frac{u_*^3}{z}, \quad \text{or} \quad \frac{L_x}{z} \gg \frac{U}{u_*}. \tag{10.28}$$

The value of U/u_* depends on height, stability, and the roughness of the surface, but typically ranges from about 10 to 30; if $\gg$ means a factor of 10 greater than, this indicates that L_x must be greater than 100 to 300 times the height. If $z = 10$ m, then L_x must be greater than 1 to 3 km. Sites this uniform can be difficult to find.

10.2.4.6 The Richardson number as an alternative stability variable

The M-O stability variable z/L involves surface fluxes, but in practice the turbulent fluxes at some height above the surface are used. The first such flux measurements were made in the 1950s, and well into the 1970s they were made only in research applications. Today, relatively inexpensive, reliable turbulent-flux instrumentation is commercially available, but stress measurements in particular are still plagued by scatter caused by the slow convergence of the time average.

The turbulence Richardson number of Chapter 9 is

$$Ri_{\mathrm{t}} = \frac{\text{buoyancy force/mass}}{\text{inertia force/mass}} \sim \frac{g\theta\ell}{\theta_0 u^2} \tag{9.6}$$

If we take $\theta \sim \ell\, \partial\Theta/\partial z$, $u \sim \ell\, \partial U/\partial z$, then this becomes

$$Ri_{\mathrm{t}} \sim \frac{\frac{g}{\theta_0}\frac{\partial\Theta}{\partial z}}{(\frac{\partial U}{\partial z})^2} \equiv Ri, \tag{10.29}$$

with Ri called the *gradient Richardson number*. It is an alternative stability variable. We can rewrite Eq. (10.29) as

$$Ri = \frac{z}{L}\frac{\phi_h}{\phi_m^2} = f(z/L), \tag{10.30}$$

which in principle can be inverted to give $z/L = f^{-1}(Ri)$. Thus, either z/L or Ri can serve as the surface-layer stability index.

The *flux Richardson number* R_{f} is defined with reference to the turbulent kinetic energy budget. In the surface layer it is

$$R_{\mathrm{f}} = \frac{\text{buoyant production rate}}{\text{shear production rate}} = \frac{\frac{g}{\theta_0}\overline{\theta w}}{\overline{uw}\frac{\partial U}{\partial z}}. \tag{10.31}$$

If we define eddy diffusivities such that

$$K_m = -\frac{\overline{uw}}{\frac{\partial U}{\partial z}}, \qquad K_h = -\frac{\overline{\theta w}}{\frac{\partial\Theta}{\partial z}}, \tag{10.32}$$

then it follows that $K_h/K_m = \phi_m/\phi_h$, and we can relate these two Richardson numbers:

$$R_{\mathrm{f}} = \frac{z}{L}\phi_m^{-1} = \frac{K_h}{K_m}Ri. \tag{10.33}$$

10.3 Asymptotic behavior of M-O similarity

Experimentalists typically achieve a range of z/L values by measuring turbulence at fixed heights (e.g., at 5.66 m, 11.3 m and 22.6 m in the Kansas experiments) and obtaining a range of L through the variations in u_* and Q_0 over the period of observations. At a given time the conditions nearest to neutral are found at the lowest observational level, and the largest departures from neutral occur at the highest level. As surface-layer conditions become very unstable, or very stable, there is reason to expect limiting behavior.

10.3.1 The asymptotic unstable state

In the surface layer the rate of buoyant production of TKE is essentially independent of height, while the rate of shear production decreases with height. Thus we might expect that above some height the direct effects of shear production are unimportant compared to buoyant production. Put another way, we might expect the *mechanical* turbulence (that due to the shear production and characterized by u_*) to become unimportant at some $-z/L$ value. We can examine the implications of the declining importance of shear production by deleting u_* from the M-O governing parameter group.

There are two ways to drop u_* as a governing M-O parameter. The first is to redo M-O similarity with only four governing parameters: g/θ_0, z, Q_0, C_0. Those plus one unknown give $m = 5$ parameters, $n = 4$ dimensions, so by the Buckingham Pi Theorem there is $m - n$ or one dimensionless parameter and it is a constant. Thus from the four governing parameters we make the scales

$$u_f = \left(\frac{g}{\theta_0} Q_0 z\right)^{1/3}, \qquad T_f = \frac{Q_0}{u_f}, \qquad c_f = \frac{C_0}{u_f}, \quad z, \tag{10.34}$$

the subscript f denoting *free convection*, or convection in zero mean wind. Then we write

$$\frac{\sigma_w}{u_f} = \frac{(\overline{w^2})^{1/2}}{u_f} = C_1, \qquad \frac{\sigma_\theta}{T_f} = C_2, \tag{10.35}$$

with C_1 and C_2 constants. This says that in very unstable conditions $\sigma_w(z)$ and $\sigma_\theta(z)$ approach their behavior in free convection.

Equation (10.35) can be rewritten in standard M-O terms:

$$\frac{\sigma_w}{u_*} = C_1 \frac{u_f}{u_*} \sim (-z/L)^{1/3}, \qquad \frac{\sigma_\theta}{T_*} = C_2 \frac{T_f}{T_*} \sim (-z/L)^{-1/3}. \tag{10.36}$$

As Figure 10.4 shows, measurements are consistent with both of these predictions.

A second way to deduce the behavior at large $-z/L$ is to require that M-O similar variables become independent of u_*. In the case of σ_w, for example, since we can write in general $\sigma_w = u_* f(z/L)$, at large $-z/L$ the similarity function f must vary as $1/u_*$ in order that $\sigma_w = f u_*$ be independent of u_*. That implies that $f(z/L) \sim (-z/L)^{1/3}$. Similarly, in order that σ_θ be independent of u_*, σ_θ / T_* must vary as $(-z/L)^{-1/3}$.

Tennekes (1970) named this *local free convection* scaling. *Local* isotropy (Part III) is isotropic behavior at large wavenumbers; *local* free convection is free-convection-like behavior at large $-z/L$.

While there are some exceptions to M-O similarity, w and θ statistics are observed to follow it well and show clear indications of local-free-convection regimes. According to Figure 10.4, these can emerge at surprisingly small $-z/L$; local free convection behavior for σ_w appears by $z/L = -2$ or so, and temperature variance

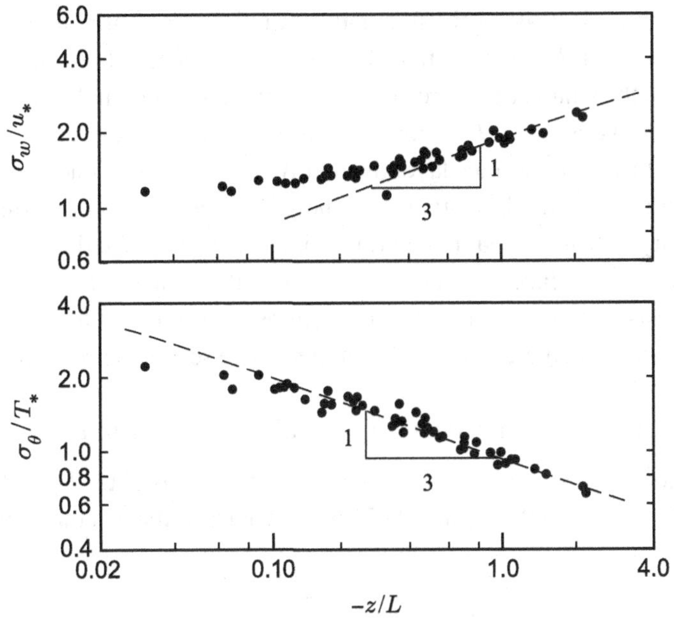

Figure 10.4 Kansas observations of the rms vertical velocity (upper) and rms temperature (lower) for unstable conditions only, plotted in M-O coordinates. The dashed lines are the asymptotic predictions for very unstable conditions, Eq. (10.36). From Wyngaard (1973).

shows it even sooner. The shear- and buoyant-production terms in the TKE budget are, in M-O terms,

$$-\overline{uw}\frac{\partial U}{\partial z} = \frac{u_*^3}{kz}\phi_m, \qquad \frac{g}{\theta_0}\overline{\theta w} = -\frac{u_*^3}{kz}\frac{z}{L}. \tag{10.37}$$

ϕ_m decreases monotonically with increasing instability; the Kansas data, for example (Figure 10.3), are fit well by the formula $\phi_m = (1 - 15z/L)^{-1/4}$ on the unstable side. At $z/L = -2$, where evidence of local-free-convection behavior of σ_w appears, this expression says that the buoyant-production rate exceeds the shear-production rate by a factor of nearly five.

10.3.2 The asymptotic stable state

The ratio of fluctuating buoyancy and inertia forces for a velocity-field eddy of size r (Chapter 2) is an "eddy Richardson number" $Ri_e(r)$,

$$Ri_e(r) = \frac{g\theta r}{\theta_0\,[u(r)]^2} = \frac{g\theta r^{1/3}}{\theta_0 \epsilon^{2/3}}, \tag{10.38}$$

with θ the amplitude scale of the fluctuations in virtual potential temperature. The $r^{1/3}$ dependence of Ri_e means that the largest eddies feel buoyancy effects the most strongly; the small eddies are dominated by inertia and (at the smallest scales) viscous forces. Thus, as z/L becomes positive from zero the largest eddies are damped first. In a stable surface layer the relative importance of buoyant destruction increases with increasing height; at some height the eddies become limited in size by the stability. Loosely speaking, at this height the turbulence does not "sense" z, the distance to the surface, as it does in an unstratified surface layer.

This suggests that under very stable conditions z drops out of the M-O governing parameter group, leaving g/θ_0, Q_0, C_0, and u_*. The scales in this limiting case are then

$$\text{velocity: } u_*, \quad \text{temperature: } T_*, \quad \text{scalar: } c_*, \quad \text{length: } L. \tag{10.39}$$

Nondimensional quantities again become universal constants. This "local z-less scaling," as it is called (Wyngaard, 1973; Chapter 12) indicates that, for example,

$$\frac{\partial U}{\partial z} \sim \frac{u_*}{L}, \qquad \frac{\partial \Theta}{\partial z} \sim \frac{T_*}{L}; \tag{10.40}$$

so that the M-O profile functions are

$$\phi_m = \frac{kz}{u_*}\frac{\partial U}{\partial z} \sim \frac{z}{L}, \qquad \phi_h = \frac{kz}{T_*}\frac{\partial \Theta}{\partial z} \sim \frac{z}{L}. \tag{10.41}$$

These predictions agree with measurements; as Figure 10.3 shows, ϕ_m and ϕ_h are essentially linear over the entire stable range of the Kansas data.

10.4 Deviations from M-O similarity

The velocity scale of the energy-containing eddies in the convective boundary layer with zero mean wind is expected to depend (at minimum) on g/θ_0, Q_0, and the boundary-layer depth z_i. These parameters define the *free-convection velocity scale* w_*:

$$w_* = \left(\frac{g}{\theta_0} Q_0 z_i\right)^{1/3}. \tag{10.42}$$

For $Q_0 = 0.2$ m K s^{-1} and $z_i = 1000$ m, typical values in fair weather over land, Eq. (10.42) gives $w_* \sim 2$ m s^{-1}.

From the definitions of w_* and L we can write $w_*/u_* = k^{-1/3}(-z_i/L)^{1/3}$. Convective boundary layers in the atmosphere have $-z_i/L$ values up to several hundred, so it is not unusual to find the free-convection-like state where $w_* \gg u_*$. Since w_* contains z_i, which is not an M-O parameter, the horizontal gusts in the surface layer due to these large eddies are not M-O similar, as pointed out by Panofsky *et al.*

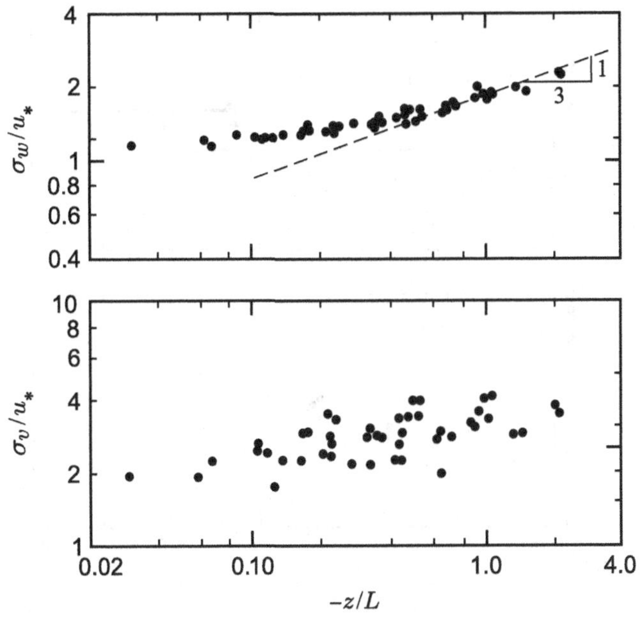

Figure 10.5 M-O plots of the rms vertical (upper) and lateral (lower) wind fluctuations at 5.7, 11.3, and 22.6 m height in the convective surface layer of the 1968 Kansas experiment. σ_w follows M-O similarity well but σ_v does not. From Wyngaard (1988).

(1977). The data in Figures 10.5–10.7 are consistent with this. But surface-layer kinematics (Appendix) suggests that $\overline{w^2}$ is only weakly influenced by the large convective eddies and to a good approximation is M-O similar, as seen in Figure 10.5.

Businger (1973) has briefly explored some of the impacts of these large convective eddies on surface-layer structure:

> Consider … a large uniform area with free convection extending over the entire planetary boundary layer up to a height h. The mean wind speed $U = 0$, consequently $u_* = 0$.… Consider now the layer close to the surface over a relatively short time period compared to the large-scale convection but relatively long compared to the time it takes to develop a local wind profile. This local temporary wind profile cannot be distinguished in characteristics from a true mean wind profile. This … leads us to believe that there is, averaged over the entire horizontal area, mean shear production of turbulence which is not related to a mean wind but to the convective circulation in the boundary layer.

He then postulated that the local friction velocity generated in this way scales with $w_* f(h/z_0)$, with f a decreasing function of h/z_0. Zilitinkevich *et al.* (2006) have since studied this problem in detail and confirmed that the surface roughness is an important parameter for surface-layer structure and heat transfer in free convection. It is conceivable that these effects extend to weak-wind, very unstable conditions, causing M-O similarity to fail there.

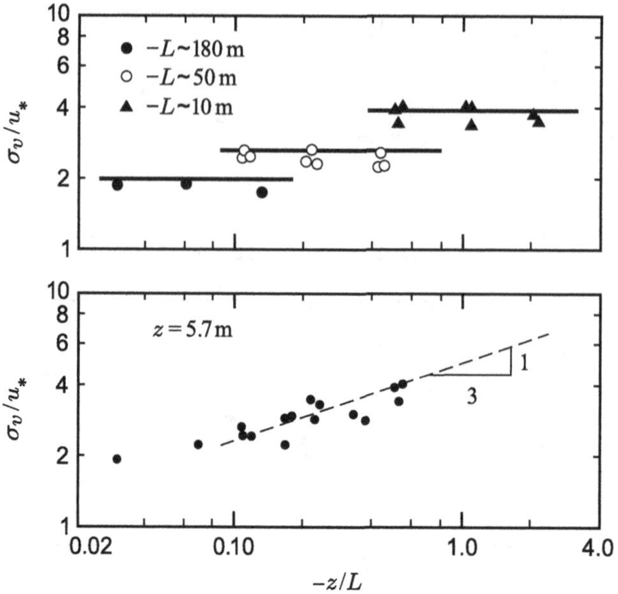

Figure 10.6 Replots of some of the σ_v data in Fig. 10.5. Lower: Data at fixed z (here 5.7 m) from three different cases appear to show a local-free-convection regime like that for σ_w, Figure 10.5. Upper: Data from the three heights in these cases suggest instead that σ_v depends only weakly on z. If so, the trend in the lower panel is likely due to the influence of boundary-layer depth, expressed through w_*, Eq. (10.42). From Wyngaard (1988).

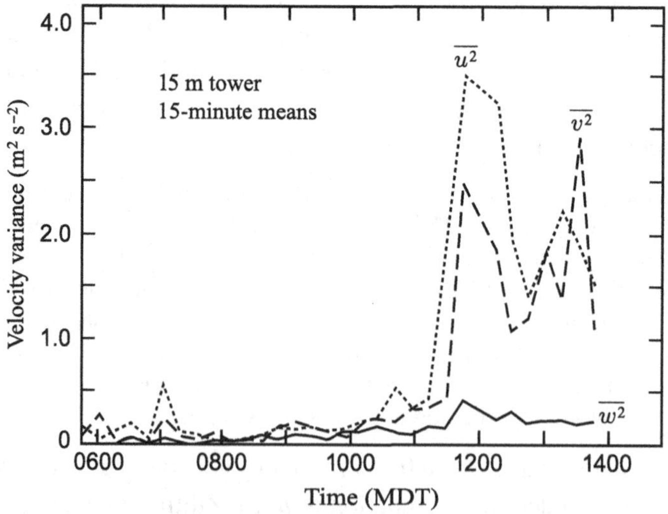

Figure 10.7 Time histories of wind variances in the convective surface layer. Shortly after 1100 local time z_i grew rapidly; the horizontal wind variances $\overline{u^2}$ and $\overline{v^2}$ increased as well, as predicted by w_* scaling, but $\overline{w^2}$, being M-O similar, did not. From Banta (1985).

Hill (1989) has pursued the implications of M-O similarity for scalars. One finding is that the correlation coefficient of any two scalars is ± 1. This is unphysical, and he postulates that here it is necessary to include the scalar physics of the surface itself in order to obtain realistic results.

10.4.1 Covariance budgets

The equations or *budgets* governing turbulent kinetic energy, temperature variance, Reynolds stress, and temperature flux had been known for some time, but the 1968 Kansas experiment provided the first comprehensive measurements of them in the atmospheric surface layer.

10.4.1.1 The TKE budget

In surface-layer coordinates this is, from Eq. (8.70),

$$\overline{uw}\frac{\partial U}{\partial z} + \frac{\partial}{\partial z}\frac{\overline{u_i u_i w}}{2} = -\frac{1}{\rho_0}\frac{\partial}{\partial z}\overline{pw} + \frac{g}{\theta_0}\overline{\theta w} - \epsilon. \tag{10.43}$$

The terms are, in order, shear production, turbulent transport, pressure transport, buoyant production, and viscous dissipation.

In the Kansas experiment all budget terms but pressure transport were measured over the M-O stability range (Wyngaard and Coté, 1971). An unexpected finding was the large rate of gain by pressure transport, which was inferred from the imbalance of the measured terms. Up to that time pressure transport had generally been assumed to be negligible, so the Kansas findings stimulated much discussion and debate in the micrometeorological community. Consensus results for the unstable side as of about 1990 are shown in Figure 10.8; overall they confirm the Kansas results. The large gain due to pressure transport approximately balances the loss to turbulent transport; thus the sum of shear and buoyant production approximately balances viscous dissipation, as also reported by Bradley *et al.* (1981a).

The TKE budget under stable conditions suggests a much more placid picture. Both turbulent and pressure transport are found to be negligible, so that shear production is essentially balanced by buoyant destruction and viscous dissipation.

10.4.2 Conserved scalar variance budgets

The quasi-steady budget of potential temperature variance is, from Eq. (8.71),

$$\overline{\theta w}\frac{\partial \Theta}{\partial z} + \frac{\partial}{\partial z}\frac{\overline{w\theta^2}}{2} = -\chi, \tag{10.44}$$

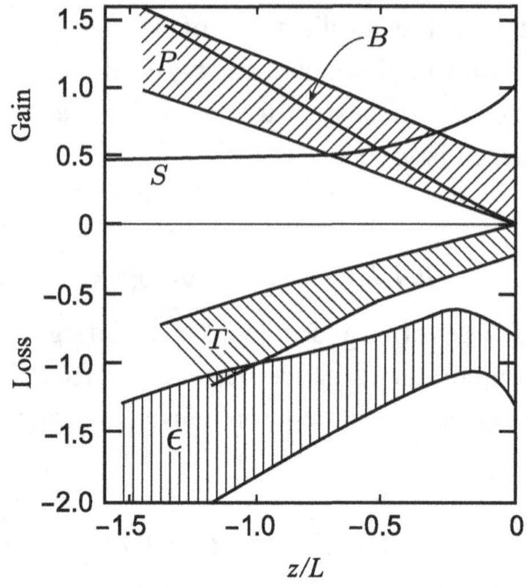

Figure 10.8 The TKE budget in the unstable surface layer. B is buoyant production; P is pressure transport; S is shear production; T is turbulent transport; ϵ is viscous dissipation. Terms are nondimensionalized with kz, u_*, and Q_0. From Wyngaard (1992). Reprinted, with permission, from *Annual Review of Fluid Mechanics*, **24**,

a balance among mean-gradient production, turbulent transport, and molecular destruction. For the entire stability range, the Kansas data indicate that mean-gradient production is the principal source term and that turbulent transport is at least an order of magnitude smaller. Therefore, in the unstable surface layer this budget is to first approximation in *local balance* – mean-gradient production balances molecular destruction. Measurements by Bradley *et al.* (1981b) confirm this.

The budget of water vapor mixing ratio (q) variance has the same form, but the difference in the buoyant production terms for $\overline{qw}$ and $\overline{\theta w}$ (Section 10.2.3) has an impact on the variances (Problem 10.20). In the surface layer the scaled variance of q is a factor of 2–3 larger, and the scaled molecular destruction rate is about 30% larger (Katul and Hseih, 1999).

10.4.3 The shear-stress budget

The quasi-steady budget of the principal component of kinematic shear stress $\overline{uw}$ is, from Eq. (8.70),

$$\overline{w^2}\frac{\partial U}{\partial z} + \frac{\partial}{\partial z}\overline{uww} = -\frac{1}{\rho_0}\left(\overline{w\frac{\partial p}{\partial x}} + \overline{u\frac{\partial p}{\partial z}}\right) + \frac{g}{\theta_0}\overline{\theta u}. \tag{10.45}$$

We have neglected the Coriolis and viscous dissipation terms, the first because it is small in the ABL and the second through local isotropy, Chapter 14.

In the unstable surface layer the Kansas results (Wyngaard *et al.*, 1971) show a three-way balance among shear production (the primary source), buoyant production (the secondary source), and the loss to the pressure-destruction term. Given the highly variable instantaneous nature of the unstable surface layer, one might have expected a significant turbulent transport term in the $\overline{uw}$ budget, but instead it is small, on the order of 10% or less of the principal terms. This reminds us that local, instantaneous structure reveals nothing about expected values, and vice versa.

To a good approximation the near-neutral $\overline{uw}$ budget is

$$\overline{w^2}\frac{\partial U}{\partial z} = \text{rate of loss to pressure term.} \tag{10.46}$$

If we write the loss term as $-\overline{uw}/\tau$, where τ is a time scale, then Eq. (10.46) reduces to the eddy-diffusivity form

$$\overline{uw} \sim -\overline{w^2}\tau\frac{\partial U}{\partial z}. \tag{10.47}$$

10.4.4 The $\overline{w\theta}$ budget

From (8.70) in quasi-steady conditions this is

$$\overline{w^2}\frac{\partial \Theta}{\partial z} + \frac{\partial}{\partial z}\overline{ww\theta} = -\frac{1}{\rho_0}\overline{\theta\frac{\partial p}{\partial z}} + \frac{g}{\theta_0}\overline{\theta^2}. \tag{10.48}$$

The Kansas results (Wyngaard *et al.*, 1971) show that in near-neutral conditions the buoyancy and transport terms are negligible. Thus, like the $\overline{uw}$ budget the near-neutral $\overline{w\theta}$ budget reduces to a balance between mean-gradient production and pressure destruction:

$$\overline{w^2}\frac{\partial \Theta}{\partial z} = -\frac{1}{\rho_0}\overline{\theta\frac{\partial p}{\partial z}}. \tag{10.49}$$

If we model the latter as $-\overline{w\theta}/\tau$, where τ is a time scale, then we have

$$\overline{w\theta} = -\overline{w^2}\tau\frac{\partial \Theta}{\partial z}, \tag{10.50}$$

as for the stress budget.

While there is overall similarity here between the the $\overline{uw}$ and $\overline{w\theta}$ budgets, the buoyancy term is relatively more important for $\overline{w\theta}$. This gives a hint that under unstable conditions the coupling between turbulent flux and mean gradient for temperature is weaker than it is for momentum. We will see in Chapter 11 that at greater heights in the convective boundary layer this coupling disappears for conserved scalars.

10.4.5 The $\overline{u\theta}$ budget

From (8.70) in quasi-steady conditions this is, using local isotropy and the Bradley *et al.* (1982) measurements to drop the molecular destruction term,

$$\overline{uw}\frac{\partial \Theta}{\partial z} + \overline{\theta w}\frac{\partial U}{\partial z} + \frac{\partial}{\partial z}\overline{wu\theta} = -\frac{1}{\rho_0}\overline{\theta \frac{\partial p}{\partial x}}. \tag{10.51}$$

The Kansas results show that in the near-neutral surface layer the turbulent transport term is small and the budget further reduces to

$$\overline{uw}\frac{\partial \Theta}{\partial z} + \overline{\theta w}\frac{\partial U}{\partial z} = -\frac{1}{\rho_0}\overline{\theta \frac{\partial p}{\partial x}}, \tag{10.52}$$

a balance between the two rates of production and the rate of destruction by turbulent pressure gradients.

On measuring turbulent pressure signals Bradshaw (1994) has written:

> Pressure fluctuations within a turbulent flow are one of the Great Unmeasureables: they are of the order of ρu^2 and so, unfortunately, are the pressure fluctuations induced on a static pressure probe by the velocity field. That is, the signal-to-noise ratio is of the order of one. To say that signals cannot be educed even with $S/N = O(1)$ is itself a fallacy, but in this case the attempts made to do so have not met with general acceptance.

Nonetheless, armed with evidence that their pressure probe could measure pressure fluctuations with a signal-to-noise ratio better than 1 (Wyngaard *et al.*, 1994), Wilczak and Bedard (2004) confirmed that the turbulent transport term in Eq. (10.51) is small, and found excellent agreement between the pressure covariance inferred from measurements of the left sides of Eqs. (10.52) and (10.51) and their direct measurements of it.

Appendix
Length scales near the surface

Since w vanishes at the surface, near the surface we can write from our zero-divergence model of the fluctuating velocity

$$w(x, y, z, t) = -\int_0^z \left(\frac{\partial u}{\partial x}(x, y, z', t) + \frac{\partial v}{\partial y}(x, y, z', t)\right) dz'. \tag{10.53}$$

We represent the fluctuating horizontal velocity vector $\mathbf{u} = (u, v)$ at a point $\mathbf{x} = (x, y)$ in the plane and at time t by a sum of Fourier components of horizontal vector wavenumber $\boldsymbol{\kappa}_n = (\kappa_{1n}, \kappa_{2n})$ with random vector coefficients $\mathbf{a}_n = (a_{1n}, a_{2n})$,

$$\mathbf{u}(\mathbf{x}, z, t; \alpha) = \sum_{n=-N}^{N} \mathbf{a}_n(\boldsymbol{\kappa}_n, z, t, \alpha) e^{i\boldsymbol{\kappa}_n \cdot \mathbf{x}}, \tag{10.54}$$

with α the realization index. It follows that $\partial u/\partial x$, for example, is

$$\frac{\partial u}{\partial x} = \sum_{n=-N}^{N} i\kappa_{1n} a_{1n} e^{i\boldsymbol{\kappa}_n \cdot \mathbf{x}}. \tag{10.55}$$

With (10.53)–(10.55) we can now write w as

$$\begin{aligned} w(\mathbf{x}, z, t; \alpha) &= -\int_0^z \left(\sum_{n=-N}^{N} i\left(\kappa_{1n} a_{1n} + \kappa_{2n} a_{2n}\right) dz' \right) e^{i\boldsymbol{\kappa}_n \cdot \mathbf{x}} \\ &= -i \sum_{n=-N}^{N} \left(\kappa_{1n} \int_0^z a_{1n} dz' + \kappa_{2n} \int_0^z a_{2n} dz' \right) e^{i\boldsymbol{\kappa}_n \cdot \mathbf{x}}. \end{aligned} \tag{10.56}$$

At horizontal scales much larger than the height z (i.e., for $\kappa_n z \ll 1$, with $\kappa_n = |\boldsymbol{\kappa}_n|$) we expect the Fourier coefficients $\mathbf{a}_n$ to vary only weakly with z because turbulent eddies with large horizontal scales (small κ) are unlikely to have small vertical scales. Thus, for $\kappa_n z \ll 1$ we can approximate the integrals in (10.56) as

$$\int_0^z a_{1n}\, dz' \simeq a_{1n} z, \qquad \int_0^z a_{2n}\, dz' \simeq a_{2n} z, \tag{10.57}$$

so that Eq. (10.56) becomes

$$\begin{aligned} w(\mathbf{x}, z, t; \alpha) &\simeq -iz \sum_{n=-N}^{N} \left(\kappa_{1n} a_{1n} + \kappa_{2n} a_{2n}\right) e^{i\boldsymbol{\kappa}_n \cdot \mathbf{x}} \\ &= -\sum_{n=-N}^{N} iz\mathbf{a}_n \cdot \boldsymbol{\kappa}_n e^{i\boldsymbol{\kappa}_n \cdot \mathbf{x}}, \quad \kappa_n z \ll 1. \end{aligned} \tag{10.58}$$

Thus, the Fourier coefficients of w are those of (u, v) dotted with $-iz\boldsymbol{\kappa}_n$. Since the spectrum at any wavenumber is proportional to the mean-square value of the Fourier coefficients at that wavenumber, it follows that

$$\text{spectrum of } w \simeq (\kappa_n z)^2 \times \text{spectrum of } u, v, \quad \kappa_n z \ll 1. \tag{10.59}$$

At horizontal scales very small compared to z (i.e., at $\kappa_n z \gg 1$) we expect the Fourier coefficients $\mathbf{a}_n$ to vary rapidly with z. Since the coefficients are random variables with zero mean, their z-integral will tend to zero. We expect the transition between the $\kappa_n z \ll 1$ and $\kappa_n z \gg 1$ regimes to occur at $\kappa_n \sim 1/z$, so the spectra of velocity in a plane in the surface layer should behave as sketched in Figure 10.2. Even though the peaks of u and v spectra occur at scales of the order of the ABL depth h (corresponding to the sweeping effect of the largest convective eddies), the peak of the w-spectrum is at much smaller scales – at horizontal wavenumbers of the order of $1/z$.

The cospectrum of the kinematic shear stress $\overline{uw}$ – the density of contributions to $\overline{uw}$ as a function of wavenumber in the horizontal plane (Chapter 15) – is at any wavenumber proportional to the mean product of the amplitudes of the Fourier coefficients of u and w (Part III). Given the wavenumber dependence of these amplitudes (Figure 10.2), we see that this stress cospectrum must peak where the w-spectrum does, at $\kappa \sim 1/z$. Thus the large-scale part of the horizontal turbulent motions in the surface layer, that with $\kappa < 1/z$, does not contribute to $\overline{uw}$. These are called *inactive motions*.

Questions on key concepts

10.1 Explain the concepts underlying the "constant-flux" layer. Why is "surface-flux" layer a better term?

10.2 Discuss how the mean-wind and stress profiles in Figure 10.1 are inferred from the mean-motion equations.

10.3 Explain the key concepts of dimensional analysis.

10.4 Of the other possible governing parameters in M-O similarity, which do you feel is the most important? Explain.

10.5 Explain some of the uses of M-O similarity.

10.6 Explain some of the deficiencies of M-O similarity.

10.7 Explain the two limiting states of M-O similarity, how their physics is simplified, and the rationale for that simplification.

10.8 Explain physically why the horizontal velocity fluctuations in Figure 10.7 suddenly intensified shortly after 1100 but the vertical fluctuations did not.

10.9 Interpret physically the pressure-transport term in the TKE budget.

10.10 There is a Lagrangian integral time scale (Chapter 4) for each of the three velocity components. How would you expect each to behave in the surface layer? Would any be M-O similar?

10.11 Explain the notions of quasi-steadiness and local homogeneity and how they simplify analysis of the surface layer.

10.12 Explain physically why turbulent pressure fluctuations are difficult to measure reliably.

10.13 Explain in your own terms the physical argument for the sensitivity of surface-layer u and v fluctuations to ABL depth but the insensitivity of w fluctuations.

10.14 Discuss the physical meaning and implications of the relative unimportance of turbulent transport in the surface-layer budgets of $\overline{uw}$ and $\overline{w\theta}$ in the unstable surface layer, despite the high turbulence level there.

10.15 Explain why the largest eddies are the most sensitive to buoyancy.

Problems

10.1 Explain why the mean wind and stress profiles in Figure 10.1 are consistent with Eqs. (10.1) and (10.2).

10.2 Discuss whether the following are M-O similar:
(a) the dissipation rate ϵ,
(b) the Kolmogorov microscale η,
(c) horizontal velocity fluctuations,
(d) the drag coefficient C_{d}.

10.3 Derive an expression like Eq. (10.21) but for the surface temperature flux.

10.4 Explain how measurements of the mean temperature difference $\Delta\Theta$ and mean wind speed difference ΔU between two heights in the surface layer can be used to infer the surface temperature and momentum fluxes.

10.5 Does M-O similarity hold in a baroclinic boundary layer? Discuss.

10.6 The temperature structure function $\overline{[\theta(\mathbf{x}) - \theta(\mathbf{x}+\mathbf{r})]^2}$ is observed to behave as $C_{T^2} r^{2/3}$ for separations $\eta \ll r \ll z$. Is C_{T^2} M-O similar? If so, write an M-O similarity expression for it.

10.7 Under the M-O hypothesis the near-surface regions of convective and stable boundary layers have the same statistical structure. Is this physically reasonable? Discuss.

10.8 In a convective boundary layer capped by an inversion having drier air, humidity fluctuations can be generated by the entrainment process at the ABL top. Could this impact the M-O similarity of humidity fluctuations? Discuss.

10.9 Discuss the role of the intercomponent-transfer terms in the u, v, and w kinetic energy equations under neutral conditions. What is their magnitude? Try to interpret them physically with sketches.

10.10 Explain why $\overline{u\theta}$ is nonzero in the surface layer. What is its production mechanism? How does it behave in free convection? Why are we generally oblivious to it?

10.11 Discuss the behavior of R_f and Ri in the asymptotic stable state. How could this state determine the depth of the nocturnal ABL?

10.12 Explain the role of the second-moment budgets in turbulent flow calculation.

10.13 Derive an expression for the eddy diffusivity for momentum in the surface layer. Discuss its asymptotic behavior.

10.14 Identify and discuss the rate of buoyant production of vertical temperature flux. What is its role in unstable conditions? In stable conditions?

10.15 Explain why large-eddy simulation generally does not perform well in the atmospheric surface layer. (Hint: consider the behavior of the length scale of the vertical velocity field near the surface.)

10.16 In large-eddy simulation of the atmospheric boundary layer the surface fluxes are typically determined from the resolved (calculated) fields through the standard surface-exchange coefficients. Is this justifiable? Discuss.

10.17 Use dimensional analysis to determine the averaging time T required to determine the time average $\overline{u}^T$ of a stationary function of time $\tilde{u}(t)$ with ensemble mean U, variance $\overline{u^2}$, and integral scale τ. Can you obtain the answer, Eq. (2.36)? Explain.

10.18 Show that Eq. (10.11) rests on good approximations.

10.19 Why is pipe length L not one of the governing parameters in pipe flow, Section 10.2.1?

10.20 Discuss the differences between the buoyant production terms for $\overline{qw}$ and $\overline{\theta w}$ and how the differences might impact their budgets.

10.21 Why does Eq. (10.38) not use $\theta(r)$?

References

Andreas, E. L., K. J. Claffey, R. E. Jordan, *et al.*, 2006: Evaluations of the von Kármán constant in the atmospheric surface layer. *J. Fluid Mech.*, **559**, 117–149.

Banta, R., 1985: Late-morning jump in TKE in the mixed layer over a mountain basin. *J. Atmos. Sci.*, **42**, 407–411.

Bradley, E. F., R. A. Antonia, and A. J. Chambers, 1981a: Turbulence Reynolds number and the turbulent kinetic energy balance in the atmospheric surface layer. *Bound.-Layer Meteor.*, **20**, 183–197.

Bradley, E. F., R. A. Antonia, and A. J. Chambers, 1981b: Temperature structure in the atmospheric surface layer. *Bound.-Layer Meteor.*, **20**, 275–292.

Bradley, E. F., R. A. Antonia, and A. J. Chambers, 1982: Streamwise heat flux budget in the atmospheric surface layer. *Bound.-Layer Meteor.*, **23**, 3–15.

Bradshaw, P., 1967: "Inactive" motion and pressure fluctuations in turbulent boundary layers. *J. Fluid Mech.*, **30**, 241–258.

Bradshaw, P., 1994: Turbulence: the chief outstanding difficulty of our subject. *Experiments in Fluids*, **16**, 203–216.

Businger, J. A., 1973: A note on free convection. *Bound.-Layer Meteor.*, **4**, 323–326.

Businger, J. A., and Yaglom, A. M., 1971: Introduction to Obukhov's paper "Turbulence in an Atmosphere with a Non-uniform Temperature." *Bound.-Layer Meteor.*, **2**, 3–6.

Businger, J. A., J. C. Wyngaard, Y. Izumi, and E. F. Bradley, 1971: Flux-profile relationships in the atmospheric surface layer. *J. Atmos. Sci.*, **28**, 181–189.

Foken, T., 2006: 50 years of Monin-Obukhov similarity theory. *Bound.-Layer Meteor.*, **119**, 431–447.

Grachev, A. A., C. W. Fairall, J. E. Hare, and J. B. Edson, 2003: Wind stress vector over ocean waves. *J. Phys. Ocean.*, **33**, 2408–2429.

Hill, R. J., 1989: Implications of Monin-Obukhov similarity theory for scalar quantities. *J. Atmos. Sci.*, **46**, 2236–2244.

Hogstrom, U., 1988: Non-dimensional wind and temperature profiles in the atmospheric surface layer. *Bound.-Layer Meteor.*, **42**, 263–270.

Horst, T. W., 1999: The footprint for estimation of atmosphere-surface exchange fluxes by profile techniques. *Bound.-Layer Meteor.*, **90**, 171–188.

Katul, G. G., and C. I. Hseih, 1999: A note on the flux-variance similarity relationships for heat and water vapour in the unstable atmospheric surface layer. *Bound.-Layer Meteor.*, **90**, 327–338.

Kolmogorov, A. N., 1941: The local structure of turbulence in incompressible viscous fluid for very large Reynolds numbers. *Doklady ANSSSR*, **30**, 301–305.

Monin, A. S., and A. M. Obukhov, 1954: Osnovnye zakonomernosti turbulentnogo peremeshivanija v prizemnom sloe atmosfery (Basic laws of turbulent mixing in the atmosphere near the ground). *Trudy geofiz. inst. AN SSSR*, **24**(151), 163–187.

Obukhov, A. M., 1946: "Turbulentnost" v temperaturnoj – neodnorodnoj atmosfere (Turbulence in an atmosphere with a non-uniform temperature). *Trudy Inst. Theor. Geofiz. AN SSSR*, **1**, 95–115.

Panofsky, H. A., H. Tennekes, D. H. Lenschow, and J. C. Wyngaard, 1977: The characteristics of turbulent velocity components in the surface layer under convective conditions. *Bound.-Layer Meteor.*, **11**, 355–361.

Tennekes, H., 1970: Free convection in the turbulent Ekman layer of the atmosphere. *J. Atmos. Sci.*, **27**, 1027–1034.

Warhaft, Z., 1976: Heat and moisture flux in the stratified boundary layer. *Quart. J. R. Meteor. Soc.*, **102**, 703–707.

Wilczak, J. M., and A. Bedard, 2004: A new turbulence microbarometer and its evaluation using the budget of horizontal heat flux. *J. Atmos. Ocean. Tech.*, **21**, 1170–1181.

Wilson, K., 2001: An alternative function for the wind and temperature gradients in unstable surface layers. *Bound.-Layer Meteor.*, **99**, 151–158.

Wyngaard, J. C., 1973: On surface-layer turbulence. *Workshop on Micrometeorology*, D. A. Haugen, Ed., American Meteorological Society, pp. 101–149.

Wyngaard, J. C., 1988: Structure of the PBL. *Lectures on Air-Pollution Modeling*, A. Venkatram and J. C. Wyngaard, Eds., American Meteorological Society, pp. 9–61.

Wyngaard, J. C., 1992: Atmospheric turbulence. *Ann. Rev. Fluid Mech.*, **24**, 205–233.

Wyngaard, J. C., and O. R. Coté, 1971: The budgets of turbulent kinetic energy and temperature variance in the atmospheric surface layer. *J. Atmos. Sci.*, **28**, 190–201.

Wyngaard, J. C., O. R. Coté, and Y. Izumi, 1971: Local free convection, similarity, and the budgets of shear stress and heat flux. *J. Atmos. Sci.*, **28**, 1171–1182.

Wyngaard, J. C., A. Siegel, and J. Wilczak, 1994: On the response of a turbulent-pressure probe and the measurement of pressure transport. *Bound.-Layer Meteorol.*, **69**, 379–396.

Zilitinkevich, S. S., J. C. R. Hunt, L. N. Esau, *et al.*, 2006: The influence of large convective eddies on the surface-layer turbulence. *Quart. J. R. Meteor. Soc.*, **132**, 1423–1456.

11

The convective boundary layer

11.1 Introduction

The mean structure of the convective boundary layer (CBL) is sketched in Figure 11.1. The *surface layer* (the lowest 10%, say) is the most accessible to observation and therefore the best understood. Above it lies the *mixed layer* (not "mixing" layer;† that is the turbulent shear layer between parallel streams of different speeds). Here the turbulent diffusivity tends to be largest and mean gradients of wind and conserved scalars smallest. The *interfacial layer* buffers the mixed layer from the free atmosphere. Its top at mean height h_2 can be thought of as the greatest height reached by the surface-driven convective elements, and its bottom at h_1 the deepest penetrations of the nonturbulent free atmosphere. The mean CBL depth z_i lies between these two; it is often taken as the height at which the vertical turbulent flux of virtual potential temperature has its negative maximum.

11.2 The mixed layer: velocity fields

11.2.1 Mixed-layer similarity

A mimimal set of governing parameters for the quasi-steady mixed layer is the M-O group u_*, z, g/θ_0, Q_0 (the surface flux of virtual temperature), $\overline{cw}_s$ (the surface flux of a conserved scalar) plus the boundary-layer depth z_i. The dimensional analysis (Chapter 10) here has $m - 1 = 6$ governing parameters and $n = 4$ dimensions, so there are $m - n = 3$ independent dimensionless quantities. One is the dimensionless dependent variable; it is conventional to take the other two as z/z_i and z_i/L. The latter is a stability parameter for the CBL as a whole.

† Tennekes (1974) admired the semantic precision of *mixed* rather than *mixing* layer.

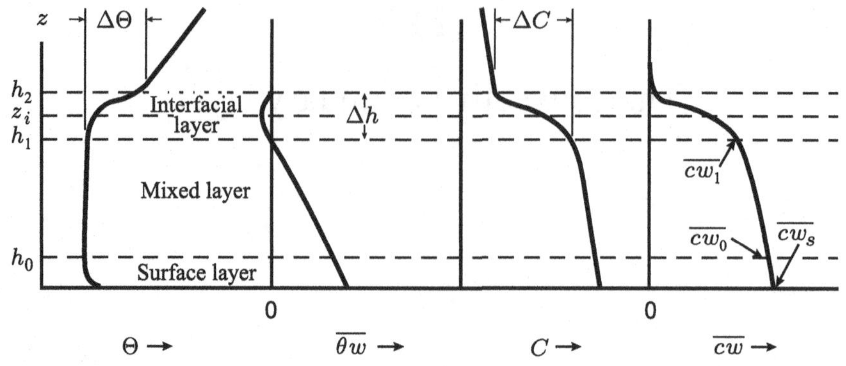

Figure 11.1 Sketches of profiles of mean quantities and their vertical fluxes in the CBL, with its layers, heights, and parameters indicated. Left pair: Virtual potential temperature and its flux. Right pair: A conserved scalar and its flux. From Deardorff (1979).

From the definition of the M-O length L it follows that the convective velocity scale w_*, Eq. (10.42), is

$$w_* = \left(\frac{g}{\theta_0} Q_0 z_i\right)^{1/3} \simeq 0.7 u_* \left(-\frac{z_i}{L}\right)^{1/3}, \tag{11.1}$$

so when the boundary-layer stability parameter $-z_i/L$ is very large w_* is much larger than the friction velocity u_*. Thus Deardorff (1970) suggested that at large $-z_i/L$ a free-convection-like state emerges. Observations and numerical simulations suggest that this state appears when $-z_i/L$ exceeds 5–10.

This asymptotic state is called *mixed-layer similarity.* Here $m - 1 = 5$ and $n = 4$, so there are two independent dimensionless quantities. The velocity scale is w_*, the length scales are z and z_i, and the intensity scale of a conserved scalar is $c_* = \overline{cw}_s/w_*$; within the mixed layer quantities nondimensionalized with these scales are predicted to be functions only of z/z_i. This can be successful for velocity statistics, as Figure 11.2 shows. But it can fail spectacularly for scalar statistics (Figure 11.2) because it neglects the entrainment-induced scalar flux at the mixed-layer top, which as pointed out by Deardorff (1972a) is an additional source of scalar fluctuations in the upper mixed layer.

Mixed-layer similarity is now known to be incorrect in other respects. For example, the lateral and streamwise integral scales (Part III), which are equal in free convection, can differ by as much as a factor of two with a mean horizontal wind. As we shall discuss, it appears that other effects – e.g., vertical variation of the mean horizontal pressure gradient (baroclinity) and z-dependent horizontal advection of mean momentum – can also influence the structure of the CBL.

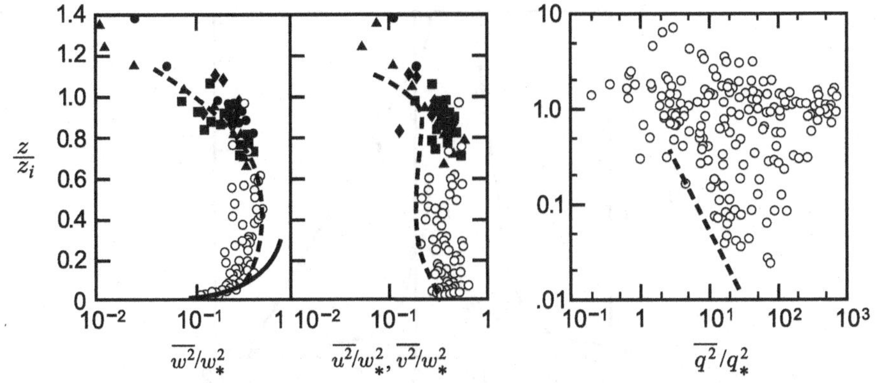

Figure 11.2 Left: Vertical and horizontal velocity variances in the CBL. Dashed lines, convection tank; solid line, asymptotic behavior of Kansas surface-layer data; open circles, Minnesota data; solid symbols, Ashchurch, England, data. From Caughey (1982). Right: Mixed-layer scaling fails for water-vapor fluctuations when their principal source is the entrainment process at CBL top. The dashed line is the observed behavior of a conserved scalar in the very unstable surface layer, Figure 10.4. From Wyngaard (1988).

11.2.2 The TKE budget

Equation (8.69) yields the TKE budget in the horizontally homogeneous ABL (Problem 11.20):

$$\frac{1}{2}\frac{\partial \overline{u_i u_i}}{\partial t} = -\left(\overline{uw}\frac{\partial U}{\partial z} + \overline{vw}\frac{\partial V}{\partial z}\right) - \frac{\partial}{\partial z}\frac{\overline{u_i u_i w}}{2} - \frac{1}{\rho_0}\frac{\partial}{\partial z}\overline{pw} + \frac{g}{\theta_0}\overline{\theta w} - \epsilon. \quad (11.2)$$

The terms on the right side are, in order, shear production (S); turbulent transport (T); pressure transport (P); buoyant production (B); and viscous dissipation.

Figure 11.3 shows its behavior within the quasi-steady mixed layer as determined from tower, balloon, and aircraft observations. The terms have been made dimensionless with the mixed-layer-similarity group $w_*^3/z_i = gQ_0/\theta_0$, which makes the dimensionless buoyant production term 1.0 at the surface. Pressure transport (which was not measured directly, but taken as the imbalance of the other terms) is a gain term in the unstable surface layer, as also shown in Figure 10.8. The turbulent-transport term also behaves as in the unstable surface layer.

11.2.3 The mean momentum balance

The steady, horizontally homogeneous mean horizontal momentum balance is

$$\frac{\partial \overline{uw}}{\partial z} = -\frac{1}{\rho_0}\frac{\partial P}{\partial x} + fV, \qquad \frac{\partial \overline{vw}}{\partial z} = -\frac{1}{\rho_0}\frac{\partial P}{\partial y} - fU. \quad (11.3)$$

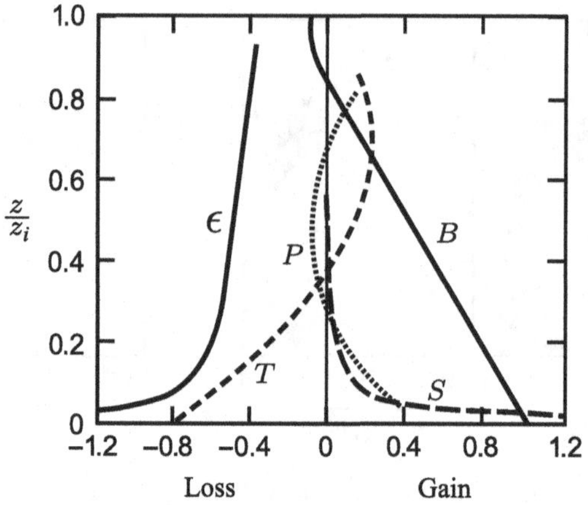

Figure 11.3 The nondimensionalized (with w_* and z_i) TKE budget in a quasi-steady, horizontally homogeneous CBL as summarized by Lenschow *et al.* (1980) from a variety of observations. ϵ, viscous dissipation; T, turbulent transport; P, pressure transport; S, shear production; B, buoyant production.

If the mean velocity scale is $\hat{U}$ and the velocity and length scales of the turbulence are u and ℓ, the ratio of the friction and Coriolis terms in Eqs. (11.3) is $u^2/f\hat{U}\ell$, a *Rossby number.* In a typical ABL with $u = 1$ m s^{-1}, $f = 10^{-4}$ s^{-1}, $\hat{U} = 10$ m s^{-1}, and $\ell = 10^3$ m this Rossby number is ~ 1, so the Coriolis terms are typically important. But the much smaller ℓ in engineering flows typically makes their Rossby number large and their Coriolis terms negligible.

In a constant-density engineering boundary layer the pressure field, and in particular the mean pressure gradient, is coupled with the velocity field through a Poisson equation (Problem 11.15). But in the ABL the mean pressure gradient is dominated by hydrostatics (Chapter 9). Thus if we convert Eq. (11.3) to a mean-shear balance by taking its z-derivative and express the z-derivative of the mean horizontal pressure gradient through Eqs. (9.29) and (9.30), we find

$$\frac{\partial^2 \overline{uw}}{\partial z^2} = -\frac{g}{T_0}\frac{\partial T}{\partial x} + f\frac{\partial V}{\partial z}, \qquad \frac{\partial^2 \overline{vw}}{\partial z^2} = -\frac{g}{T_0}\frac{\partial T}{\partial y} - f\frac{\partial U}{\partial z}, \tag{11.4}$$

We'll see shortly that Eqs. (11.4) can imply strongly curved stress profiles in an ABL with a modest horizontal temperature gradient.

11.2.3.1 The barotropic case

With horizontally uniform mean temperature the mean shear budgets (11.4) reduce, after scaling with u_* and z_i, to

$$\frac{\partial^2 (\overline{uw}/u_*^2)}{\partial (z/z_i)^2} = \left(\frac{f z_i}{u_*}\right)\frac{z_i}{u_*}\frac{\partial V}{\partial z}, \qquad \frac{\partial^2 (\overline{vw}/u_*^2)}{\partial (z/z_i)^2} = -\left(\frac{f z_i}{u_*}\right)\frac{z_i}{u_*}\frac{\partial U}{\partial z}. \tag{11.5}$$

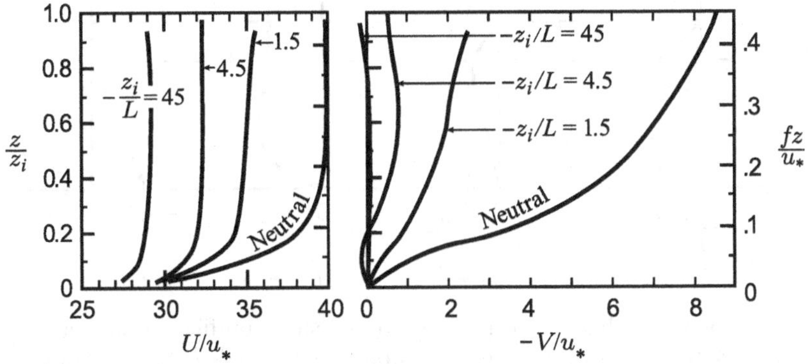

Figure 11.4 Mean wind profiles in a 45 north latitude ABL from the large-eddy simulations of Deardorff (1972a). The mean wind shear above the surface layer decreases sharply as $-z_i/L$ increases. Neutral curves are read on the right axis, unstable curves on the left.

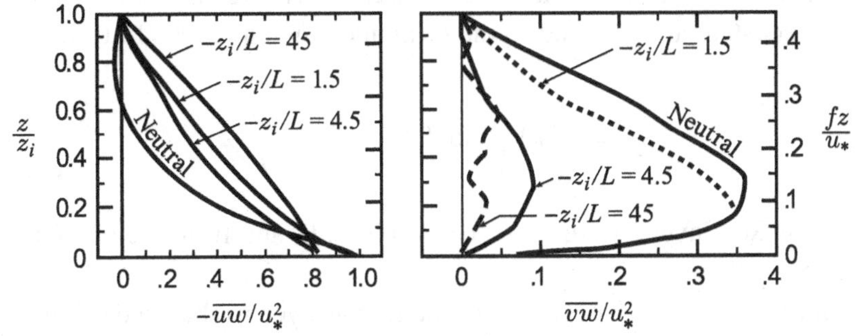

Figure 11.5 Streamwise (left) and lateral (right) turbulent stress profiles in the ABL as calculated through large-eddy simulation by Deardorff (1972a). Their curvature is significant in the neutral case but decreases to nearly zero at $-z_i/L = 45$. Neutral curves are read on the right axis, unstable curves on the left.

The dimensionless parameter fz_i/u_* is typically $\sim$1 or smaller (with $f = 10^{-4}$ s^{-1}, $z_i = 1000$ m, and $u_* = 0.3$ m s^{-1} it is 1/3). Under increasingly convective conditions (i.e., as $-z_i/L$ increases) we expect decreasing mean wind shear, so from Eqs. (11.5) we expect decreasing curvature of the stress profiles as well. Figures 11.4 and 11.5 nicely illustrate this tendency in Deardorff's (1972a) large-eddy simulations.

Figure 11.6 is an idealized model of stress and mean wind profiles in the barotropic CBL. We can sketch out its vertical structure using surface-layer coordinates and the momentum equations (11.3) written in the form

$$\frac{\partial \overline{uw}}{\partial z} = f(V - V_g), \qquad \frac{\partial \overline{vw}}{\partial z} = f(U_g - U). \tag{11.6}$$

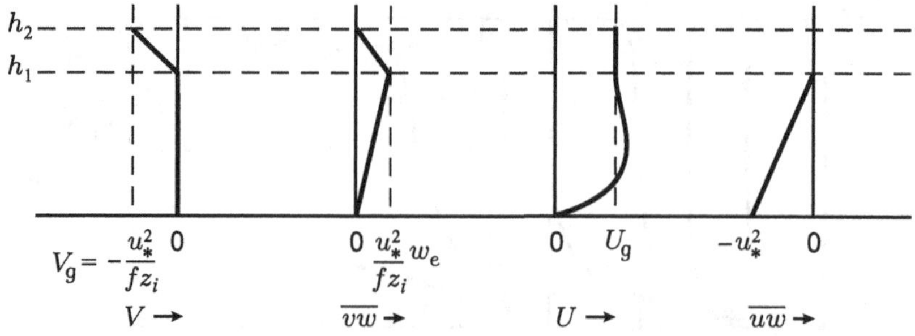

Figure 11.6 An idealization of mean wind and stress profiles in the barotropic CBL. w_e is the entrainment velocity, Section 11.4.2; x and U are in the direction of the mean wind in the surface layer.

Starting at the surface, U increases with z and V is zero. In the mixed layer U is constant and V remains zero. As $\overline{uw}$ decreases linearly to zero at the ABL top its gradient is constant at u_*^2/z_i and so the U-momentum equation (11.6) is

$$\frac{u_*^2}{z_i} = -fV_g, \text{ so that } V_g = -\frac{u_*^2}{fz_i}. \tag{11.7}$$

Within the mixed layer $\partial V/\partial z \simeq 0$ so that $\overline{vw} \simeq 0$, and from the second of Eqs. (11.6) $U \simeq U_g$ there.

The adjustment $V \rightarrow V_g$ occurs in the interfacial layer, as sketched in Figure 11.6. If the wind turning angle is α, then $\tan\alpha = V_g/U_g \simeq -u_*^2/fUz_i$. This is typically of the order of u_*/U, which is small, so $\tan\alpha \simeq \alpha \simeq -u_*^2/fUz_i$. Thus, the wind turning angle can be only a few degrees, much smaller than typical values in neutral or stable ABLs. Finally, as discussed in Section 11.4.1 entrainment of the inversion at velocity w_e in the presence of a mean velocity jump ΔV generates a lateral stress $-w_e\Delta V \simeq w_e\alpha U \simeq w_e u_*^2/fz_i$ at the inversion base, as sketched.

11.2.3.2 The baroclinic case

The mean wind and stress profiles sketched in Figure 11.6 are not typically observed in the CBL because naturally occurring horizontal temperature gradients are apt to make it baroclinic. Equation (11.4) says baroclinity is balanced in steady conditions by a Coriolis term involving mean wind shear and a term involving the curvature of the stress profile. A natural question then is whether turbulent mixing can minimize the mean wind shear so that the baroclinity is balanced mainly by the latter. Figure 11.7, which shows a 180 degree change in mean wind direction across a very unstable ($-z_i/L = 250$) baroclinic CBL over land, gives little support to this notion.

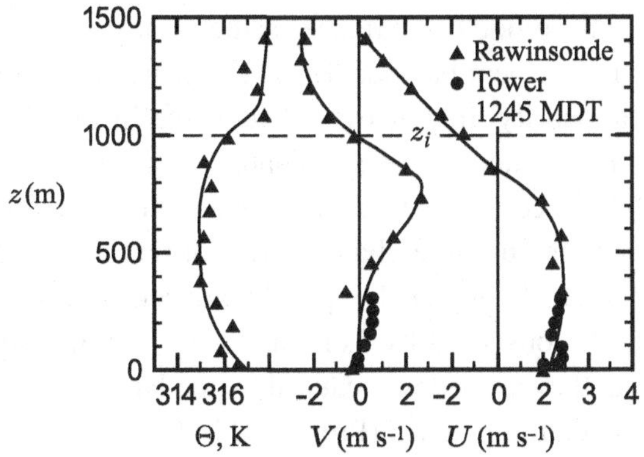

Figure 11.7 Mean wind profiles in a baroclinic CBL with $-z_i/L \sim 250$ at the Boulder Atmospheric Observatory. Despite the very unstable conditions the mean wind direction changes by 180 degrees between the surface and the inversion base. From Wyngaard (1985).

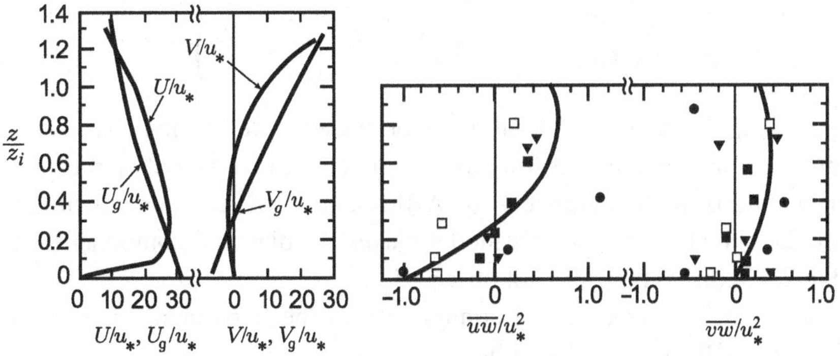

Figure 11.8 Mean profiles measured in the baroclinic CBL in AMTEX. Left: Mean wind and geostrophic wind; from Garratt *et al.* (1982). Right: Turbulent kinematic stress; from Lenschow *et al.* (1980).

Figure 11.8 shows profiles of mean wind, geostrophic wind, and stress measured during wintertime cold-air outbreaks over the East China Sea in the AMTEX experiment. In interpreting them we'll write the mean shear balances (11.4) as

$$\frac{\partial^2 \overline{uw}}{\partial z^2} = f\left(\frac{\partial V}{\partial z} - \frac{\partial V_g}{\partial z}\right), \qquad \frac{\partial^2 \overline{vw}}{\partial z^2} = f\left(\frac{\partial U_g}{\partial z} - \frac{\partial U}{\partial z}\right). \tag{11.8}$$

The curvature of the AMTEX stress profiles, Figure 11.8, indicates that the left sides of Eqs. (11.8) are negative, as are the right sides in at least the lower part of the mixed layer. But as is typical, the data do not allow a definitive test of Eqs. (11.8).

The simplest mean momentum balance in the steady, barotropic mixed layer, Eqs. (11.3), and its mean shear balance, Eqs. (11.8), are easily upset by horizontal advection or nonstationarity. If the stress divergence in the CBL is of the order of $u_*^2/z_i \simeq 10^{-4}$ m s^{-2}, then a 1 m s^{-1} change in a 10 m s^{-1} mean wind over 100 km horizontal distance and a 0.4 m s^{-1} change in mean wind speed per hour give horizontal advection and time-change terms of that magnitude. Likewise, the mean shear equations (11.8) are easily upset by advection and time change. Some of the baroclinic CBLs observed by Lemone *et al.* (1999) displayed substantial mean wind shear that they argued was caused by the large time-change term in the mean horizontal momentum balance (Problem 11.19).

11.2.3.3 Direct testing of eddy-diffusivity models

The simplest turbulence closure assumes the *deviatoric* turbulent kinematic stress, the difference between the turbulent kinematic stress $\overline{u_i u_j}$ and its isotropic form $\overline{u_k u_k}\delta_{ij}/3 = 2e\delta_{ij}/3$ (Chapter 14), is proportional to the mean strain-rate tensor:

$$-\left(\overline{u_i u_j} - \frac{2}{3}\delta_{ij}e\right) = K_m\left(\frac{\partial U_i}{\partial x_j} + \frac{\partial U_j}{\partial x_i}\right), \tag{11.9}$$

with $e = \overline{u_i u_i}/2$. The proportionality factor, the eddy diffusivity,† has a subscript m, signifying momentum, to distinguish it from the eddy diffusivity for a conserved scalar. In a horizontally homogeneous ABL the diagonal components of the right side of Eq. (11.9) vanish, so this model makes the diagonal components of $\overline{u_i u_j}$ equal to $2e/3$; this is typically not observed.

For the principal off-diagonal components in the horizontally homogeneous, quasi-steady ABL, Eq. (11.9) implies

$$\overline{uw} = -K_m\frac{\partial U}{\partial z}, \qquad \overline{vw} = -K_m\frac{\partial V}{\partial z}. \tag{11.10}$$

In the surface layer the set (11.10) serves as a way of readily inferring $K_m(z)$ from measurements of $\partial U/\partial z$ and u_*^2:

$$K_m(z) = -\frac{\overline{uw}}{\partial U/\partial z} \simeq \frac{\tau_0/\rho_0}{\partial U/\partial z} = \frac{u_*^2}{\partial U/\partial z}. \tag{11.11}$$

$\partial U/\partial z$ in the surface layer has been well documented and expressed through the Monin–Obukhov similarity function ϕ_m (Chapter 10). The ϕ_m form (10.17)

† Turbulence theoreticians have traditionally criticized the representation of a turbulent flux as $-K$ times a mean gradient, but the late Les Kovasznay (Johns Hopkins University) reportedly enjoyed pointing out that it was his civil right to *define* such an eddy diffusivity. We discuss it here in that spirit.

of Wilson (2001), for example, has a local-free-convection limit in which K_m behaves as

$$K_m \sim u_f z \sim \left(\frac{g}{\theta_0} Q_0\right)^{1/3} z^{4/3}. \tag{11.12}$$

The K_m defined in Eq. (11.10) has not been documented extensively above the surface layer, perhaps in part due to the inherently large scatter. To illustrate, consider a mid-layer kinematic stress $\overline{uw}$ of the order of $-u_*^2$, so that if $K_m \sim 0.05 w_* z_i$ (which we'll see shortly is a reasonable estimate) the mean wind shear is

$$\frac{\partial U}{\partial z} = \frac{-\overline{uw}}{K_m} \simeq \frac{u_*^2}{0.05 w_* z_i}, \tag{11.13}$$

For $z_i = 1500$ m, $u_* = 0.3$ m s^{-1}, and $w_* = 2$ m s^{-1}, all typical of the CBL in clear weather over land, this gives $\partial U/\partial z \simeq 6 \times 10^{-4}$ s^{-1} – a mean wind difference of 0.6 m s^{-1} per 1000 m height. The uncertainty in a mean wind difference due to finite averaging time scales with the rms wind fluctuation (Chapter 2); this is of the order of w_*, which here is three times the mean wind difference over a height of 1000 m. Likewise, the error in time-averaged stress scales with w_*^2, which here is *two orders of magnitude* larger than the mean stress. Thus reliably determining the K_m defined in Eq. (11.10) in the mixed layer with conventional observations can require unachievable averaging times.

11.2.3.4 Inferences from the stress budgets

The Reynolds-stress budgets (8.69) provide another path to diagnosing the behavior of K_m. As discussed in Chapter 9, their Coriolis terms are typically small in the ABL and in steady, locally homogeneous conditions they reduce to a balance among shear production (S), turbulent transport (T), buoyant production (B), and pressure covariance (P) terms:

$$\frac{\partial}{\partial t}\overline{uw} = 0 = \underset{S}{-\overline{w^2}\frac{\partial U}{\partial z}} - \underset{T}{\frac{\partial}{\partial z}\overline{uww}} + \underset{B}{\frac{g}{\theta_0}\overline{\theta u}} - \underset{P}{\frac{1}{\rho_0}\left(\overline{w\frac{\partial p}{\partial x}} + \overline{u\frac{\partial p}{\partial z}}\right)}, \tag{11.14}$$

$$\frac{\partial}{\partial t}\overline{vw} = 0 = -\overline{w^2}\frac{\partial V}{\partial z} - \frac{\partial}{\partial z}\overline{vww} + \frac{g}{\theta_0}\overline{\theta v} - \frac{1}{\rho_0}\left(\overline{w\frac{\partial p}{\partial y}} + \overline{v\frac{\partial p}{\partial z}}\right). \tag{11.15}$$

We'll continue to take x as the direction of the mean wind in the surface layer. Wyngaard (1984) found two runs from the Minnesota experiment (Kaimal *et al.*, 1976) with enough lateral mean wind shear to allow the $\overline{vw}$ budget to be resolved, and combined four runs from AMTEX (Lenschow *et al.*, 1980) to produce a $\overline{uw}$

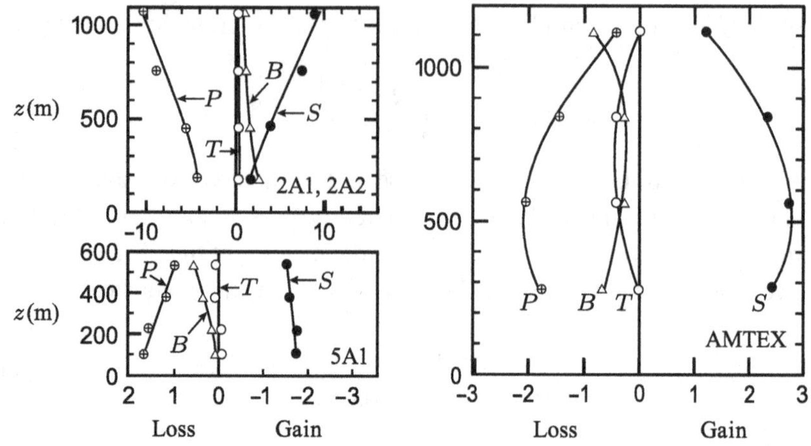

Figure 11.9 Observations of the budgets of kinematic stress, Eqs. (11.14) and (11.15), in the CBL. S is shear production; T, turbulent transport; B, buoyant production; P, pressure destruction. The units are 10^{-3} m^2 s^{-3}. Left: $\overline{vw}$ budgets from the Minnesota experiment. Right: $\overline{uw}$ budget from AMTEX. From Wyngaard (1984).

budget. From these data the S, T, and B terms were calculated directly, allowing P to be obtained from Eqs. (11.14) and (11.15) by difference. The results, shown in Figure 11.9, indicate that in each budget the principal gain term is shear production and the principal loss term is the pressure covariance.[†]

We can define Rotta (1951) time scales for the pressure-covariance terms, allowing for different values for the two components:

$$-\frac{1}{\rho_0}\left(\overline{w\frac{\partial p}{\partial x}}+\overline{u\frac{\partial p}{\partial z}}\right)=-\frac{\overline{uw}}{\tau^s},\qquad -\frac{1}{\rho_0}\left(\overline{w\frac{\partial p}{\partial y}}+\overline{v\frac{\partial p}{\partial z}}\right)=-\frac{\overline{vw}}{\tau^\ell}. \tag{11.16}$$

Figure 11.10 (left) reveals no strong differences between the τ^s and τ^ℓ values deduced from the measured stress budgets in Figure 11.9. Thus we'll write them simply as τ, and the stress budgets (11.14) and (11.15) can then be written as expressions for eddy diffusivities,

$$\overline{uw}\simeq-\overline{w^2}\tau\left(1+\frac{T+B}{S}\right)_{\overline{uw}}\frac{\partial U}{\partial z}=-K_m^s\frac{\partial U}{\partial z}, \tag{11.17}$$

$$\overline{vw}\simeq-\overline{w^2}\tau\left(1+\frac{T+B}{S}\right)_{\overline{vw}}\frac{\partial V}{\partial z}=-K_m^\ell\frac{\partial V}{\partial z}. \tag{11.18}$$

Figures 11.2 and 11.10 suggest that $\overline{w^2}\tau\simeq 0.05w_*z_i$ in mid-CBL.

† A *gain* term on the right side of Eq. (11.14) or (11.15) is one that produces stress of the observed sign. In runs 2A1 and 2A2 in Figure 11.9, for example, $\overline{vw}$ is positive, so gain terms are positive. But in run 5A1 $\overline{vw}$ is negative and therefore gain terms are negative.

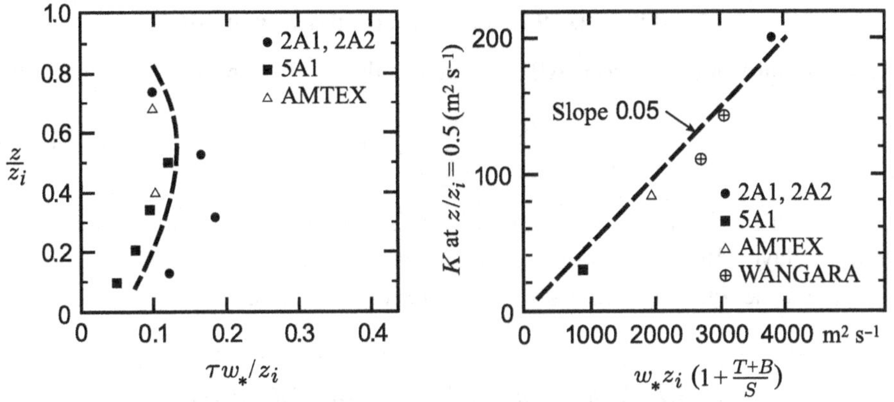

Figure 11.10 Left: Observed values of the time scales τ^s and τ^ℓ in the model (11.16) of the pressure-destruction term in the stress budgets. Right: The observed mid-CBL eddy diffusivity (ordinate) versus that predicted from the models (11.17) and (11.18). From Wyngaard (1984).

Figure 11.10 (right) is a plot of K_m^s and K_m^ℓ calculated from Eqs. (11.17) and (11.18) at $z \simeq 0.5z_i$ in the Minnesota and AMTEX experiments, as functions of the parameter $w_* z_i[1 + (T + B)/S]$. This plot also includes midlayer K_m^s and K_m^ℓ values estimated from Deardorff's (1974) LES results† for hours 14 and 15, respectively, of day 33 of the Wangara experiment. This plot shows no significant difference between K_m^s and K_m^ℓ and suggests that $K_m \simeq 0.05 w_* z_i$ in mid-CBL. The factor $1 + (T + B)/S$ varied from about 0.6 to 1.25 in mid-CBL, suggesting the range of variation of that K_m value.

As we discussed in Chapter 10, Subsection 10.2.4.5, if we write the time-change and horizontal advection terms in the $\overline{uw}$ budget, for example, as $\overline{uw}/\tau_u$ and $U\overline{uw}/L_x$, with τ_u and L_x the scales of the unsteadiness and inhomogeneity, then if $\tau_u \gg \tau$ and $L_x \gg \tau U$ the quasi-steady, locally homogeneous assumptions underlying Eq. (11.17) are valid, and likewise for Eq. (11.18) (Problem 11.18). Using $\tau = 0.1 z_i/w_*$ (Figure 11.10), $z_i = 1000$ m, $U = 10$ m s^{-1}, and $w_* = 1$ m s^{-1}, these are $\tau_u \gg 100$ s, $L_x \gg 1000$ m. In many situations these criteria should not be difficult to satisfy. Then if in addition the stress budget behavior on which Eqs. (11.17) and (11.18) are based holds over a reasonable range of $-z_i/L$ values, these eddy-diffusivity expressions could be useful in applications.

† Deardorff's LES calculation, which used 40^3 grid points on a CDC 6600 "mainframe" computer, ran a factor of 10 slower than real time. In late 2007 a comparable 40^3 LES code on a dedicated, single processor of an IBM SP5 ran 16 times faster than real time – a factor of 160 faster than Deardorff's code. (Peter Sullivan, personal communication.)

11.3 The mixed layer: conserved-scalar fields

In a horizontally homogeneous ABL the mean value C of a conserved scalar satisfies

$$\frac{\partial C}{\partial t} + \frac{\partial \overline{cw}}{\partial z} = 0. \tag{11.19}$$

As indicated in Figure 11.1, in general $\overline{cw}$ is nonzero at the bottom and top of the mixed layer. In a horizontally homogeneous situation where $\partial C/\partial z$ does not depend on time, $\overline{cw}$ varies linearly with z (Problem 11.14).

11.3.1 "Top-down" and "bottom-up" diffusion

Since a conserved scalar constituent $\tilde{c}$ is governed by a linear differential equation, we can superpose its solutions in the same velocity field. Thus we can consider separately the C profiles that result from the scalar fluxes at the top and bottom of the mixed layer. We'll call these individual processes *top-down* and *bottom-up* diffusion. These labels refer not to the direction of the scalar flux, but rather to where the flux is applied.[†]

We'll consider dynamically passive scalars – ones that do not affect the velocity field – in a quasi-steady, horizontally homogeneous CBL. Temperature and water vapor are not passive, for they induce buoyancy forces. But we'll assume that two conserved scalars with the same boundary and initial conditions and in the same turbulence field diffuse identically, so that one can infer the diffusion properties of temperature and water vapor in a given velocity field from those of passive, conserved tracers diffusing in that velocity field.

We write the scalar flux profile as the sum of two linear sub-profiles, each with one nonzero boundary flux:

$$\overline{cw}(z) = \overline{cw}_0 + (\overline{cw}_1 - \overline{cw}_0)\frac{z}{z_i} = \overline{cw}_0\left(1 - \frac{z}{z_i}\right) + \overline{cw}_1\left(\frac{z}{z_i}\right). \tag{11.20}$$

The right panel of Figure 11.11 sketches an example of top-down diffusion of a trace constituent $\tilde{c}$. There is a positive flux of constituent at the top of the mixed layer and zero flux at the surface. Here the top flux is due to the entrainment of air with lower constituent concentration (c negative) in downward (w negative) turbulent motion and higher-concentration, boundary-layer air in upward motion. In a quasi-steady state the $\overline{cw}$ profile is linear, with $\partial\overline{cw}/\partial z$ positive so that $\partial C/\partial t$ is negative, as sketched.

A bottom-up case is sketched in the left panel of Figure 11.11. Here the flux divergence is negative so that C increases with time. Here the entrainment at the

[†] In upper-ocean applications the *top* is the thermocline and the *bottom* is the surface.

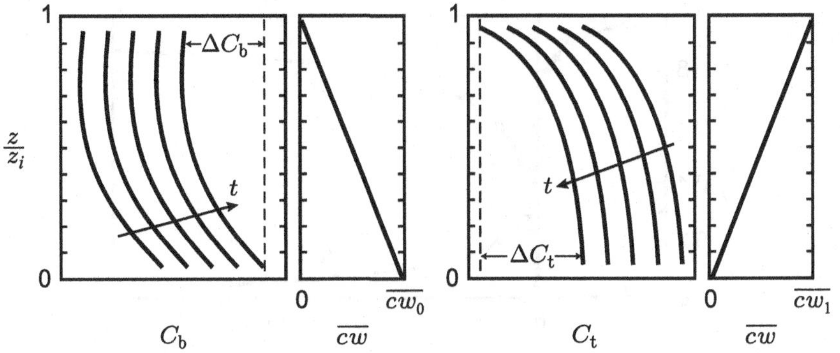

Figure 11.11 Sketches of profiles of the mean value and vertical flux of a conserved scalar in bottom-up (left) and top-down (right) diffusion. From Wyngaard (1987).

mixed-layer top, in conjunction with the change in C that develops across the interfacial layer, will generate top-down diffusion as well. Thus the bottom-up process, unlike the top-down one, is unlikely to exist alone.

A natural question is whether top-down and bottom-up diffusion have the same properties. This would be the case in a flow with statistical symmetry about its horizontal midplane – for example, convection between parallel plates, the bottom plate heated and the top plate cooled, the two surface heat fluxes being equal in magnitude. Here the profiles of the eddy diffusivities K_b and K_t are symmetric about the midplane: $K_t(1 - z/z_i) = K_b(z/z_i)$. But since the ABL is not symmetric about its midplane, top-down and bottom-up diffusion in the ABL should also lack that symmetry.†

To examine this asymmetry we use the linearity of its conservation equation to represent a conserved scalar field as the sum of top-down and bottom-up parts coexisting in the velocity field:

$$\tilde{c} = \tilde{c}_b + \tilde{c}_t, \quad C = C_t + C_b, \quad c = c_t + c_b. \tag{11.21}$$

The simplest mixed-layer similarity hypothesis is that the mean scalar gradient in each process depends only on the boundary flux, the mixed-layer scales w_* and z_i, and z:

$$\frac{\partial C_t}{\partial z} = -\frac{\overline{cw_1}}{w_* z_i} g_t(z/z_i), \qquad \frac{\partial C_b}{\partial z} = -\frac{\overline{cw_0}}{w_* z_i} g_b(z/z_i), \tag{11.22}$$

where g_t and g_b are dimensionless functions. Then the symmetry question is whether $g_b(z/z_i) = g_t(1 - z/z_i)$.

† In fact, one expects that turbulent boundary layers in general lack that symmetry.

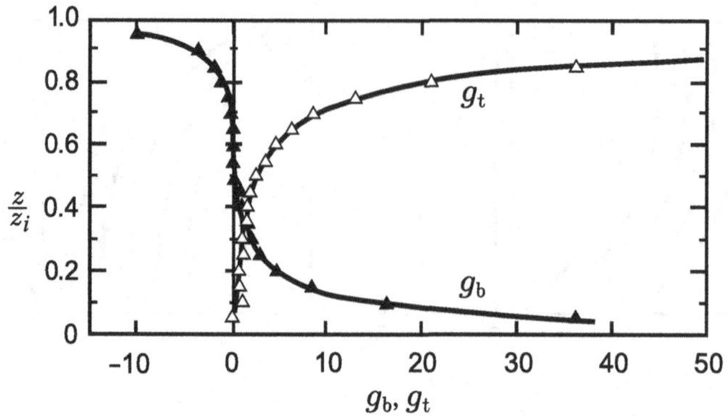

Figure 11.12 LES calculations of the dimensionless mean-gradient functions g_b and g_t defined in Eq. (11.22). From Moeng and Wyngaard (1984).

11.3.1.1 Early LES results

Moeng and Wyngaard (1984, 1986a, 1986b) used LES with 40^3 grid points to simulate the diffusion of passive, conserved "dyes" in a CBL with $-z_i/L \simeq 10$. A blue dye was emitted continuously from the surface; in quasi-steady state it experienced both bottom-up and top-down diffusion. A red dye continuously entrained into the CBL from above experienced only top-down diffusion. The function g_t, Eq. (11.22), was determined directly from the red-dye field:

$$g_t = -\frac{w_* z_i}{\overline{cw}_1} \frac{\partial C_t}{\partial z}. \tag{11.23}$$

From the decomposition $C = C_b + C_t$ and the definitions (11.22) of the gradient functions we can write

$$g_b = -\frac{w_* z_i}{\overline{cw}_0} \left(\frac{\partial C}{\partial z} + \frac{\overline{cw}_1}{w_* z_i} g_t \right). \tag{11.24}$$

which with g_t now known allowed g_b to be evaluated from the statistics of the blue dye field.

The resulting dimensionless mean-gradient functions g_b and g_t are shown in Figure 11.12. The sign change in g_b at $z \simeq 0.6 z_i$ is caused by a sign change in the mean gradient $\partial C_b/\partial z$; since the bottom-up flux is nonzero there, this implies a singularity in the bottom-up eddy diffusivity K_b at that point. K_t is well behaved. Thus the eddy diffusivities for the two processes are indeed not symmetric.

Based on LES studies Patton *et al.* (2003) proposed modified forms of g_t and g_b over a plant canopy. Wang *et al.* (2007) have attempted to determine them from long-term, well-calibrated, point measurements of carbon dioxide mixing ratio over a

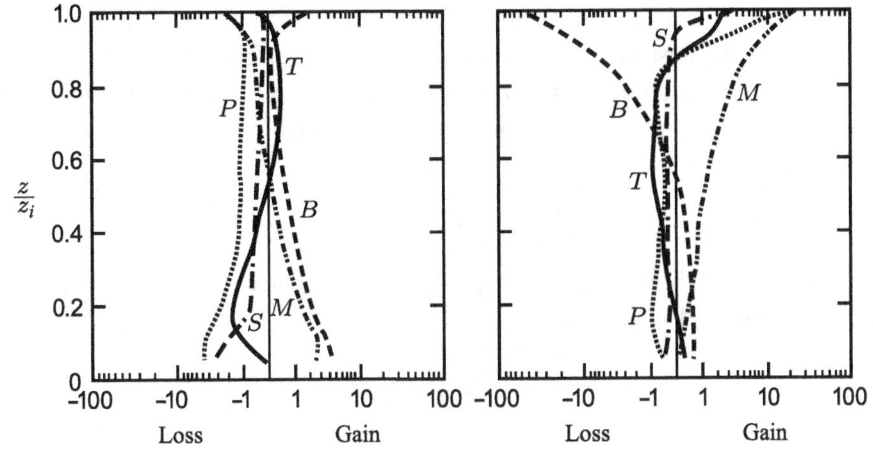

Figure 11.13 LES calculations of the budgets (11.26) of the bottom-up flux (left) and (11.25) of the top-down flux (right) of a conserved scalar. The terms are nondimensionalized with w_*, z_i, and the boundary flux. The scale between -1 and 1 on the abscissa is linear. M, mean-gradient production; T, turbulent transport; B, buoyant production; P, pressure destruction; S, subgrid term. From Moeng and Wyngaard (1986a).

forested site. Previous attempts had been unsuccessful, in their view largely because of the difficulty of making these measurements.[†] The Wang *et al.* (2007) results were qualitatively consistent with those from LES but not all differences between them could be explained.

The turbulent-flux budget (8.71) can give insight into the behavior of eddy diffusivity. In horizontally homogeneous conditions it becomes, for the top-down and bottom-up scalar cases,

$$\frac{\partial \overline{c_t w}}{\partial t} = 0 = -\overline{w^2}\frac{\partial C_t}{\partial z} - \frac{\partial \overline{c_t w^2}}{\partial z} + \frac{g}{\theta_0}\overline{c_t \theta} - \frac{1}{\rho_0}\left(\overline{c_t \frac{\partial p}{\partial z}}\right), \tag{11.25}$$

$$0 = \quad M \quad + \quad T \quad + \quad B \quad + \quad P$$

$$\frac{\partial \overline{c_b w}}{\partial t} = 0 = -\overline{w^2}\frac{\partial C_b}{\partial z} - \frac{\partial \overline{c_b w^2}}{\partial z} + \frac{g}{\theta_0}\overline{c_b \theta} - \frac{1}{\rho_0}\left(\overline{c_b \frac{\partial p}{\partial z}}\right). \tag{11.26}$$

The terms on the right side are M, mean-gradient production; T, turbulent transport; B, buoyant production; and P, pressure destruction.

Figure 11.13 shows these budgets as calculated with 40^3 LES by Moeng and Wyngaard (1986a). They are not only asymmetric, but also quite different. The top-down budget is dominated by the gain through mean-gradient production;

† As discussed in Chapter 2, the variance of the difference between a time-averaged point measurement and the ensemble mean scales with the integral scale/averaging length, whereas the corresponding quantity for area-averaged LES fields scales with the square of that ratio. As a result, time-averaged single-point measurements tend to have much more scatter than area-averaged LES results.

buoyant production is generally smaller, and the loss terms are turbulent transport and pressure destruction. In the bottom-up budget buoyant production, not mean-gradient production, is the principal source; mean-gradient production and turbulent transport are generally smaller, and the principal loss term is pressure destruction.

11.3.1.2 Scalar variances

With the top-down, bottom-up decomposition (11.21) we can write the variance of a passive, conserved scalar c as

$$\overline{c^2} = \overline{c_t^2} + 2\overline{c_t c_b} + \overline{c_b^2}. \tag{11.27}$$

The simplest scaling hypothesis that accounts for the scalar flux at mixed-layer top takes the governing parameters for c_t statistics to be w_*, z_i, and $\overline{cw}_1$; those for c_b statistics to be w_*, z_i, and $\overline{cw}_0$; and those for their joint statistics to be w_*, z_i, $\overline{cw}_0$, and $\overline{cw}_1$. Thus from Eq. (11.27) Moeng and Wyngaard (1984) wrote

$$\overline{c^2} = \left(\frac{\overline{cw}_1}{w_*}\right)^2 f_t(z/z_i) + 2\left(\frac{\overline{cw}_0\overline{cw}_1}{w_*^2}\right) f_{tb}(z/z_i) + \left(\frac{\overline{cw}_0}{w_*}\right)^2 f_b(z/z_i). \tag{11.28}$$

They found that in midlayer $f_b \sim 1,\ f_{tb} \sim 1,\ f_t \sim 6$.

The correlation coefficient of the top-down and bottom-up scalar fields is

$$r_{tb} = \frac{\overline{c_t c_b}}{\left(\overline{c_t^2}\ \overline{c_b^2}\right)^{1/2}} = \frac{f_{tb}}{(f_t f_b)^{1/2}}\mathrm{sgn}(\overline{cw}_0\ \overline{cw}_1), \tag{11.29}$$

where sgn means "the sign of." Moeng and Wyngaard (1984) found that $|r_{tb}| \sim 0.5$ in mid-CBL. The positive sign of f_{tb} has interesting implications. For example, if two species are diffusing into the mixed layer, one from above and one from below, then $\overline{cw}_0$ is positive, $\overline{cw}_1$ is negative, and in midlayer $r_{tb} \simeq -0.5$. If the two species undergo a binary reaction, the mean reaction rate is

$$\text{mean reaction rate} \propto \overline{\tilde{c}_1 \tilde{c}_2} = C_1 C_2 + \overline{c_1 c_2}. \tag{11.30}$$

Because Eq. (11.29) says that $\overline{c_1 c_2}$ is negative here, the mean reaction rate is less than the mean concentrations would indicate.

We saw in Figure 11.13 that the top-down and bottom-up scalar flux budgets have significant buoyant-production terms. For $\overline{w\theta}$ and $\overline{wc}$ these terms are proportional to $\overline{\theta^2}$ and $\overline{\theta c}$, respectively. By using the top-down and bottom-up decomposition for θ and c we can derive, using the approach in Eqs. (11.27) and (11.28) (Problem 11.16), the following expressions for the profiles of $\overline{\theta c}$ and $\overline{\theta^2}$:

$$\overline{\theta c} = \frac{\overline{w\theta}_0\, \overline{wc}_0}{w_*^2}\left[R_\theta R_c f_{\mathrm{t}} + (R_\theta + R_c) f_{\mathrm{tb}} + f_{\mathrm{b}}\right], \quad R_\theta = \frac{\overline{w\theta}_1}{\overline{w\theta}_0}, \quad R_c = \frac{\overline{wc}_1}{\overline{wc}_0}, \tag{11.31}$$

$$\overline{\theta^2} = \frac{(\overline{w\theta}_0)^2}{w_*^2}\left(R_\theta^2 f_{\mathrm{t}} + 2R_\theta f_{\mathrm{tb}} + f_{\mathrm{b}}\right). \tag{11.32}$$

Equations (11.31) and (11.32) show that unless $R_c = R_\theta$, i.e., unless the top/bottom flux ratio R is the same for the scalar c and θ, the buoyant production term in the dimensionless budgets of $\overline{w\theta}$ and $\overline{wc}$ will not be the same. The nature of the CBL makes R_θ negative, but R_c can be positive, as for water vapor in a CBL over an evaporating surface with drier air in the capping inversion. Thus the two flux budgets, and the two eddy diffusivities, are apt to differ.

11.3.2 Generalizations of K-closure for scalars

We have seen that in top-down diffusion (where the flux of c-stuff is nonzero at the top of the mixed layer and zero at the bottom) K appears to be well behaved, but in bottom-up diffusion K is significantly larger and has a midlayer singularity.

This latter phenomenon was characterized earlier by Deardorff (1966), for example, as "the existence of upward heat flux, $\overline{w\theta} > 0$, with vanishing or counter (positive) potential temperature gradient, $\partial\Theta/\partial z \geq 0$." He cited several studies in which this had been observed, beginning with the Great Plains Turbulence Field Program in 1957. It is also evident in high-resolution LES (Moeng and Wyngaard, 1989). Here the mean-gradient production term in the temperature variance budget (10.44) is negligible or even a slight loss term; the gain is by turbulent transport of variance from below.

Because the simplicity of K-closure is attractive in numerical modeling, this evidence of its misbehavior stimulated early efforts to develop improved versions. The simplest is the modified form suggested by Deardorff (1972b),

$$\overline{w\theta} = -K\left(\frac{\partial\Theta}{\partial z} - \gamma_\theta\right), \tag{11.33}$$

with a simple model of the $\overline{w\theta}$ conservation equation (10.48) used to estimate γ_θ as $g\overline{\theta^2}/\theta_0\overline{w^2}$. Holtslag and Moeng (1991) obtained another estimate through a different model of that equation.

Using results of the Moeng and Wyngaard (1989) 96^3 LES runs, Holtslag and Moeng (1991) also derived an expression for K in the flux-gradient form (11.33):

$$K = \frac{(1 - z/z_i + R_c z/z_i)\tilde{K}_b \tilde{K}_t}{(1 - z/z_i)\tilde{K}_t + R_c(z/z_i)\tilde{K}_b}, \qquad R_c = \overline{cw}_1/\overline{cw}_0. \tag{11.34}$$

Here $\tilde{K}_t$ and $\tilde{K}_b$ are the top-down and bottom-up eddy diffusivities K_t and K_b modified to be compatible with the modified flux-gradient expression (11.33):

$$\frac{\tilde{K}_b}{w_* z_i} = \left(\frac{z}{z_i}\right)^{4/3}\left(1 - \frac{z}{z_i}\right)^2, \quad \frac{\tilde{K}_t}{w_* z_i} = 7\left(\frac{z}{z_i}\right)^2\left(1 - \frac{z}{z_i}\right)^3. \tag{11.35}$$

Fiedler (1984) discussed the generalization of the eddy diffusivity to the integral form

$$\overline{wc}(z,t) = \int D(z,z')\frac{\partial}{\partial z'}C(z',t)dz', \tag{11.36}$$

which is the physical-space version of the "spectral diffusivity" model of Berkowicz and Prahm (1979). Stull's (1984) "transilient turbulence" closure is also of this general form. Hamba's (1993) two-term eddy diffusivity expression for conserved scalars,

$$\overline{wc} = -K\frac{\partial C}{\partial z} + K_2\frac{\partial^2 C}{\partial z^2}, \tag{11.37}$$

can reproduce the Moeng and Wyngaard (1984) results for top-down and bottom-up diffusion.

11.3.3 Generalized mixed-layer similarity

Equations (11.21) and (11.22) generalize the mixed-layer similarity statement for a mean scalar gradient in the quasi-steady, horizontally homogeneous mixed layer to include the flux of the scalar at the CBL top:

$$\frac{\partial C}{\partial z} = -\frac{\overline{cw}_1}{w_* z_i}g_t(z/z_i) - \frac{\overline{cw}_0}{w_* z_i}g_b(z/z_i). \tag{11.38}$$

A typical profile of vertical flux of virtual potential temperature in the CBL is sketched in Figure 11.1. When Eq. (11.38) is used for Θ the negative flux of virtual temperature at the top, caused by entrainment of the capping inversion, makes the two terms on its rhs of opposite signs. The resulting Θ profile can be nearly uniform in the CBL. This is no doubt the origin of the term *mixed layer* and the erroneous notion that within it conserved scalars are "well mixed" (constant).

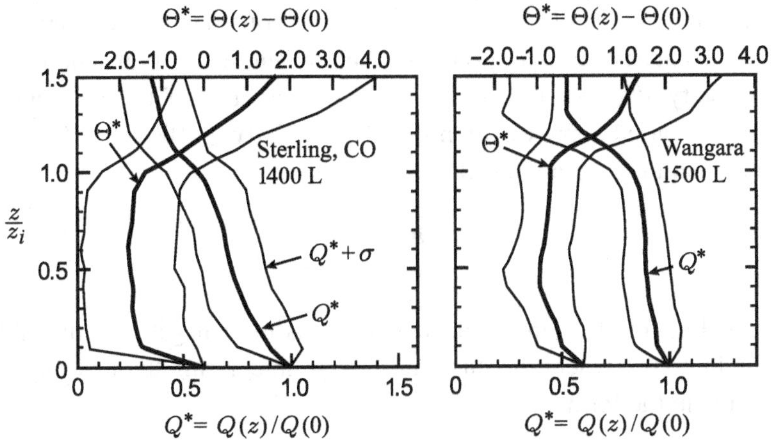

Figure 11.14 Mean radiosonde profiles of potential temperature and specific humidity measured in the midday, clear-weather CBL in Sterling, Colorado (left) and Wangara (right). The center, heavier lines are the mean over 33 days (Sterling) and 9 days (Wangara), and the thinner lines represent ± 1 standard deviation. From Mahrt (1976).

Water vapor mixing ratio in the CBL over an evaporating surface provides an everyday counterexample of an "unmixed" mean scalar profile. A typical water-vapor flux profile, with the top flux also positive due to the entrainment of drier air aloft, is sketched in Figure 11.1. Here the two terms on the rhs of Eq. (11.38) are of the same sign, which typically gives a detectable mean gradient of water-vapor mixing ratio in the mixed layer.

Figure 11.14 illustrates this point through mean profiles of specific humidity (a conserved water-vapor variable) and potential temperature in the mixed layer. The averaging of 33 radiosonde profiles has removed much of the instantaneous detail, revealing mean profiles that suggest a significant vertical gradient of mean specific humidity but an essentially zero gradient of mean potential temperature.

11.4 The interfacial layer

11.4.1 The entrainment flux

The top of the CBL is locally and instantaneously quite thin; averaging over time, space, or an ensemble produces a much thicker *interfacial layer*. Its thickness $\Delta h = h_2 - h_1$ (Figure 11.1) is a good fraction of z_i, according to the convection-tank experiments of Willis and Deardorff (1974). It maintains the entrainment fluxes of momentum, potential temperature, water vapor, and trace constituents at the top of the mixed layer.

We'll illustrate with the entrainment flux of potential temperature, relaxing slightly our usual assumption of horizontal homogeneity to allow a nonzero mean vertical velocity W. This implies a nonzero divergence of the horizontal wind field, since to a good approximation $\partial U/\partial x + \partial V/\partial y + \partial W/\partial z = 0$. The mean potential temperature equation is then

$$\frac{\partial \Theta}{\partial t} + W\frac{\partial \Theta}{\partial z} + \frac{\partial \overline{w\theta}}{\partial z} = 0. \tag{11.39}$$

We integrate Eq. (11.39) from z_i, the height of maximum negative temperature flux (Figure 11.1), to h_2, where the flux vanishes. In integrating the first term we use Leibnitz' rule in the form

$$\frac{\partial}{\partial t}\int_{z_i(t)}^{h_2(t)} \Theta\, dz = \Theta(h_2)\frac{\partial h_2}{\partial t} - \Theta(z_i)\frac{\partial z_i}{\partial t} + \int_{z_i}^{h_2} \frac{\partial \Theta}{\partial t} dz. \tag{11.40}$$

For the second term we use the mean-value theorem in the form

$$\int_{z_i}^{h_2} W\frac{\partial \Theta}{\partial z} dz = W(h_m)[\Theta(h_2) - \Theta(z_i)], \qquad z_i \le h_m \le h_2. \tag{11.41}$$

Since $\overline{w\theta}$ vanishes at h_2, the third term integrates to $-\overline{w\theta}(z_i)$.

The interfacial layer thickness typically is in the range 0.2–0.6 z_i, but nonetheless it is conventional here to take the limit as $h_2 - z_i \to 0$, giving a *jump model* of the interfacial layer. In this zero-thickness limit several terms in the integral of Eq. (11.39) vanish (Problem 11.21). The resulting expression is

$$\overline{w\theta}(z_i) \equiv \overline{w\theta}_1 = -\left(\frac{\partial z_i}{\partial t} - W(z_i)\right)\left[\Theta(h_2) - \Theta(z_i)\right] = -w_e\Delta\Theta. \tag{11.42}$$

w_e, the *entrainment velocity*, is the mean rate of erosion of the nonturbulent fluid by the turbulent fluid. If $w_e = -W$ the rate of entrainment is balanced by the mean subsidence, so $\partial z_i/\partial t = 0$. This happens often in the marine boundary layer under steady synoptic conditions. Over land, w_e usually exceeds $|W|$ in the morning and early afternoon hours, allowing z_i to grow. Under a high-pressure system Ekman pumping (Chapter 9) causes W to be negative, suppressing the growth of z_i and thereby limiting the formation of the small cumulus at the top of the mixed layer. We often have clear skies in high-pressure areas.

The jump equations for the fluxes of momentum and conserved scalar c are

$$\overline{uw}_1 = -w_e\Delta U, \qquad \overline{vw}_1 = -w_e\Delta V, \qquad \overline{wc}_1 = -w_e\Delta C. \tag{11.43}$$

Clearly, the entrainment velocity w_e is an important parameter.

11.4.2 The entrainment velocity

The mixed layer over land typically deepens at a few tens of meters to perhaps 100 m per hour ($\sim$ 1–3 cm s^{-1}). But the convective velocity scale w_* is typically at least 1 m s^{-1} and often larger. Why is $w_e \ll w_*$?

The answer lies in the effects of stability on turbulence. In the interfacial layer the buoyant production term represents a rate of TKE loss because $\overline{w\theta}$ is negative there. Thus the turbulent energy budget under quasi-steady conditions says:

$$\text{rate of TKE gain from shear production, turbulent and pressure transport}$$
$$= \text{rate of TKE loss to viscous dissipation and buoyancy.} \qquad (11.44)$$

If the underlying boundary layer is in free convection the shear production term vanishes, and it can also be quite small in a barotropic CBL. In such cases the only source term in the TKE budget is turbulent and pressure transport from below.

In the interfacial layer capping a convective ABL we can estimate this transport term as of order $w_*^3/\Delta h$, with Δh the thickness of the interfacial layer (Figure 11.1). If Δh scales with z_i, this is of order w_*^3/z_i. Therefore a flux Richardson number characteristic of interfacial-layer turbulence is

$$R_f = \frac{\text{rate of energy loss to buoyancy}}{\text{rate of energy gain by transport}} \sim \frac{\frac{g}{\theta_0} w_e \Delta\Theta}{w_*^3/z_i}. \qquad (11.45)$$

Observations suggest that R_f in steady turbulence cannot exceed a relatively small value, say 0.2–0.3; at larger values the turbulence is extinguished.

We can look at this another way. If in the interfacial layer $R_f \to$ constant $= a$, then a crude statement of the TKE budget there is

$$-\frac{g}{\theta_0}\overline{w\theta}_1 = a\frac{w_*^3}{z_i}. \qquad (11.46)$$

It then follows from the definition of w_* that $-\overline{w\theta}_1/Q_0 = a$. This is a common closure in CBL modeling – that the entrainment flux of temperature is a constant negative fraction of the surface flux.

Another Richardson number, one characteristic of CBL structure (Problem 11.21), is (Deardorff and Willis, 1985)

$$R_* = \frac{\frac{g}{\theta_0} z_i \Delta\Theta}{w_*^2}. \qquad (11.47)$$

R_* is typically large; for example, if z_i = 1 km, $\Delta\Theta$ = 1 K, w_* = 1 m s^{-1}, then $R_* = 30$. The definitions of R_f and R_*, Eqs. (11.45) and (11.47), imply that they are related by

$$\frac{R_{\rm f}}{R_*} = \frac{w_e}{w_*}. \tag{11.48}$$

Thus with $R_* = 30$ and $R_{\rm f} \simeq 0.3$, for example, $w_{\rm e} \sim 0.01 w_*$. This is very small indeed.

In the interfacial layer the intensity of turbulent fluctuations decreases from the mixed-layer values to the much lower values typical of the essentially nonturbulent free atmosphere above. Thus, we expect the eddy diffusivity in the interfacial layer to be less than in the mixed layer, and vertical mean gradients to be larger. If we know the value of a mean property – mean velocity, temperature, water vapor mixing ratio, say – in the free atmosphere just above the ABL and the value at the surface, it is not clear a priori how the changes in this mean property are distributed across the surface layer, the mixed layer, and the interfacial layer. One can model each layer, however, and estimate the profile from the resulting coupled set of equations.

In this way one finds that the baroclinic, convective ABL tends to concentrate its mean wind shear within the interfacial layer. The "jump conditions" of Eq. (11.43) imply that the turbulent kinetic energy budget there will have positive-definite shear production terms,

$$-\overline{uw}\frac{\partial U}{\partial z} - \overline{vw}\frac{\partial V}{\partial z} \simeq \frac{w_{\rm e}}{\Delta h}\left[(\Delta U)^2 + (\Delta V)^2\right]. \tag{11.49}$$

Even in barotropic cases we saw there is a mean wind jump across the interfacial layer in order to meet the geostrophic upper boundary condition on the mean wind. We conclude that there typically is non-negligible shear production in the interfacial layer. If so, the concept that $R_{\rm f} \to a$ in the interfacial layer implies that the ratio of the negative flux of entrainment and the surface flux can exceed a.

Questions on key concepts

11.1 Discuss the three-layer structure of the CBL.

11.2 Discuss the notion of mixed-layer similarity, giving examples of statistics that do and do not follow it.

11.3 Discuss the behavior of the TKE budget in the CBL.

11.4 What is K-closure? Why is it so difficult to assess its reliability in the mixed layer? Why could it be easier to evaluate with numerically computed fields?

11.5 Explain the shape of the mean profiles in the CBL as represented in Figure 11.6.

11.6 Explain why it is so difficult to evaluate the mean-momentum balance from observations in the ABL.

11.7 Explain how the budgets of Reynolds stress and scalar flux can be used to produce expressions for eddy diffusivities.

11.8 Explain what is meant by top-down and bottom-up diffusion. Why can the first exist alone, but not the second?
11.9 Explain the physical meaning of the finding that the correlation coefficient of top-down and bottom-up scalar fields is negative when both boundary fluxes are into the CBL.
11.10 Explain why the CBL is often "well mixed" in Θ but not in Q.
11.11 Explain the concept of *entrainment velocity*.
11.12 Explain physically why the entrainment velocity is typically much smaller than the turbulence velocity scale u.
11.13 Explain why the ABL is apt to be baroclinic. Why is this not true of engineering boundary layers?
11.14 Explain physically why the vertical flux of temperature is a source of TKE in unstable stratification and a sink in stable stratification.

Problems

11.1 Explain why the concept of mixed-layer similarity is consistent with that of local free convection.
11.2 Develop a criterion for the negligibility of the effects of horizontal inhomogeneity and time changes on mixed-layer similarity.
11.3 Discuss the averaging-time requirement for stress measurements in the CBL. Does it explain why we know so little about stress profiles there?
11.4 Discuss the failure of mixed-layer similarity for scalars in the context of the right panel of Figure 11.2. What does that figure suggest also about the M-O similarity of scalars?
11.5 Examine the potential importance of the term that horizontal advection of mean momentum produces in the mean-shear equation. Is it likely to be as important as baroclinity?
11.6 Interpret physically the reason why the correlation coefficient r_{tb} in Eq. (11.29) is nonzero. What sign do you expect it to have?
11.7 Generalize Eq. (11.42) to the more realistic case of finite depth of the interfacial layer. Does it have an appreciable effect on the result?
11.8 Interpret in the context of the mixing-length model the failure of the eddy-diffusivity model for diffusion of a conserved scalar in the CBL.
11.9 Stress measurements in the CBL are plagued by scatter, as Figure 11.8 indicates. How would area averaging help this? Use the expression for convergence of a spatial average to the ensemble average.
11.10 Derive the conservation equation for $\overline{w^3}$ in a CBL. Simplify it as much as possible. What is its principal source term? Does it explain why $\overline{w^3}$ is positive?

11.11 Derive the conservation equation for $\overline{\theta^3}$ in a CBL. Simplify it as much as possible. What is its principal source term? Use it to predict the sign of $\overline{\theta^3}$.

11.12 What do the results of Section 11.3 imply about the vertical diffusion of trace species in cloud layers?

11.13 Discuss the influence of the parameter $m = f z_i / w_*$ on the mean wind profile in a baroclinic CBL. Can you interpret m physically?

11.14 Use Eq. (11.19) to show that the scalar flux profile is linear in a quasi-steady, horizontally homogeneous ABL.

11.15 Derive the Poisson equation for pressure by taking the divergence of the Navier–Stokes equation. How does it indicate that the pressure field in the ABL is fundamentally different from that in a constant-density engineering flow?

11.16 Using the approach in Eqs. (11.27) and (11.28), derive Eqs. (11.31) and (11.32). Assume that any two c_t, and any two c_b, are perfectly correlated. (Is that a reasonable assumption? Discuss.)

11.17 Derive Deardorff's expression for γ_θ in Eq. (11.33). Use the scalar flux conservation equation (10.48), assume the turbulent transport term is negligible, and model the pressure-destruction term with a Rotta time scale.

11.18 Show that if $\tau_u \gg \tau$ and $L_x \gg \tau U$ the time-change and horizontal advection terms in the stress budgets Eqs. (11.14) and (11.15) are negligible.

11.19 Develop a criterion for the negligibility of the time-change term in the mean-momentum balance (11.3). Then show that this criterion can be difficult to meet in practice.

11.20 Show that Eq. (8.69) does yield the TKE budget (11.2) for a horizontally homogeneous ABL. Why does the Coriolis term vanish?

11.21 Explain why R_*, Eq. (11.47), is characteristic of CBL structure.

References

Berkowicz, R., and L. P. Prahm, 1979: Generalization of K-theory for turbulent diffusion. Part I: Spectral turbulent diffusivity concept. *J. Appl. Meteor.*, **18**, 266–272.

Caughey, S. J., 1982: Observed characteristics of the atmospheric boundary layer. In *Atmospheric Turbulence and Air Pollution Modelling*, F. T. M. Nieuwstadt and H. Van Dop, Eds., Dordrecht: Reidel, pp. 107–158.

Deardorff, J. W., 1966: The counter-gradient heat flux in the atmosphere and in the laboratory. *J. Atmos. Sci.*, **23**, 503–506.

Deardorff, J. W., 1970: Convective velocity and temperature scales for the unstable planetary boundary layer and for Rayleigh convection. *J. Atmos. Sci.*, **27**, 1211–1213.

Deardorff, J. W., 1972a: Numerical investigation of neutral and unstable planetary boundary layers. *J. Atmos. Sci.*, **29**, 91–115.

Deardorff, J. W., 1972b: Theoretical expression for the counter-gradient vertical heat flux. *J. Geophys. Res.*, **77**, 5900–5904.

Deardorff, J. W., 1974: Three-dimensional numerical study of the height and mean structure of a heated planetary boundary layer. *Bound.-Layer Meteor.*, **7**, 81–106.

Deardorff, J. W., 1979: Prediction of convective mixed-layer entrainment for realistic capping inversion structure. *J. Atmos. Sci.*, **36**, 424–436.

Deardorff, J. W., and G. E. Willis, 1985: Further results from a laboratory model of the convective planetary boundary layer. *Bound.-Layer Meteor.*, **32**, 205–236.

Fiedler, B. H., 1984: An integral closure model for the vertical turbulent flux of a scalar in a mixed layer. *J. Atmos. Sci.*, **41**, 674–680.

Garratt, J. R., J. C. Wyngaard, and R. J. Francey, 1982: Winds in the atmospheric boundary layer – prediction and observation. *J. Atmos. Sci.*, **39**, 1307–1316.

Hamba, F., 1993: A modified first-order model for scalar diffusion in the convective boundary layer. *J. Atmos. Sci.*, **50**, 2800–2810.

Holtslag, A. A. M., and C.-H. Moeng, 1991: Eddy diffusivity and countergradient transport in the convective atmospheric boundary layer. *J. Atmos. Sci.*, **48**, 1690–1698.

Kaimal, J. C., J. C. Wyngaard, D. A. Haugen, *et al.*, 1976: Turbulence structure in the convective boundary layer. *J. Atmos. Sci.*, **33**, 2152–2169.

Lemone, M. A., Mingyou Zhou, C.-H. Moeng, *et al.*, 1999: An observational study of wind profiles in the baroclinic convective mixed layer. *Bound.-Layer Meteor.*, **90**, 47–82.

Lenschow, D. H., J. C. Wyngaard, and W. T. Pennell, 1980: Mean-field and second-moment budgets in a baroclinic, convective boundary layer. *J. Atmos. Sci.*, **37**, 1313–1326.

Mahrt, L., 1976: Mixed layer moisture structure. *Mon. Wea. Rev.*, **104**, 1403–1407.

Moeng, C.-H., and J. C. Wyngaard, 1984: Statistics of conservative scalars in the convective boundary layer. *J. Atmos. Sci.*, **41**, 3161–3169.

Moeng, C.-H., and J. C. Wyngaard, 1986a: Recalculation of the pressure-gradient/scalar covariance in top-down and bottom-up diffusion. *J. Atmos. Sci.*, **43**, 1182–1183.

Moeng, C.-H., and J. C. Wyngaard, 1986b: An analysis of closures for pressure-scalar covariances in the convective boundary layer. *J. Atmos. Sci.*, **43**, 2499–2513.

Moeng, C.-H., and J. C. Wyngaard, 1989: Evaluation of turbulent transport and dissipation closures in second-order modeling. *J. Atmos. Sci.*, **46**, 2311–2330.

Patton, E. G., P. P. Sullivan, and K. J. Davis, 2003: The influence of a forest canopy on top-down and bottom-up diffusion in the planetary boundary layer. *Quart. J. R. Meteor. Soc.*, **129**, 1415–1434.

Rotta, J. C., 1951: Statistiche theorie nichthomogener turbulenz. *Z. Phys.*, **129**, 547–572.

Stull, R. B., 1984: Transilient turbulence theory, Part 1: The concept of eddy mixing across finite distances. *J. Atmos. Sci.*, **41**, 3351–3367.

Tennekes, H., 1974: The atmospheric boundary layer. *Phys. Today*, **27**, 52–63.

Wang, W., K. J. Davis, C. Yi, *et al.*, 2007: A note on the top-down and bottom-up gradient functions over a forested site. *Bound.-Layer Meteor.*, **124**, 305–314.

Willis, G. E., and J. W. Deardorff, 1974: A laboratory model of the unstable planetary boundary layer. *J. Atmos. Sci.*, **31**, 1297–1307.

Wilson, D. K., 2001: An alternative function for the wind and temperature gradients in unstable surface layers. *Bound.-Layer Meteor.*, **99**, 151–158.

Wyngaard, J. C., 1984: The mean wind structure of the baroclinic, convective boundary layer. *Proceedings of the First Sino-American Workshop on Mountain Meteorology*, E. Reiter, Z. Baozhen, and Q. Yongfu, Eds., American Meteorological Society, pp. 371–396.

Wyngaard, J. C., 1985: Structure of the planetary boundary layer and implications for its modeling. *J. Climate Appl. Meteor.*, **24**, 1131–1142.
Wyngaard, J. C., 1987: A physical mechanism for the asymmetry in top-down and bottom-up diffusion. *J. Atmos. Sci.*, **44**, 1083–1087.
Wyngaard, J. C., 1988: Structure of the PBL. *Lectures on Air-Pollution Modeling*, A. Venkatram and J. C. Wyngaard, Eds., American Meteorological Society, pp. 9–61.

12
The stable boundary layer

12.1 Introduction

The stable boundary layer (SBL) is as different from the convective boundary layer (CBL) as night is from day. The SBL is typically much thinner and much less diffusive; chimney plumes in the stable air just after sunrise can travel intact for long distances, quite unlike those in midday. The decrease in surface wind speed around sunset on a clear day, which is prominent in wind climatologies (Arya, 2001), is also evident to casual observers. Figure 12.1 shows the root cause: in a given mean horizontal pressure gradient the surface mean wind speed is lower in stable stratification than in neutral or unstable stratification. Thus the lower nocturnal wind speeds tend to persist until the initiation of a CBL at sunrise.

We gave examples of SBLs in Chapter 9. The two sketched in Figure 9.7 are made stably stratified by flow over a cooler surface and by entrainment of warmer air aloft. Perhaps the most common example is the nocturnal SBL over land in fair weather. Another is the "long-lived" SBL at the South Pole; it is caused by a combination of a temperature inversion over a cooled, sloped surface and the typically downslope orientation of the mean pressure gradient (Neff *et al.*, 2008).

As we'll see, turbulence in the SBL tends to be in a delicate dynamical balance, and so SBL structure tends to be more difficult to study and to parameterize than that in the CBL. But through analytical, computational, and observational studies it is gradually yielding to understanding.

The material in this chapter is arranged as follows. We'll begin with an overview of the energetics of stably stratified turbulence and the notion of a degree of stability past which turbulence cannot exist. We'll then discuss some applications of second-order closure and LES to the SBL. Section 12.2 surveys the familiar late-afternoon transition of the ABL over land – the response of the flow near the surface, the inertial oscillation aloft in what was the previous afternoon's CBL, and treating the effects of sloping terrain and gravity waves. In Section 12.3 we'll discuss some

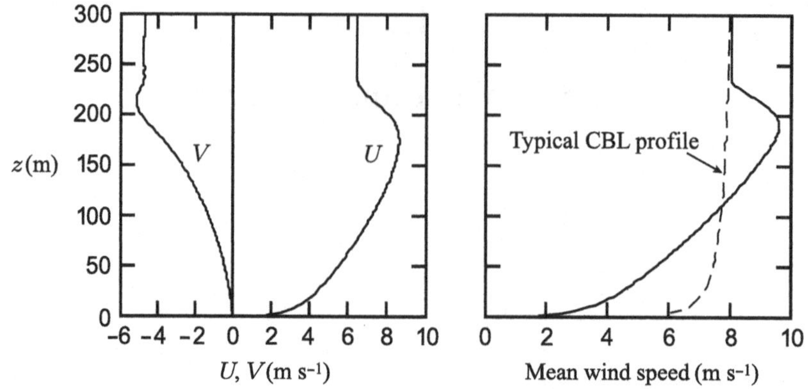

Figure 12.1 Left: Profiles of mean wind components in an SBL, in coordinates aligned with the surface wind, as calculated through LES. Right: The profile of mean wind speed, and for comparison a CBL mean speed profile calculated from Wilson's formula, Eq. (10.18), with $L = -15$ m, $z_0 = 0.01$ m, $u_* = 0.46$ m s^{-1}, and the same mean horizontal pressure gradient. The contrast in surface-layer mean wind speeds between the CBL and SBL cases is striking. LES data courtesy Peter Sullivan, NCAR.

of the important features of the quasi-steady SBL – its limiting structure near a cooled surface and below a capping inversion, its equilibrium height formula, and its limiting surface heat flux; the generalization of its height formula to a range of neutral and stable ABLs; and LES-based insights into SBL structure. In Section 12.4 we'll cover some of the important dynamical and structural features of the evolving SBL, and in Section 12.5 we'll discuss models of equilibrium SBL depth.

12.1.1 Energetics of stably stratified turbulence

The interfacial layer that caps the CBL (Chapter 11) is stably stratified, but its TKE budget has two gain terms: shear production and turbulent transport from the CBL below. The first involves the Reynolds stress, which can be extinguished by sufficiently strong stable stratification. But the roots of the transport term lie outside the interfacial layer, so we would not expect it to be sensitive to this stable stratification. By contrast the SBL has TKE gain only through shear production, and so its turbulence, unlike that in the interfacial layer, tends to be quite sensitive to stable stratification. Figures 12.2 and 12.3 show the rapid decay of turbulence parameters after the onset of surface cooling and stable stratification.

The dynamical balance of shear-driven turbulence is not considered delicate. Why does stable stratification change this?

The answer lies in the different natures of the TKE loss to viscous and buoyancy forces. Viscous forces directly impact only the smallest eddies; in equilibrium their

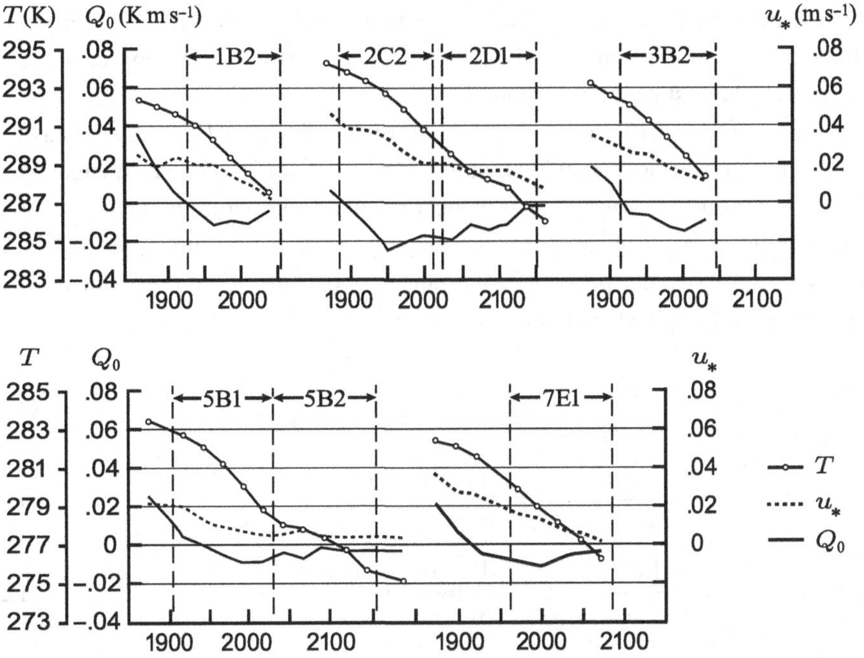

Figure 12.2 Time histories of 1-m temperature, friction velocity u_*, and surface temperature flux Q_0 measured in the 1973 Minnesota experiment. From Caughey *et al.* (1979).

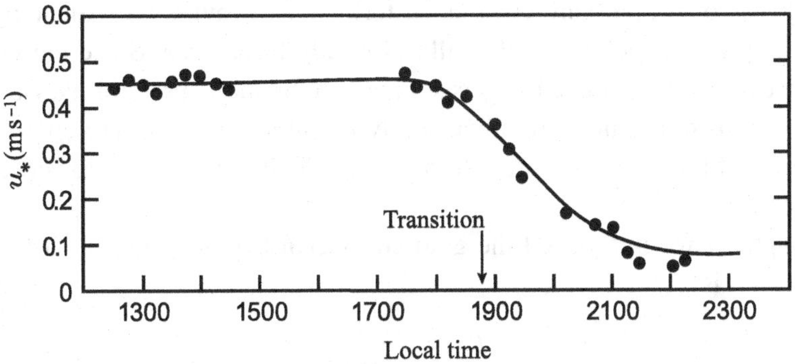

Figure 12.3 The late-afternoon decay of the friction velocity u_* in Run 2 of the 1973 Minnesota experiment. The solid line is the prediction of a second-order-closure model. From Wyngaard (1975).

mean rate of energy loss through viscous dissipation is balanced by their mean rate of gain through the Kolmogorov energy cascade (Chapter 6). If the rate of viscous dissipation were larger than the cascade rate, the kinetic energy of the dissipative eddies would decrease, decreasing the dissipation rate – and vice versa.

But buoyancy directly impacts only the energy-containing eddies significantly; excessive stability extinguishes them, which shuts down the energy cascade and extinguishes the rest of the turbulence.

We'll illustrate with the component kinetic energy budgets, Eq. (8.69), for a quasi-steady, horizontally homogeneous, stably stratified surface layer. Wyngaard and Coté (1971) found that turbulent transport is small compared to shear production and dissipation. The Kansas observations suggest that pressure transport is also small, so we rewrite the pressure covariances and express the component energy budgets to a first approximation as

$$\frac{1}{2}\frac{\partial \overline{u^2}}{\partial t} = 0 = -\overline{uw}\frac{\partial U}{\partial z} + \frac{1}{\rho_0}\overline{p\frac{\partial u}{\partial x}} - \frac{\epsilon}{3}, \tag{12.1}$$

$$\frac{1}{2}\frac{\partial \overline{v^2}}{\partial t} = 0 = \frac{1}{\rho_0}\overline{p\frac{\partial v}{\partial y}} - \frac{\epsilon}{3}, \tag{12.2}$$

$$\frac{1}{2}\frac{\partial \overline{w^2}}{\partial t} = 0 = \frac{1}{\rho_0}\overline{p\frac{\partial w}{\partial z}} + \frac{g}{\theta_0}\overline{\theta w} - \frac{\epsilon}{3}. \tag{12.3}$$

The buoyancy term in Eq. (12.3), now being negative, represents the mean rate of loss of $\overline{w^2}/2$ by working against buoyancy forces (Chapter 9). The spectral dynamics (Chapter 16) indicates that this rate of loss occurs in the energy-containing range of the w field. The viscous dissipation term in Eq. (12.3) is another mean rate of loss, so the pressure-covariance term must provide the balancing mean rate of input of kinetic energy that allows steady turbulence. Since the pressure-covariance terms in the three budgets sum to zero by incompressibility, we see that $\overline{w^2}$ is maintained at a nonzero value solely by *intercomponent energy transfer* – i.e., transfer of kinetic energy from horizontal velocity fluctuations to vertical ones (Chapter 5).

In Chapter 9 we introduced the gradient Richardson number; in surface-layer coordinates it is

$$Ri = \frac{\frac{g}{\Theta_0}\frac{\partial \Theta}{\partial z}}{\left(\frac{\partial U}{\partial z}\right)^2}. \tag{12.4}$$

The surface-layer observations of Kondo *et al.* (1978) show that temperature and velocity signals tend to be fully turbulent for $Ri < 0.2$, but for more stable conditions intermittent turbulence appears, especially in the temperature field. As Ri increases further the fluctuations become more intermittent, weaker, and finally disappear. This intermittency makes the measurement, understanding, and parameterization of the nocturnal SBL particularly challenging.

Nieuwstadt (2005) used direct numerical simulation of channel flow to study the maintenance of turbulence in stable stratification. The flow was horizontally homogeneous and driven by a pressure gradient; it had a no-slip boundary condition on the lower wall and a free-slip (stress free) wall at the top. He computed neutral turbulent flow as an initial condition and then continued the computations with a constant, negative heat flux imposed at the bottom wall. He found that the flow continued to be turbulent only for $h/L < 0.5$.†

Nieuwstadt also found that weak stable stratification increased the rate of viscous dissipation, which in equilibrium required an increase in the rate of shear production. The interpretation is that the stable stratification, whose damping effects are strongest for the largest eddies, Eq. (10.38), decreases ℓ and hence increases $\epsilon \sim u^3/\ell$. The resulting increase in the rate of shear production was larger than the rate of loss to buoyant destruction.

12.1.2 Second-order-closure modeling of the SBL

Numerical modeling has been applied to the ABL since the 1960s, and "second-order closure" came into use in the 1970s. As in the engineering community (Chapter 5), as experience was gained with the models the hopes for their universality dimmed. Fine-mesh LES "data bases" facilitated the testing of models in atmospheric applications, and in this way Moeng and Wyngaard (1989) concluded that typical shear-flow closures for dissipation rate and turbulent transport performed poorly in the CBL.

Second-order models seemed to perform better in the SBL, where the turbulent transport terms in second-moment budgets (Chapter 5) are much less important than in the CBL and simple closures for pressure covariances appeared to suffice. Delage (1974) used an eddy-diffusivity model with a rate equation for TKE to explore SBL structure with results that seemed quite physical. Many such studies followed, including the second-order closure calculation of Wyngaard (1975) that was "tuned" in part with the Kansas observations. It reproduced well the evolving surface-layer properties measured in the 1973 Minnesota evening runs.

This model system spawned the much simpler SBL model of Brost and Wyngaard (1978). By neglecting the turbulent-transport, time-change, and Coriolis terms in the second-moment equations and replacing a highly parameterized ϵ rate equation with $\epsilon \simeq u^3/\ell$ and a model for ℓ, they converted the set of eleven partial differential equations to eight algebraic equations for second moments (Fitzjarrald, 1979). The solutions of the BW set were quite similar to those of the original model system.

Nieuwstadt (1984) then showed that the BW set of algebraic equations, when nondimensionalized with the local kinematic stress magnitude $\tau(z) = [(\overline{uw})^2 + (\overline{vw})^2]^{1/2}$ and the local vertical temperature flux $\overline{w\theta}(z)$, has solutions that depend only on the local Monin–Obukhov length

† 1.25 in Nieuwstadt's convention, in which L is defined without the von Karman constant.

$$\Lambda(z) = -\frac{\tau^{3/2}}{k(g/\theta_0)\overline{\theta w}}. \tag{12.5}$$

Nieuwstadt named this property *local scaling.* It implies that a turbulence quantity in this set, made dimensionless with the local fluxes of temperature and momentum, is a universal function of $z/\Lambda(z)$. Nieuwstadt viewed local scaling as "an extension of Monin–Obukhov similarity to the whole stable boundary layer."

Nieuwstadt also pointed out that for $z \gg \Lambda$ the set should display the "z-less scaling" observed in the very stable surface layer (Chapter 10). Physically, this means that under very stable conditions the length scale of the turbulence is determined by the local length scale Λ rather than by z.

Nieuwstadt presented data taken under stable conditions along a meteorological mast at Cabauw, the Netherlands. The results in "local-similarity" coordinates – variables measured at a given height z, made dimensionless with the fluxes at that height, and plotted against z/Λ – supported the local-scaling hypothesis and the concept of a z-less limit.

The Monin–Obukhov length L is independent of height, so as we saw in Chapter 10 we can use $\phi_m(z/L)$, τ_0, and Q_0 to calculate $U(z)$, for example. But in local scaling $\Lambda = \Lambda(z)$, so to obtain vertical structure one needs $\tau(z)$ and $\overline{w\theta}(z)$. With the closure assumption that both the flux and gradient Richardson numbers are constant at 0.2, Nieuwstadt (1984) found analytical solutions for these flux profiles in stationary conditions:

$$\overline{w\theta} = Q_0(1 - z/h), \qquad \tau = u_*^2(1 - z/h)^{3/2}. \tag{12.6}$$

12.1.3 Large-eddy simulation of the SBL

Among the first applications of large-eddy simulation (LES) to the SBL was that of Mason and Derbyshire (1990). They used a grid of $40 \times 32 \times 62$ points in a domain 1000 m deep over uniform, flat terrain; the horizontal resolution was about 12 m. Finding great difficulty in starting runs from stable conditions (the turbulence tended to decay), they began with a neutral turbulent boundary layer and then applied cooling to the lower surface. Their three stably stratified cases ran for about two hours after the onset of cooling. Case B had a rather small surface temperature flux (-0.01 m s^{-1} K), as in some of the Minnesota runs (Figure 12.2); Case C had a flux of three times that; and D had a constant cooling rate.

Their results did show the rapid decay of the friction velocity, as in Figure 12.3, and the establishment of a thin, quasi-equilibrium, stably stratified boundary layer within two hours. Case B had an equilibrium boundary layer depth of about 200 m; the flux and gradient Richardson numbers increased monotonically to about 0.2 at the boundary-layer top. In their conclusions they pointed to the broad agreement with the models of Nieuwstadt (1984), as adapted by Derbyshire (1990), and of Brost and Wyngaard (1978).

The subgrid model used by Mason and Derbyshire is the standard one (Chapter 6) in which the deviatoric subgrid stress is taken to be proportional to the resolved strain rate. This makes the rate of transfer of kinetic energy from resolved to subgrid motions, which is the subgrid stress contracted with this resolved strain rate (Chapter 6), positive definite. But this rate of transfer is known to be instantaneously of either sign in turbulent flows – i.e., to or from the resolved motions. Subsequently Kosovic and Curry (2000), contending that the standard subgrid model with one-way energy transfer may not be appropriate for SBLs because it can cause spurious laminarization of the flow, used LES with such two-way energy transfer to study the SBL. Their results also agreed well with observations and with Nieuwstadt's analytical model.

12.2 The late-afternoon ABL transition over land

In clear weather, the net shortwave radiative flux at the surface (Figure 9.4) approaches zero as the sun nears the horizon in late afternoon. The rate of energy loss through longwave emission becomes the dominant term in the surface energy budget, changing its net effect from surface warming to surface cooling. This changes the sign of the vertical gradient of temperature at the surface, which changes the direction of the surface heat flux (Problem 12.3). Above the diffusive sublayer on the surface (Chapter 1) this now-downward heat flux continues to be carried by the turbulence. Its divergence, supplemented by the divergence of radiative heat flux (Garratt and Brost, 1981), causes the near-surface air to cool.

As time proceeds, turbulence diffuses this cooling upward in the boundary layer. We'll see that the effects of the resulting stable stratification increase with increasing distance from the surface. Meanwhile the CBL turbulence well aloft decays as its rate of input through buoyant production decreases to zero. This decay attenuates the turbulent stress divergence term in the mean horizontal momentum budget, which, as we'll see, triggers an inertial oscillation in the horizontal wind field aloft.

12.2.1 The near-surface response

We'll define *transition* as the time when the surface temperature flux Q_0 (Figure 9.5) changes sign. After transition the large, CBL-spanning eddies begin to decay, and the emerging stable stratification also contributes to the decrease in surface stress and near-surface mean wind speed (Problems 12.10, 12.11). Figures 12.2

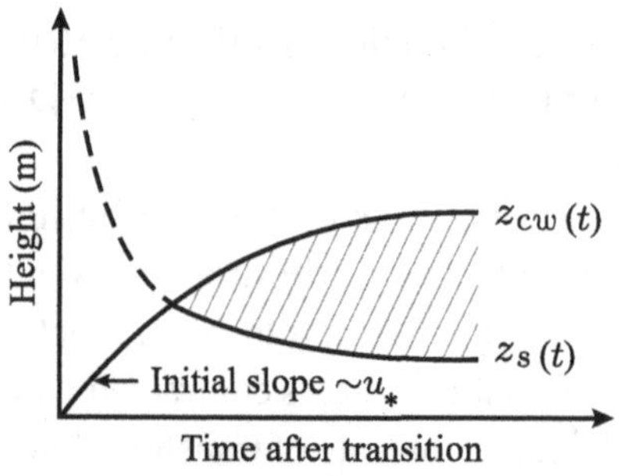

Figure 12.4 A schematic of the upward diffusion of stable stratification in the young nocturnal SBL. The curve labeled $z_{cw}(t)$ is the mean upper edge of the diffusing cooling wave; the solid portion of the curve labeled $z_s(t)$ is the mean height above which the effects of this cooling are dynamically significant. The shaded region between the curves is the region influenced by the stable stratification.

and 12.3 show time histories of near-surface temperature, friction velocity u_*, and Q_0 near transition in the Minnesota experiment.

Observational data on the late-afternoon CBL decay process are scarce, but LES studies by Nieuwstadt and Brost (1986) gave some insight. They focused on a somewhat different problem, the response of a quasi-steady CBL to the abrupt zeroing of its surface heat flux. The decay of TKE began about one z_i/w_* time after transition, the time scale of the decay being of order $\ell/u \sim z_i/w_*$. The decay of temperature variance began sooner – perhaps because in a CBL with a negligible or even slightly positive mean potential temperature gradient it has only one source, turbulent transport from below.

A plausible model of the mean evolution of stable stratification in the nocturnal SBL is the following. After transition $z_{cw}(t)$, the mean height of the top of the surface-based cooling layer, Figure 12.4, moves upward through turbulent diffusion. We'll take the initial surface-layer stratification as near-neutral, the initial vertical velocity v_{cw} of this mean cooling wave as $\sim u_*$, and the initial time trajectory of z_{cw} as

$$z_{cw} = v_{cw}t \sim u_* t. \tag{12.7}$$

At some height $z_s < z_{cw}$ the rate of buoyant destruction of ambient TKE caused by the upward-diffusing stable stratification becomes dynamically important. We'll define z_s as the height where the rate of buoyant destruction of TKE is a given fraction $a < 1$ of the rate of shear production:

$$\frac{g}{\theta_0}\overline{w\theta}(z_s) = a\,\overline{uw}(z_s)\frac{\partial U}{\partial z}(z_s). \tag{12.8}$$

We require that z_s not exceed the height z_{cw} to which the cooling wave has diffused:

$$z_s(t) \le z_{cw}(t). \tag{12.9}$$

We'll use neutral surface-layer scaling to approximate Eq. (12.8) as

$$\frac{g}{\theta_0} Q_0 = -a\frac{u_*^3}{kz_s}, \quad z_s(t) = a\frac{-u_*^3}{k\frac{g}{\theta_0}Q_0} \sim aL(t), \tag{12.10}$$

with $L(t)$ the evolving M-O length. With the constraint (12.9) this gives

$$aL(t) \le z_s(t) \le z_{cw}(t). \tag{12.11}$$

The shaded region $z_s \le z \le z_{cw}$ sketched in Figure 12.4 is influenced by the growing stable stratification induced by surface cooling. Within an hour or so there can be a shallow (no more than a few hundreds of meters deep) nocturnal SBL in place.

12.2.2 The inertial oscillation aloft

As we discussed in Chapter 11, in the quasi-steady, horizontally homogeneous mixed layer the mean horizontal momentum equations are

$$\begin{aligned} \frac{\partial U}{\partial t} &= -\frac{\partial \overline{uw}}{\partial z} + f(V - V_g) \simeq 0, \\ \frac{\partial V}{\partial t} &= -\frac{\partial \overline{vw}}{\partial z} + f(U_g - U) \simeq 0. \end{aligned} \tag{12.12}$$

In the late-afternoon transition the turbulence aloft, which is supported largely by buoyant production, begins to decay, causing the stress-divergence terms in these equations to decay. Within a few large-eddy turnover times the mean momentum equations have lost important terms and so the mean wind components begin to evolve in time:

$$\frac{\partial U}{\partial t} = f(V - V_g), \qquad \frac{\partial V}{\partial t} = f(U_g - U). \tag{12.13}$$

Solution of the coupled equations (12.13) is facilitated by defining a complex mean horizontal velocity $W(z,t) = U(z,t) + iV(z,t)$.† If the corresponding complex geostrophic wind depends only on z, i.e., $W_g(z) = U_g(z) + iV_g(z)$, the equation for $\Delta W \equiv W - W_g$ is

$$\frac{\partial \Delta W}{\partial t} = -if\Delta W, \tag{12.14}$$

† This is the conventional notation. The reader should not confuse W with the mean vertical velocity, which vanishes here.

whose solution is (Problem 12.13)

$$\Delta W(t) = \Delta W(t_0)\, e^{-if(t-t_0)} = \Delta W(t_0)\,[\cos f(t-t_0) - i \sin f(t-t_0)], \tag{12.15}$$

with t_0 the time at which the stress-divergence terms in Eqs. 12.12 have decayed to zero.

Equation (12.15) can have important consequences for the clear-weather, night-time behavior of winds at heights that on the previous afternoon were in the mixed layer of the CBL. These follow from multiplying Eq. (12.15) by its complex conjugate (indicated by an asterisk):

$$\Delta W(t)\, \Delta W^*(t) = \left|\Delta W(t)\right|^2 = \Delta W(t_0)\, \Delta W^*(t_0) = \text{constant}. \tag{12.16}$$

For the physical components of velocity this implies (Problem 12.14)

$$\begin{aligned}(\Delta U)^2 + (\Delta V)^2 &\equiv [U(z,t) - U_g(z)]^2 + [V(z,t) - V_g(z)]^2 \\ &= [U(z,t_0) - U_g(z)]^2 + [V(z,t_0) - V_g(z)]^2 = \text{constant}.\end{aligned} \tag{12.17}$$

Equation (12.17) is a statement of kinetic energy conservation for a time-dependent, mean horizontal flow having only pressure-gradient and Coriolis forces – a flow with neither turbulent friction nor mean advection. When the flow has decayed to a nonturbulent state we can drop the modifier *mean* and interpret Eq. (12.17) as saying that the squared magnitude of the difference between the velocity and the geostrophic velocity is constant in time. The direction of the horizontal velocity changes with time, but its energy constraint (12.17) causes the tip of its vector to trace out a circular path in the U, V plane. As sketched in Figure 12.5, the center of the circle is (U_g, V_g); its squared radius is $R^2 = \left[U(t_0) - U_g\right]^2 + \left[V(t_0) - V_g\right]^2$, and its angular frequency around this circle is f. This phenomenon, discussed by Blackadar (1957), is one cause of the *nocturnal* or *low-level jet.*

We'll consider the cases where the initial state is a barotropic CBL (Figure 11.6) and a baroclinic CBL (Figure 11.8). In the barotropic case from Eqs. (11.6), Figure 11.5, and Figure 11.6 we estimate

$$\text{barotropic:}\quad \Delta U \ll \Delta V \simeq u_*^2/f z_i \simeq 5u_*, \quad (\Delta U)^2 + (\Delta V)^2 \simeq 25u_*^2. \tag{12.18}$$

In the baroclinic case, from Figure 11.8 we take

$$\text{baroclinic:}\quad \Delta U \simeq 5u_*, \qquad \Delta V \simeq 10u_*, \qquad (\Delta U)^2 + (\Delta V)^2 \simeq 125u_*^2. \tag{12.19}$$

These show that the intensity of this inertial oscillation can be strongly affected by baroclinity. In combination with the direction change due to the mean wind spiral

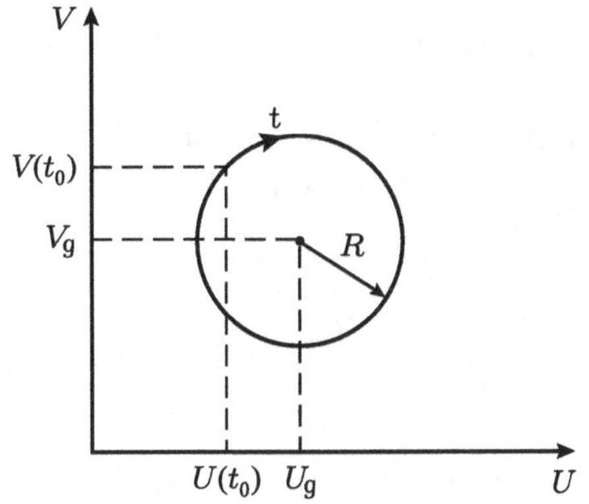

Figure 12.5 A schematic of the nocturnal oscillation of the horizontal wind vector that can exist when turbulence above the SBL has decayed. The evolving wind vector, with initial value $[U(t_0), V(t_0)]$, traces a clockwise (in the northern hemisphere) path at angular frequency f on the circle of squared radius $R^2 = [U(t_0) - U_g]^2 + [V(t_0) - V_g]^2$.

and the stability-depressed mean wind speeds in the surface layer, within a few hours this can create a substantial difference between the mean wind speeds at the surface and at a few hundred meters height in the nocturnal boundary layer.

As discussed by Stull (1988), low-level jets are quite common and originate from several types of forcing in addition to the one we discussed. Such jets can lead to strong differences in nocturnal dispersion of effluents from surface and elevated sources, and create both opportunities and hazards for wind-power generation.

Even though over land there is apt to be an SBL every night in clear weather, its structure and dynamics are less well understood than those of the convective case. We'll cover some of the reasons for this.

12.2.3 The effects of sloping terrain

The downslope buoyancy forces experienced by cooler, and therefore denser, air over sloping terrain can be quite important in the mean momentum balance in the SBL. We'll analyze flow over flat terrain having a downslope of small angle β to the horizontal by tilting our coordinates by β so the x and y axes are parallel to the surface. With this tilting operation the horizontally homogeneous momentum equations parallel to the surface (Caughey *et al.*, 1979) are

$$\begin{aligned} \frac{\partial U}{\partial t} + \frac{\partial \overline{uw}}{\partial z} &= f(V - V_g) - \frac{g}{\theta_0}\overline{\tilde{\theta}'}\beta_x, \\ \frac{\partial V}{\partial t} + \frac{\partial \overline{vw}}{\partial z} &= f(U_g - U) - \frac{g}{\theta_0}\overline{\tilde{\theta}'}\beta_y, \end{aligned} \tag{12.20}$$

with β_x and β_y the components of the downslope angle β and $\overline{\tilde{\theta}'}$ the deviation from θ_0.

If we define the mean wind direction as $\alpha = \tan^{-1}(V/U)$, then from the evolution equations (12.20) for U and V we can derive that for α (Problem 12.18):

$$\frac{d\alpha}{dt} = \frac{1}{S^2}\left(V\frac{\partial \overline{uw}}{\partial z} - U\frac{\partial \overline{vw}}{\partial z}\right) + f\left(\frac{G}{S}\cos\gamma - 1\right) + \frac{g}{\theta_0}\frac{\overline{\tilde{\theta}'}}{S^2}(V\beta_x - U\beta_y). \quad (12.21)$$

Here γ is the angle between the geostrophic and mean winds. In the Minnesota experiment the terrain-slope term in Eq. (12.21) had become important at $z = 4$ m by the mid-stages of each of the runs in Figure 12.2. In five of the seven runs its magnitude reached f, which in mid-latitudes corresponds to a rate of turning of the mean wind direction of 20 degrees per hour (Problem 12.19).

Given that few land surfaces are as level as the Minnesota site, we conclude that drainage forces can be very important in the nocturnal ABL; they can foster slope-driven local flows that evolve in time and interact in complicated ways.

12.2.4 An approach to treating gravity waves

The nocturnal ABL is also a rich medium for the growth and propagation of internal gravity waves, which interact with turbulence. Coulter (1990) has reported acoustic sounder measurements in which Kelvin–Helmholtz waves and instabilities above the nocturnal SBL induced a factor of two to four modulation in rms vertical velocity and other turbulence quantities within the boundary layer. Figure 12.6 shows an acoustic sounder record from a late-night SBL with its characteristic evidence of strong gravity waves.

Finnigan *et al.* (1984) have used the Reynolds and Hussein (1972) three-part decomposition of flow variables, mean + turbulence + wave, to isolate the wave-induced and turbulent motions in the nocturnal ABL and study their interactions. That decomposition generalizes the ensemble-mean plus fluctuation decomposition of velocity, pressure, and conserved scalar fields to

$$\tilde{u}_i = U_i + u_i^{\mathrm{w}} + u_i^{\mathrm{t}}, \quad \tilde{p} = P + p^{\mathrm{w}} + p^{\mathrm{t}}, \quad \tilde{c} = C + c^{\mathrm{w}} + c^{\mathrm{t}}. \quad (12.22)$$

As usual the tilde indicates the full variable and a capital denotes the ensemble-mean value, but now the superscripts w and t indicate the wave and turbulent parts of the fluctuation about the ensemble mean.

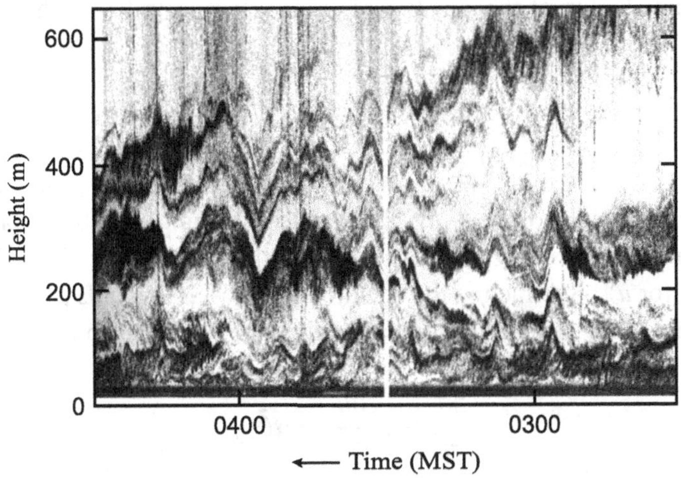

Figure 12.6 An acoustic sounder record from a nocturnal SBL with strong gravity waves. Photo courtesy W. D. Neff. From Wyngaard (1988).

The decomposition of a variable $\tilde{a} = A + a^{\mathrm{w}} + a^{\mathrm{t}}$ involves both phase and ensemble averaging. The phase averaging is defined through

$$\tilde{a}^{\mathrm{p}} = \lim_{N\to\infty}\left(\frac{1}{N}\sum_{n=1}^{N}\tilde{a}(t+n\tau)\right) = A + a^{\mathrm{w}}, \tag{12.23}$$

since the phase average of a^{t} is zero. Under stationary conditions a time average converges to the ensemble average, so in practice one determines the mean part A through

$$A = \lim_{T\to\infty}\frac{1}{T}\int_0^T \tilde{a}(t)\,dt. \tag{12.24}$$

Thus, taking the time average of $\tilde{a}$ gives its mean part A; subtracting the mean A from the phase average of $\tilde{a}$ gives the wave part a^{w}; and subtracting the phase average $A + a^{\mathrm{w}}$ from $\tilde{a}$ gives the turbulent part a^{t}.

The averaging rules (Finnigan *et al.*, 1984) are

$$(a^{\mathrm{t}})^{\mathrm{p}} = 0, \qquad \overline{a^{\mathrm{w}}} = 0, \qquad \overline{a^{\mathrm{t}}} = 0,$$

$$\overline{\bar{a}b} = \bar{a}\bar{b}, \qquad (a^{\mathrm{w}}b)^{\mathrm{p}} = a^{\mathrm{w}}b^{\mathrm{p}}, \qquad (\bar{a}b)^{\mathrm{p}} = \bar{a}b^{\mathrm{p}},$$

$$\overline{a}^{\mathrm{p}} = \bar{a}, \qquad \overline{a^{\mathrm{p}}} = \bar{a}, \qquad \overline{a^{\mathrm{w}}b^{\mathrm{t}}} = (\overline{a^{\mathrm{w}}b^{\mathrm{t}}})^{\mathrm{p}} = 0. \tag{12.25}$$

The decomposition of the equations is carried out as follows. Phase averaging the momentum equation,

$$\frac{\partial \tilde{u}_i}{\partial t} + \tilde{u}_j \frac{\partial \tilde{u}_i}{\partial x_j} = -\frac{1}{\rho_0}\frac{\partial \tilde{p}'}{\partial x_i} - 2\epsilon_{ijk}\Omega_j \tilde{u}_k + \frac{g}{\theta_0}\tilde{\theta}'\delta_{3i} + \nu \nabla^2 \tilde{u}_i, \tag{8.57}$$

produces an equation for $U_i + u_i^{\mathrm{w}}$. Ensemble averaging Eq. (8.57) produces an equation for U_i. Subtracting that equation from the equation for $U_i + u_i^{\mathrm{w}}$ produces an equation for u_i^{w}. Subtracting the equation for $U_i + u_i^{\mathrm{w}}$ from Eq. (8.57) gives an equation for u_i^{t}. The same sequence is applied to the potential temperature equation

$$\frac{\partial \tilde{\theta}}{\partial t} + \tilde{u}_i \frac{\partial \tilde{\theta}}{\partial x_i} = \alpha \tilde{\theta}_{,jj}. \tag{12.26}$$

The equations for the wave and turbulent components of the fluctuating velocity field are

$$u^{\mathrm{w}}_{i,t} + u^{\mathrm{w}}_{i,j}U_j + U_{i,j}u^{\mathrm{w}}_j + r^{\mathrm{w}}_{ij,j} + \left(u^{\mathrm{w}}_i u^{\mathrm{w}}_j - \overline{u^{\mathrm{w}}_i u^{\mathrm{w}}_j}\right)_{,j} = -\frac{1}{\rho_0}p^{\mathrm{w}}_{,i} + \frac{g}{\theta_0}\theta^{\mathrm{w}}\delta_{3i}, \tag{12.27}$$

$$u^{\mathrm{t}}_{i,t} + u^{\mathrm{t}}_{i,j}U_j + U_{i,j}u^{\mathrm{t}}_j - r^{\mathrm{w}}_{ij,j} + \left(u^{\mathrm{t}}_i u^{\mathrm{t}}_j - \overline{u^{\mathrm{t}}_i u^{\mathrm{t}}_j} + u^{\mathrm{t}}_i u^{\mathrm{w}}_j + u^{\mathrm{w}}_i u^{\mathrm{t}}_j\right)_{,j}$$
$$= -\frac{1}{\rho_0}p^{\mathrm{t}}_{,i} + \frac{g}{\theta_0}\theta^{\mathrm{t}}\delta_{3i} + \nu u^{\mathrm{t}}_{i,jj}. \tag{12.28}$$

Here

$$r^{\mathrm{w}}_{ij} = \left[(u^{\mathrm{t}}_i u^{\mathrm{t}}_j)^{\mathrm{p}} - \overline{u^{\mathrm{t}}_i u^{\mathrm{t}}_j}\right]. \tag{12.29}$$

We have neglected the molecular diffusion term in the equation for the wave component. Equations (12.27) and (12.28) sum to the usual equation for fluctuating velocity $u_i = \tilde{u}_i - U_i$, as required:

$$u_{i,t}+u_{i,j}U_j+U_{i,j}u_j+(u_i u_j)_{,j}-(\overline{u_i u_j})_{,j} = -\frac{1}{\rho_0}p_{,i}+\frac{g}{\theta_0}\theta\delta_{3i}+\nu u_{i,jj}. \tag{12.30}$$

The wave and turbulent parts of θ satisfy

$$\theta^{\mathrm{w}}_{,t} + \theta^{\mathrm{w}}_{,j}U_j + \Theta_{,j}u^{\mathrm{w}}_j + r^{\mathrm{w}}_{j\theta,j} + \left(u^{\mathrm{w}}_j\theta^{\mathrm{w}} - \overline{u^{\mathrm{w}}_j\theta^{\mathrm{w}}}\right)_{,j} = 0, \tag{12.31}$$

$$\theta^{\mathrm{t}}_{,t} + \theta^{\mathrm{t}}_{,j}U_j + \Theta_{,j}u^{\mathrm{t}}_j - r^{\mathrm{w}}_{j\theta,j} + \left(u^{\mathrm{t}}_j\theta^{\mathrm{t}} - \overline{u^{\mathrm{t}}_j\theta^{\mathrm{t}}} + u^{\mathrm{w}}_j\theta^{\mathrm{t}} + u^{\mathrm{t}}_j\theta^{\mathrm{w}}\right)_{,j} = \gamma\theta^{\mathrm{t}}_{,jj}. \tag{12.32}$$

Here we define

$$r^{\mathrm{w}}_{j\theta} = \left[(u^{\mathrm{t}}_j\theta^{\mathrm{t}})^{\mathrm{p}} - \overline{u^{\mathrm{t}}_j\theta^{\mathrm{t}}}\right]. \tag{12.33}$$

Their sum is the equation for the temperature fluctuation:

$$\theta_{,t} + \theta_{,j}U_j + \Theta_{,j}u_j + \left(u_j\theta - \overline{u_j\theta}\right)_{,j} = \gamma\theta_{,jj}. \tag{12.34}$$

By constructing kinetic energy budgets of the wave and turbulence components Finnigan *et al.* (1984) found that kinetic energy flows from wave motions to the turbulence and that the mechanics of this energy flow depends on the nonlinear character of the wave field. The TKE so produced is dissipated primarily by working against buoyancy rather than viscous forces, which they point out is consistent with the quasi two-dimensional character of the turbulence imposed by the strong stability of their flow. Finnigan (1988) later generalized the approach to unsteady waves and studied several wave–turbulence interaction events in detail.

12.3 The quasi-steady SBL

We have seen in Chapter 10 that M-O similarity has given remarkable order to turbulence statistics in the stable surface layer. Furthermore, acoustic sounding (Neff and Coulter, 1986; Neff *et al.*, 2008) has provided evidence of quasi-steady SBLs of both the nocturnal and the long-lived variety. These developments have stimulated analytical work on the SBL.

12.3.1 The stable surface layer limit

The behavior of the mean vertical gradients of wind and potential temperature in very stable conditions, Figure 10.3, indicates that

$$\phi_m = \frac{kz}{u_*}\frac{\partial U}{\partial z} \sim \frac{z}{L}, \quad \text{so that} \quad \frac{\partial U}{\partial z} \sim \frac{u_*}{L}, \tag{12.35}$$

and similarly for $\partial\Theta/\partial z$. This reflects the emergence of *z-less scaling* in very stable conditions.

We can also view this stable limit through the flux Richardson number, the ratio of the rates of buoyant destruction and shear production of TKE:

$$R_{\mathrm{f}} = \frac{\frac{g}{\theta_0}\overline{\theta w}}{\overline{uw}\frac{\partial U}{\partial z}}. \tag{12.36}$$

This can be expressed in M-O similarity terms as (Problem 12.17)

$$R_{\mathrm{f}} = \frac{z}{L}\frac{1}{\phi_m}. \tag{12.37}$$

Since $\phi_m = 1 + \beta z/L$ (Figure 10.3), where $\beta \simeq 5$, R_f is

$$R_f = \frac{z/L}{1 + \beta z/L}. \tag{12.38}$$

Thus in the very stable surface layer $R_f \to 1/\beta \simeq 0.20$.

12.3.2 Entrainment-induced stratification

As shown in Figure 9.7, an ABL can be made stably stratified by the entrainment of warmer air aloft. This is sometimes called the "inversion-capped neutral" ABL. Here we can revisit the "eddy Richardson number" Ri_e for an eddy of size r, introduced in Chapter 10:

$$Ri_e(r) = \frac{g\theta r}{\theta_0 [u(r)]^2} = \frac{g\theta r^{1/3}}{\theta_0 \epsilon^{2/3}}. \tag{10.38}$$

$Ri_e(r)$ is the ratio of fluctuating buoyancy and inertia forces on an energy-containing-range eddy of size r, with θ the amplitude of the fluctuations in potential temperature. We interpreted Eq. (10.38) as saying that the largest eddies feel the strongest buoyancy effects. If we write $\theta \sim \ell \partial\Theta/\partial z$, then

$$Ri_e(r) = \left(\frac{g}{\theta_0}\frac{\partial\Theta}{\partial z}\right)\frac{\ell r^{1/3}}{\epsilon^{2/3}} = N^2 \frac{\ell r^{1/3}}{\epsilon^{2/3}}, \tag{12.39}$$

with N the *Brunt–Vaisala frequency*. If we also write $\epsilon \sim \sigma_w^3/\ell$, where σ_w is the rms vertical velocity fluctuation, and take $r = \ell$ then Eq. (12.39) is

$$Ri_e(\ell) = \frac{N^2\ell^2}{\sigma_w^2}. \tag{12.40}$$

We might expect $Ri_e(\ell)$, like the flux Richardson number, to approach a constant under very stable stratification so that the turbulence scale approaches the *buoyancy length scale* ℓ_b,

$$\ell_b \sim \frac{\sigma_w}{N} \sim \frac{u}{N}. \tag{12.41}$$

As we shall discuss, u/N has been used as the length scale of the energy-containing eddies in the stably stratified, equilibrium limit.

12.3.3 Equilibrium height of the nocturnal SBL

Derbyshire (1990) determined analytically some of the properties of the equilibrium nocturnal SBL implied by Nieuwstadt's (1984) local-scaling model. An underlying assumption here is that the flux Richardson number R_f is constant. This is perhaps a fair assumption in the global sense, since Eq. (12.38) shows that the region of appreciable variation in R_f is confined to $z \leq L$, say.

In a quasi-steady state the mean potential temperature gradient is independent of time,

$$\frac{\partial}{\partial t}\frac{\partial \Theta}{\partial z} = 0 = -\frac{\partial^2 \overline{\theta w}}{\partial z^2}, \tag{12.42}$$

so as indicated in (12.6) the $\overline{\theta w}$ profile is linear: $\overline{\theta w} = Q_0(1 - z/h)$, with h the SBL depth. The surface temperature flux Q_0 is negative, directed from the air to the surface. From the definition of the flux Richardson number R_f and the linear $\overline{\theta w}$ profile,

$$R_f = \frac{\frac{g}{\theta_0}\overline{w\theta}}{\overline{uw}\frac{\partial U}{\partial z} + \overline{vw}\frac{\partial V}{\partial z}} = \frac{\frac{g}{\theta_0}Q_0(1 - z/h)}{\overline{uw}\frac{\partial U}{\partial z} + \overline{vw}\frac{\partial V}{\partial z}}. \tag{12.43}$$

It follows that

$$\overline{uw}\frac{\partial U}{\partial z} + \overline{vw}\frac{\partial V}{\partial z} = \frac{g}{\theta_0}\frac{Q_0}{R_f}\left(1 - \frac{z}{h}\right). \tag{12.44}$$

In terms of the complex covariance $T = \overline{uw} + i\overline{vw}$ and the complex mean wind $W = U + iV$, if T and dW/dz are parallel Eq. (12.44) can be written

$$T^*\frac{dW}{dz} = \frac{g}{\theta_0}\frac{Q_0}{R_f}\left(1 - \frac{z}{h}\right), \tag{12.45}$$

where * denotes complex conjugate.

The quasi-steady mean horizontal momentum balance is, in complex notation,

$$\frac{dT}{dz} = -if\left(W - W_g\right), \tag{12.46}$$

where $W_g = U_g + iV_g$. Differentiating (12.46), assuming $W_g \neq W_g(z)$, and multiplying by T^* yields, using Eq. (12.45),

$$T^*\frac{d^2T}{dz^2} = -if\frac{g}{\theta_0}\frac{Q_0}{R_f}\left(1 - \frac{z}{h}\right). \tag{12.47}$$

A solution for the T profile is

$$T = -u_*^2\left(1 - \frac{z}{h}\right)^{\alpha}, \tag{12.48}$$

with $\alpha = \alpha^r + i\alpha^i$ a complex constant to be determined. Substituting Eq. (12.48) into (12.47) and equating the powers of $(1 - z/h)$ on each side of the resulting equation yields $\alpha^* + \alpha - 2 = 1$, which implies that $\alpha^r = 3/2$. Requiring that the left side of the equation be imaginary, as the right side is, gives $\alpha^i = \sqrt{3}/2\ \mathrm{sgn(f)}$, so that $\alpha = 3/2 + (\sqrt{3}/2)i\ \mathrm{sgn(f)}$.

With α determined Eq. (12.47) then yields an expression for boundary-layer depth h:

$$h^2 = -\sqrt{3}R_f \frac{u_*^4}{\frac{g}{\theta_0}Q_0|f|} = \sqrt{3}kR_f\left(\frac{u_*L}{|f|}\right). \tag{12.49}$$

For $R_f \simeq 0.20$–0.25 this yields

$$h \simeq 0.4\left(\frac{u_*L}{|f|}\right)^{1/2}, \quad \frac{h|f|}{u_*} \simeq 0.4\left(\frac{u_*}{|f|L}\right)^{-1/2}. \tag{12.50}$$

This was perhaps the first analytical derivation of the power law that was originally derived by Zilitinkevich (1972) on dimensional grounds.

Model calculations (Businger and Arya, 1974; Brost and Wyngaard, 1978) and numerical simulations (Zilitinkevich *et al.*, 2007) support Eq. (12.50), showing that if the SBL reaches a quasi-steady state its depth h follows the predicted $(u_*L/|f|)^{1/2}$ dependence closely. The implied value of h can be quite small; for example, if $u_* = 0.1$ m s^{-1}, $L = 100$ m, $f = 10^{-4}$ s^{-1}, and the proportionality factor = 0.4, then from Eq. (12.50) $h \sim 100$ m.

12.3.4 A constraint on the maintainence of turbulence in the SBL

Derbyshire (1990) used Eq. (12.48) for the T profile to write Eq. (12.45) as

$$\frac{dW}{dz} = -\frac{g}{\theta_0}\frac{Q_0}{u_*^2 R_f}(1 - z/h)^{1-\alpha^*}. \tag{12.51}$$

Integrating this from the surface to height z and evaluating the result at $z = h$ gives the geostrophic drag law (Problem 12.22)

$$\frac{G}{u_*} = \frac{1}{kR_f}\frac{h}{L}. \tag{12.52}$$

This drag law, with Eq. (12.49) and the identity $gQ_0/\theta_0 = -u_*^3/(kL)$, implies that the surface buoyancy flux is

$$\frac{g}{\theta_0}Q_0 = -\frac{R_f}{\sqrt{3}}G^2|f|. \tag{12.53}$$

In Derbyshire's analytical model the parameters on the right side are all constant, so this equation says the surface buoyancy flux is also constant. This clearly needs interpretation.

Derbyshire (1990) interprets Eq. (12.53) as giving the maximum surface buoyancy flux the steady SBL can have while remaining fully turbulent – i.e., without the onset of turbulence intermittency and extinction at $Ri \simeq 0.25$ (Subsection 12.1.1). We can support that interpretation by rewriting Eq. (12.53) as

$$R_{\mathrm{f}} = \frac{-\sqrt{3}\frac{g}{\theta_0} Q_0}{G^2 |f|} \simeq \text{constant}. \tag{12.54}$$

The numerator of the fraction is proportional to the SBL-averaged rate of buoyant destruction of TKE. Since $|f|G$ is the magnitude of the mean horizontal pressure gradient, the MKE balance (Chapter 5, Subsection 5.5.2) shows that the denominator is proportional to the rate of production of mean flow kinetic energy (MKE). If the MKE balance is in equilibrium, the denominator is also proportional to the SBL-averaged rate of shear production of TKE. Thus as Derbyshire suggests, Eq. (12.54) can be interpreted as an estimate of the maximum "global" flux Richardson number for the SBL – i.e., that beyond which it cannot support turbulence:

$$\begin{aligned} R_{\mathrm{f}}(\max) &= \text{constant} \\ &\sim \frac{\text{SBL-averaged rate of buoyant destruction of TKE}}{\text{SBL-averaged rate of shear production of TKE}}. \end{aligned} \tag{12.55}$$

Nieuwstadt (2005) provided some intriguing support for Derbyshire's notion of a maximum flux Richardson number, Eq. (12.55), through DNS of stably stratified turbulent channel flow. He applied cooling at the bottom and allowed a surface stress. At the channel top he used a "free slip" condition that caused the stress and buoyancy flux to vanish there. Thus the flow was qualitatively like an SBL without significant Coriolis effects. He found that the flow ceased to be turbulent when the Monin–Obukhov length L became less than twice the channel depth. This can be expressed as a constraint like Eq. (12.55) on the maximum global flux Richardson number (Problem 12.23).

The surface heat flux in the SBL over land is determined by the surface energy balance, which in clear weather is dominated by radiative effects that are independent of SBL dynamics. It appears that such SBLs are prone to such strong surface cooling that excessive values of the global flux Richardson number are generated and the turbulence is extinguished.

12.3.5 Insights from LES

Calculations of SBL structure with several different LES codes (Beare *et al.*, 2006) showed broad agreement at sufficiently high resolution, i.e., a minimum grid dimension of 3 m and less. The test case was an SBL previously simulated by Kosovic and

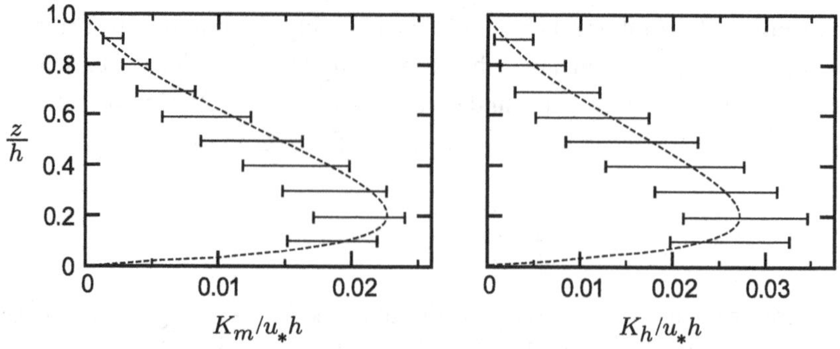

Figure 12.7 Calculated profiles of eddy diffusivities in the SBL. The horizontal lines indicate the range of values calculated by a suite of 11 different LES codes with 6 m vertical resolution; the dashed curves are calculated from the Brost–Wyngaard (1978) algebraic model. Adapted from Beare *et al.* (2006).

Curry (2000). The initial potential temperature profile was constant up to 100 m and increased at 0.01 K m^{-1} above that. The geostrophic wind speed was 8 m s^{-1}. A surface cooling rate of 0.25 K h^{-1} was applied for 9 h to give a quasi-steady, moderately stable SBL with $h/L \simeq 2$.

An interesting feature of the mean wind speed profile (Figure 12.1) is the "jet" – wind speed exceeding the geostrophic value by about 20% – at the SBL top (Problem 12.20). When coupled with the inertial oscillation following the afternoon transition (Subsection 12.2.2) this could give quite significant nocturnal wind maxima at the SBL top. Figure 12.7 shows the profiles of eddy diffusivities K_m and K_h; the Brost–Wyngaard (1978) algebraic model gives eddy-diffusivity profiles within the scatter of the LES results.

12.4 The evolving SBL

12.4.1 Structure

In their study of nocturnal SBL behavior over a grassland site, Mahrt *et al.* (1998) identify two limiting regimes that emerge after transition. The fully turbulent *weakly stable* regime is typically either an early, quasi-steady phase of the nocturnal SBL or one present under strong winds and/or cloud cover. The *very stable* regime occurs under clear skies; it has a larger surface cooling rate and lower mean wind speed. Its turbulence is weak and intermittent, even near the surface. Presumably its surface buoyancy flux can approach the limiting value discussed in Subsection 12.3.4.

Figure 12.8 shows vertical profiles of turbulence variances and covariances measured in the clear, early evening runs in the 1973 Minnesota experiments. The mid-run times averaged only one hour after transition, but the collapse to a similarity

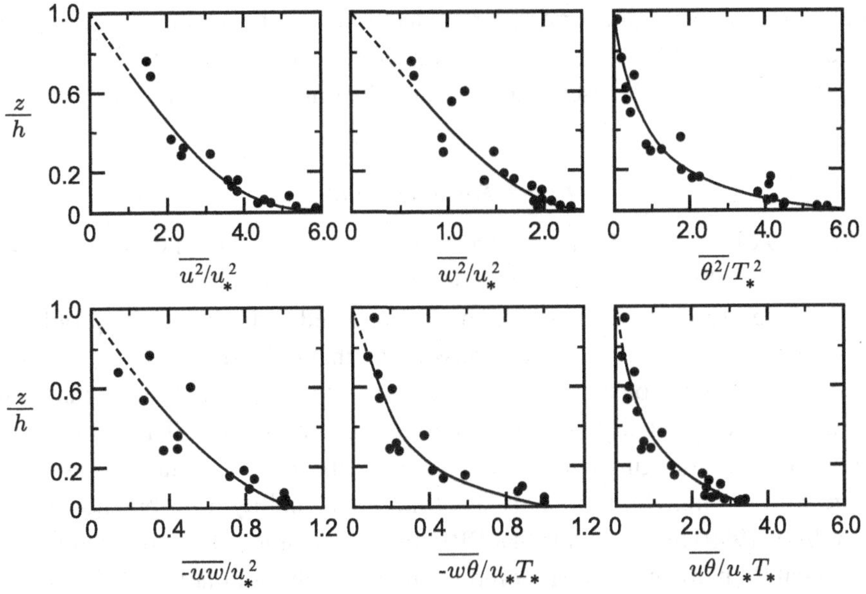

Figure 12.8 Vertical profiles of dimensionless variances and covariances measured in the early-evening Minnesota runs shown in Figure 12.2. The SBL depth h ranged from 30 m to 400 m over the seven runs. The curves are visual fits to the data. From Caughey *et al.* (1979).

structure is evident. The boundary-layer depth followed the equilibrium expression Eq. (12.50), but with a proportionality constant about double the equilibrium value (Caughey *et al.*, 1979).

This state was transient, however, and the runs were terminated about five hours after transition. In some there was an extreme shift in mean wind direction caused by the very slight (0.0014) terrain slope or by baroclinity; others entered the very stable regime and turbulence at 4 m height was extinguished.

In the very stable regime an internal intermittency mechanism discussed by Businger (1973) can modulate the turbulence. The extinguishing of the turbulence eliminates the turbulent friction term in the momentum equation, and the unbalanced horizontal pressure gradient then accelerates the flow until the Richardson number decreases enough to regenerate turbulence. Van de Wiel *et al.* (2002) show that surface vegetation can have a strong influence on this intermittency dynamics.

Nocturnal SBLs can differ from those we have discussed here. Mahrt and Vickers (2003) have described the contrasting vertical structure of nocturnal boundary layers observed in the CASES99 experiment in relatively flat grassland in the US Midwest. In addition to the "traditional" stable ABLs of varying depth, they observed what they call an "upside-down" variety in which the TKE increases with height and turbulent transport is a local rate of gain due to import from above. The principal source of TKE in this case is shear production associated with the nocturnal jet aloft. Banta *et al.* (2002) discuss this nocturnal jet in some detail.

12.4.2 Parameterization

A test of 19 different SBL parameterizations used in weather and climate models (Cuxart *et al.*, 2006), with high-resolution LES being the standard, showed a wide variation in performance. The scatter in the predicted eddy-diffusivity profiles, for example, was an order of magnitude larger than that in Figure 12.7.

SBL parameterizations for such models are typically one-dimensional, since SBL depth is very much less than horizontal grid mesh size. Their formulation is particularly challenging because the nature of the effects of stable stratification change with its strength, ultimately determining the ability of turbulence to exist. Derbyshire (1999) recommends that SBL parameterizations be expressed in terms of the local Richardson number, pointing out that such formulations can mimic the behavior of the very stable boundary layer.

12.5 Modeling the equilibrium height of neutral and stable ABLs

Zilitinkevich *et al.* (2007) suggest that the equilibrium height h_E of neutral and stable ABLs can be represented through the "Ekman formula" $h_E \sim (K/|f|)^{1/2}$, with f the Coriolis parameter and K a characteristic eddy viscosity $\sim u_T \ell_T$, the product of the velocity and length scales of the ABL turbulence. They parameterize three ABL states: the "truly neutral" ABL with zero surface heat flux and neutral stratification aloft; the "conventionally neutral" ABL (our "inversion-capped neutral" case, Figure 9.7) with zero surface heat flux and stable stratification of strength N aloft; and the "nocturnal stable" ABL with a negative surface temperature flux and neutral stratification aloft. The $u_T \ell_T$ values are chosen as:

- "Truly neutral" $(Q_0 = 0,\ N = 0)$ $u_T \ell_T \sim u_* h_E$.
- "Conventionally neutral" $(Q_0 = 0,\ N > 0)$ $u_T \ell_T \sim u_*(u_* N^{-1})$.
- "Nocturnal stable" $(Q_0 < 0,\ N = 0)$ $u_T \ell_T \sim u_* L$.

The turbulence length scale in the conventionally neutral case, $\ell_T \sim u_* N^{-1}$, was introduced by Kitaigorodskii and Joffre (1988); it is like our ℓ_b of Eq. (12.41). That in the nocturnal stable case, $\ell_T \sim L$, is our Eq. (12.35).

This analysis provides the ABL height expressions

$$h_E \sim \frac{u_*}{|f|}, \quad \text{truly neutral ABL;}$$

$$h_{\mathrm{E}} \sim \frac{u_*}{|fN|^{1/2}}, \quad \text{conventionally neutral ABL;} \tag{12.56}$$

$$h_{\mathrm{E}} \sim \left(\frac{u_* L}{|f|}\right)^{1/2}, \quad \text{nocturnal stable ABL.}$$

The first of Eqs. (12.56) is the Rossby–Montgomery (1935) expression for the equilibrium depth of a neutral ABL; the second is due to Pollard *et al.* (1973); the third is the Zilitinkevich (1972) form, Eq. (12.50).

Zilitinkevich *et al.* (2007) then suggest the ABL height h_{E} across this neutral and stable regime can be obtained from the interpolation formula

$$\frac{1}{h_{\mathrm{E}}^2} = \frac{f^2}{(C_{\mathrm{R}} u_*)^2} + \frac{N|f|}{(C_{\mathrm{CN}} u_*)^2} + \frac{|f|}{C_{\mathrm{NS}}^2 k u_* L}. \tag{12.57}$$

Here C_{R}, C_{CN}, and C_{NS} are constants associated with the truly neutral, conventionally neutral, and nocturnal stable states. From a variety of observations and LES results they suggest that $C_{\mathrm{R}} \simeq 0.6$, $C_{\mathrm{CN}} \simeq 1.36$, and $C_{\mathrm{NS}} \simeq 0.51$. The last implies the constant in the Zilitinkevich expression (12.50) is $\simeq 0.3$.

Questions on key concepts

12.1 Explain some of the prominent differences between the SBL and the CBL. What is their common, root cause?

12.2 Explain why and how the late-afternoon transition process depends on clear weather.

12.3 Sketch the dynamics of the low-level jet. Explain why it can be most intense in a clear-weather, baroclinic situation.

12.4 Discuss why geophysical turbulence is more sensitive to stable stratification than is engineering turbulence.

12.5 What are the two broad classes of SBL? How are they different?

12.6 Discuss how stable stratification of the ABL develops in late afternoon and early evening in clear weather.

12.7 Discuss the positive and negative impacts of the low-level jet on wind-power generation.

12.8 Explain why and how the dynamics of stably stratified turbulence is "delicate."

12.9 Discuss and interpret the sensitivity of the SBL to terrain slope.

12.10 Discuss the generalization of the traditional ensemble-mean-plus-fluctuation decomposition of turbulent fields to include a wave component.

12.11 Interpret the buoyancy length scale (12.41) physically.

12.12 Explain why the nocturnal SBL is unlikely to achieve a quasi-steady state.

12.13 Explain why second-order closure can be more reliable in the SBL than the CBL.

Problems

12.1 The surface energy balance has a term involving heat transfer at the top of the surface slab. Discuss its nature and explain why this is sometimes referred to as *convective* heat transfer.

12.2 A surface is cooler than the air above and is evaporating water so that the vertical flux of water vapor is positive. The virtual temperature flux is zero. What is the stability index z/L? Using M-O similarity, write the expression for the vertical gradient of potential temperature.

12.3 Explain why the late-afternoon change in direction of the surface heat flux occurs before sunset.

12.4 Explain and discuss why the nocturnal ABL is less likely to provide quasi-steady conditions than its daytime counterpart.

12.5 Under what conditions would you expect to find an inertial subrange in stably stratified turbulence?

12.6 Sketch the vertical profile of the length scale of the energetic eddies of the w-field in the inversion-capped neutral ABL of Figure 9.7.

12.7 The change in mean wind direction across the ABL tends to be least in the convective case and greatest in the stable case. Why? (Hint: Consider the streamwise mean equation of motion.)

12.8 Explain why the nocturnal jet can be particularly strong following the late-afternoon decay of a baroclinic convective ABL.

12.9 Discuss some of the differences between the daytime and nocturnal ABLs as potential sources of wind power.

12.10 Explain why the mean wind speed in the surface layer typically decreases in late afternoon in clear weather. (Hint: use the forms of the M-O functions for dimensionless mean wind shear, and assume the mean horizontal pressure gradient does not change.)

12.11 Which term in the horizontal equation of mean motion also contributes to the post-transition decrease in mean wind speed in the surface layer? Interpret the mechanism physically.

12.12 Show that the indicated solution to Eq. (12.48) for the complex stress profile is consistent with Eq. (12.6).

12.13 Derive Eq. (12.14) and the solution (12.15).

12.14 Derive Eqs. (12.16) and (12.17).

12.15 Explain the nature of the opportunities and hazards that the nocturnal jet poses for wind power.

12.16 Discuss the important similarities and differences of the SBL and the interfacial layer of the CBL.

12.17 Show that Eq. (12.37) for R_f follows from Eq. (12.36).

12.18 Derive Eq. (12.21).

12.19 Show that the Coriolis term in Eq. (12.21) is equivalent to a wind-turning rate of 20 degrees per hour in mid-latitudes.

12.20 Explain why the jet exists in the U-component wind profile of Figure 12.1. (Hint: in the coordinates used for that figure, integrate the lateral equation of mean motion across the boundary layer.)

12.21 Use the equations of motion to discuss the mechanism through which in clear weather the surface wind speed typically decreases as the boundary layer transitions in late afternoon. What can you infer about the mechanism?

12.22 Show that integrating Eq. (12.51) from the surface to height z and evaluating the result at $z = h$ gives the geostrophic drag law, Eq. (12.52).

12.23 Nieuwstadt (2005) found his stably stratified, horizontally homogeneous channel flow (Subsection 12.3.4) to be turbulent only for $L \geq 2h$, with L the M-O length and h the channel depth.

(a) Assuming that the temperature flux decreases linearly from Q_0 at the surface to zero at the channel top, calculate the depth-averaged rate of buoyant destruction of TKE per unit mass.

(b) Assume the MKE equation (5.51) reduces to a balance between the rate of production of MKE by flow down the mean pressure gradient and the rate of loss of MKE through shear production of turbulence. Recall from Example 3.2.1, Chapter 3, that the mean pressure gradient here does not depend on z. Use this balance to express the depth-averaged rate of shear production of TKE as the product of depth averaged mean velocity, U_{ave}, and the mean kinematic pressure gradient.

(c) Use the streamwise mean momentum balance, Eq. (3.10), the zero-stress condition at $z = h$, and a high-Re approximation to neglect its viscous term and express the mean kinematic pressure gradient in terms of u_*^2 and h.

(d) Use a friction factor f (as in the Moody chart, Figure 1.2) to express U_{ave} in terms of u_*.

(e) Write the global flux Richardson number, ratio of the depth-integrated rates of buoyant production and shear production of TKE, in terms of h/L. Relate your result to Nieuwstadt's finding that turbulence cannot be supported for $h/L \geq 0.5$.

References

Arya, S. P., 2001: *Introduction to Micrometeorology*. San Diego: Academic Press.

Banta, R. M., R. K. Newsom, J. K. Lundquist, *et al.*, 2002: Nocturnal low-level jet characteristics over Kansas during CASES-99. *Bound.-Layer Meteor.*, **105**, 221–252.

Beare, R. J., and 16 others, 2006: An intercomparison of large-eddy simulations of the stable boundary layer. *Bound.-Layer Meteor.*, **118**, 247–272.

Blackadar, A. K., 1957: Boundary layer wind maxima and their significance for the growth of nocturnal inversions. *Bull. Amer. Meteor. Soc.*, **38**, 283–290.

Brost, R. A., and J. C. Wyngaard, 1978: A model study of the stably stratified planetary boundary layer. *J. Atmos. Sci.*, **35**, 1427–1440.

Businger, J. A., 1973: Turbulent transfer in the atmospheric surface layer. *Workshop on Micrometeorology*, D. A. Haugen, Ed., American Meteorological Society, pp. 67–98.

Businger, J. A., and Arya, S. P. S., 1974: Height of the mixed layer in a stably stratified planetary boundary layer. *Adv. Geophys.*, **18A**, 73–92.

Caughey, S. J., J. C. Wyngaard, and J. C. Kaimal, 1979: Turbulence in the evolving nocturnal boundary layer. *J. Atmos. Sci.*, **36**, 1041–1052.

Coulter, R. L., 1990: A case study of turbulence in the stable nocturnal boundary layer. *Bound.-Layer Meteor.*, **52**, 75–91.

Cuxart, J., and 23 others, 2006: Single-column model intercomparison for a stably stratified atmospheric boundary layer. *Bound.-Layer Meteor.*, **118**, 273–303.

Delage, Y., 1974: A numerical study of the nocturnal atmospheric boundary layer. *Quart. J. Roy. Meteor. Soc.*, **100**, 351–364.

Derbyshire, S. H., 1990: Nieuwstadt's stable boundary layer revisited. *Quart. J. R. Meteor. Soc.*, **116**, 127–158.

Derbyshire, S. H., 1999: Stable boundary-layer modelling: established approaches and beyond. *Bound.-Layer Meteor.*, **90**, 423–446.

Finnigan, J. J., 1988: Kinetic energy transfer between internal gravity waves and turbulence. *J. Atmos. Sci.*, **45**, 486–505.

Finnigan, J. J., F. Einaudi, and D. Fua, 1984: The interaction between an internal gravity wave and turbulence in the stably stratified nocturnal boundary layer. *J. Atmos. Sci.*, **41**, 2409–2436.

Fitzjarrald, D. E., 1979: On using a simplified turbulence model to calculate eddy diffusivities. *J. Atmos. Sci.*, **36**, 1817–1820.

Garratt, J. R., and R. A. Brost, 1981: Radiative cooling effects within and above the nocturnal boundary layer. *J. Atmos. Sci.*, **38**, 2730–2746.

Kitaigorodskii, S. A., and Joffre, S. M., 1988: In search of simple scaling for the heights of the stratified atmospheric boundary layer. *Tellus*, **40A**, 419–433.

Kondo, J., O. Kanechika, and N. Yasuda, 1978: Heat and momentum transfers under strong stability in the atmospheric surface layer. *J. Atmos. Sci.*, **35**, 1012–1021.

Kosovich, B., and J. A. Curry, 2000: A large-eddy simulation study of a quasi-steady, stably stratified atmospheric boundary layer. *J. Atmos. Sci.*, **57**, 1052–1068.

Mahrt, L., and D. Vickers, 2003: Contrasting vertical structures of nocturnal boundary layers. *Bound.-Layer Meteor.*, **105**, 351–363.

Mahrt, L., J. Sun, W. Blumen, T. Delaney, and S. Oncley, 1998: Nocturnal boundary-layer regimes. *Bound.-Layer Meteor.*, **88**, 255–278.

Mason, P. J., and S. H. Derbyshire, 1990: Large-eddy simulation of the stably-stratified atmospheric boundary layer. *Bound.-Layer Meteor.*, **53**, 117–162.

Moeng, C.-H., and J. C. Wyngaard, 1989: Evaluation of turbulent transport and dissipation closures in second-order modeling. *J. Atmos. Sci.*, **46**, 2311–2330.

Neff, W. D., and R. L. Coulter, 1986: Acoustic remote sensing. In *Probing the Atmospheric Boundary Layer*, D. H. Lenschow, Ed., Boston: American Meteorological Society.
Neff, W., D. Helmig, A. Grachev, and D. Davis, 2008: A study of boundary layer behavior associated with high NO concentrations at the South Pole using a minisodar, tethered balloon, and sonic anemometer. *Atmos. Env.*, **42**, 2762–2779.
Nieuwstadt, F. T. M., 1984: The turbulent structure of the stable, nocturnal boundary layer. *J. Atmos. Sci.*, **41**, 2202–2216.
Nieuwstadt, F. T. M., 2005: Direct numerical simulation of stable channel flow at large stability. *Bound.-Layer Meteorol.*, **116**, 277–299.
Nieuwstadt, F. T. M., and R. A. Brost, 1986: The decay of convective turbulence. *J. Atmos. Sci.*, **43**, 532–546.
Pollard, R. T., P. B. Rhines, and R. Thompson, 1973: The deepening of the wind-mixed layer. *Geophys. Fluid Dyn.*, **3**, 381–404.
Reynolds, W. C., and A. K. M. F. Hussein, 1972: The mechanics of an organized wave in turbulent shear flow. Part 3. Theoretical models and comparisons with experiments. *J. Fluid Mech.*, **54**, 263–288.
Rossby, C. G., and R. B. Montgomery, 1935: The layer of frictional influence in wind and ocean currents. *Pap Phys. Oceanogr. Meteorol. (MIT and Woods Hole Oceanogr. Inst.)* **3**, 1–101.
Stull, Roland B., 1988: *An Introduction to Boundary Layer Meteorology.* Kluwer.
Van de Wiel, B. J. H., R. J. Ronda, A. F. Moene, H. A. R. De Bruin, and A. A. M. Holtslag, 2002: Intermittent turbulence and oscillations in the stable boundary layer over land. Part 1: A bulk model. *J. Atmos. Sci.*, **59**, 942–958.
Wyngaard, J. C., 1975: Modeling the planetary boundary layer – extension to the stable case. *Bound.-Layer Meteor.*, **9**, 441–460.
Wyngaard, J. C., 1988: Structure of the PBL. *Lectures on Air-Pollution Modeling*, A. Venkatram and J. C. Wyngaard, Eds., American Meteorological Society, pp. 9–61.
Wyngaard, J. C., and O. R. Coté, 1971: The budgets of turbulent kinetic energy and temperature variance in the atmospheric surface layer. *J. Atmos. Sci.*, **28**, 190–201.
Zilitinkevich, S. S., 1972: On the determination of the height of the Ekman boundary layer. *Bound.-Layer Meteor.*, **3**, 141–145.
Zilitinkevich, S. S., Esau, I., and Bakalov, A., 2007: Further comments on the equilibrium height of neutral and stable planetary boundary layers. *Quart. J. R. Meteor. Soc.*, **133**, 265–271.

Part III

Statistical representation of turbulence

13

Probability densities and distributions

13.1 Introduction

Since Osborne Reynolds' time it has been accepted that turbulence is a stochastic phenomenon that must be analyzed statistically, and averaging remains central to those analyses. Reynolds used a volume average, but the properties he attributed to it are those of the ensemble average; space averages have somewhat different properties (Chapter 3). The ensemble average is the natural choice for statistical analyses of turbulence. Furthermore, the concept of an ensemble of realizations of a flow has new roots today in the sensitive dependence of turbulence on its initial conditions.

In Part I we introduced these concepts, writing the velocity field in realization α of an experiment, a function of spatial position $\mathbf{x}$ and time t, as $\tilde{u}_i(\mathbf{x}, t; \alpha)$. We defined its ensemble average as

$$\overline{\tilde{u}_i}(\mathbf{x}, t) = \lim_{N \to \infty} \frac{\tilde{u}_i(\mathbf{x}, t; \alpha_1) + \cdots + \tilde{u}_i(\mathbf{x}, t; \alpha_N)}{N}. \tag{13.1}$$

In the *Reynolds convention* we write the variable as the sum of its ensemble average and a deviation from it,

$$\tilde{u}_i(\mathbf{x}, t; \alpha) = \overline{\tilde{u}_i}(\mathbf{x}, t) + u_i(\mathbf{x}, t; \alpha), \tag{13.2}$$

the average of the deviation (or fluctuation, as we have come to call it) being zero. Here we have indicated the dependence on the realization, but in practice we often suppress this.

In Parts I and II we used the Reynolds convention in the equations of motion to examine the structure and dynamics of turbulent flows. Here we begin our more detailed discussions of the statistical representation of turbulence.†

† The material in the following section is adapted from Lumley and Panofsky (1964).

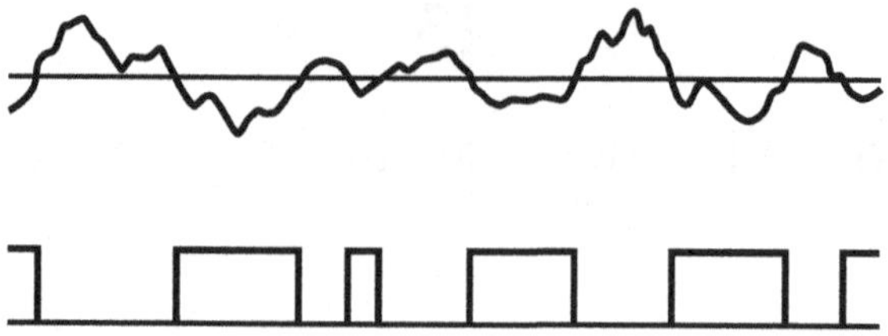

Figure 13.1 Upper: A random, stochastic scalar function $u(t)$ in one realization. The horizontal line represents a value of u^*. Lower: The indicator function ϕ for that value of u^*. Adapted from Lumley and Panofsky (1964).

13.2 Probability statistics of scalar functions of a single variable

13.2.1 Probability densities and distributions

We'll consider first an ensemble of random (different in each realization), stochastic (irregular), zero-mean, scalar functions of one independent variable, say time t. Such an ensemble is sometimes called a random *process*. We'll denote these functions as $u(t; \alpha)$, with α the realization index. In each member of the ensemble we introduce an *indicator function* $\phi(u^*; t)$. It is so-named because it indicates when $u(t) < u^*$ (Figure 13.1):

$$\begin{aligned} \phi(u^*; t) &= 1 \text{ if } u(t) < u^*; \\ \phi(u^*; t) &= 0 \text{ if } u(t) \geq u^*. \end{aligned} \tag{13.3}$$

For any given t this set of indicator functions in effect counts the ensemble members in which $u(t) < u$. Put another way, its ensemble mean $\overline{\phi}(u; t)$ is the fraction of the realizations in which $u(t) < u$. This is defined as the *probability distribution function* $P(u; t)$:

$$P(u; t) = \overline{\phi}(u; t). \tag{13.4}$$

It should be evident that $P(u; t)$ has the properties

$$\begin{aligned} P(u_1; t) &\leq P(u_2; t), \qquad u_1 \leq u_2, \\ \lim_{u \to \infty} P(u; t) &= 1, \\ \lim_{u \to -\infty} P(u; t) &= 0. \end{aligned} \tag{13.5}$$

The derivative of the probability distribution with respect to amplitude is called the *probability density*:

$$\beta(u; t) \equiv \frac{\partial P(u; t)}{\partial u}. \tag{13.6}$$

Since from the definition of the probability distribution we can write

$$P(u_2;t) - P(u_1;t) = \Pr\{u_1 \le u(t) < u_2\} = \frac{P(u_2;t) - P(u_1;t)}{u_2 - u_1}(u_2 - u_1) \cong \beta(u_1;t)\Delta u, \quad (13.7)$$

for $\Delta u \to 0$ we can interpret $\beta(u;t)\Delta u$ as the probability that values of $u(t)$ occur in a narrow band of width Δu lying above u. Thus β has the properties

$$\beta(u;t) \ge 0, \quad \int_{-\infty}^{+\infty} \beta(u;t)\, du = 1. \quad (13.8)$$

13.2.2 Moments and characteristic functions

The probability density $\beta(u;t)$ yields the *moments* of $u(t)$ – the mean values of powers of $u(t)$:

$$\overline{u^n}(t) = \int_{-\infty}^{+\infty} u^n \beta(u;t)\, du. \quad (13.9)$$

This multiplies each value of u^n by the probability of occurrence of that value and sums the products. In the general case when $\overline{u} \neq 0$ using $(u - \overline{u})^n$ in the integrand yields *central moments*, or *moments about the mean* (Problem 13.15).

If the probability density is symmetric about the origin, all odd moments are zero. Thus odd moments, suitably nondimensionalized, are measures of the asymmetry of the probability density. The third moment, nondimensionalized with the variance, is called the *skewness*:

$$S = \frac{\overline{u^3}(t)}{(\overline{u^2}(t))^{3/2}}. \quad (13.10)$$

The second moment is called the *variance*; the fourth moment, when made dimensionless with the square of the second, is called the *kurtosis*† or *flatness factor* F:

$$F = \frac{\overline{u^4}(t)}{(\overline{u^2}(t))^2}. \quad (13.11)$$

It describes the shape of the probability density. For a Gaussian distribution $F = 3$.

The Fourier transform of $\beta(u;t)$ is called the characteristic function $f(\kappa;t)$:

$$\int_{-\infty}^{+\infty} e^{i\kappa u}\beta(u;t)du = f(\kappa;t),$$
$$\beta(u;t) = \frac{1}{2\pi}\int_{-\infty}^{+\infty} e^{-i\kappa u} f(\kappa;t)\, d\kappa. \quad (13.12)$$

† A function with a peaked probability density is sometimes referred to as *kurtic*.

In view of Eq. (13.9) we can interpret the first of Eqs. (13.12) as

$$f(\kappa; t) = \overline{e^{i\kappa u(t)}}. \tag{13.13}$$

Taking the n-th derivative of the first of Eqs. (13.12) with respect to $i\kappa$ and evaluating the derivatives at the origin yields the moments of u:

$$\left.\frac{\partial^n f(\kappa; t)}{\partial(i\kappa)^n}\right|_{\kappa=0} = \overline{u^n}(t). \tag{13.14}$$

These moments determine the series expansion of f:

$$f(\kappa; t) = \sum_{n=0}^{\infty} \frac{1}{n!}\overline{u^n}(t)(i\kappa)^n. \tag{13.15}$$

Thus knowledge of all the moments, or knowledge of f or of β, are equivalent.

13.2.3 Joint probability densities and distributions

The probability density $\beta(u; t)$ gives information about $u(t)$ at any given time but not at two or more times. But we can use the indicator function $\phi(u; t)$ to write

$$\text{if } u(t_1) < u_1, \quad \phi(u_1; t_1) = 1; \qquad \text{if } u(t_1) \geq u_1, \quad \phi(u_1; t_1) = 0,$$

and it follows that the probability that $u(t_1) < u_1$ and $u(t_2) < u_2$ is given by the average of $\phi(u_1; t_1)\phi(u_2; t_2)$:

$$\overline{\phi(u_1; t_1)\phi(u_2; t_2)} = P_2(u_1, u_2; t_1, t_2). \tag{13.16}$$

This is called a *joint probability distribution function.* Its second derivative with respect to amplitudes is a *joint probability density*:

$$\beta_2(u_1, u_2; t_1, t_2) = \frac{\partial^2}{\partial u_1 \partial u_2} P_2(u_1, u_2; t_1, t_2). \tag{13.17}$$

This can be interpreted as a probability:

$$\begin{aligned}&\beta_2(u_1, u_2; t_1, t_2)\Delta u_1 \Delta u_2 \\ &\quad = \Pr\{u_1 \leq u(t_1) < u_1 + \Delta u_1, \;\; u_2 \leq u(t_2) < u_2 + \Delta u_2\}.\end{aligned} \tag{13.18}$$

This has the normalization properties

$$\int\!\!\int_{-\infty}^{+\infty} \beta_2(u_1, u_2; t_1, t_2)\, du_1\, du_2 = 1, \tag{13.19}$$

which means that there is exactly one pair of values which is taken on by any particular $u(t)$ at t_1 and t_2. Further, we see that (from (13.16) and (13.17))

$$\int_{-\infty}^{+\infty} \beta_2(u_1, u_2; t_1, t_2)\, du_2 = \beta(u_1; t_1), \tag{13.20}$$

and correspondingly for the integral over u_1. Its Fourier transform, called a characteristic function, is

$$\begin{aligned} f_2(\kappa_1, \kappa_2; t_1, t_2) &= \int\int_{-\infty}^{+\infty} e^{i\kappa_1 u_1 + i\kappa_2 u_2} \beta_2(u_1, u_2; t_1, t_2)\, du_1\, du_2 \\ &= \overline{e^{i\kappa_1 u_1 + i\kappa_2 u_2}}. \end{aligned} \tag{13.21}$$

Through the property (13.20) all the moments at t_1 or t_2 only can be obtained, e.g., $\overline{u^n}(t_1)$. There are also joint moments (also called cross moments):

$$\overline{u^n(t_1)u^m(t_2)} = \int\int_{-\infty}^{+\infty} u_1^n u_2^m \beta_2(u_1, u_2; t_1, t_2)\, du_1\, du_2. \tag{13.22}$$

The complete set of these moments formally represents f_2,

$$f_2 = \sum_{n,m=0}^{\infty} \frac{1}{n!m!} \overline{u^n(t_1)u^m(t_2)}(i\kappa_1)^n (i\kappa_2)^m, \tag{13.23}$$

and hence β_2.

Rather than considering $u(t)$ at two different times, we could have considered two different processes, say $u(t)$ and $v(t)$, at the same time – two components of velocity, for example, or velocity and a scalar. We can also consider products similar to (13.16) of any number of ϕ's, giving simultaneous information at arbitrarily many times, for several different processes, if desired. For a single process, for example, we could define

$$\beta_n(u_1, ..., u_n; t_1, ..., t_n), \quad n = 1, 2, \ldots, \tag{13.24}$$

with the interpretation paralleling that given before. The entire set of such densities, for all n, gives all statistical information about $u(t)$.

13.2.4 Stationarity

A random process is *stationary* when β is not a function of time, and $\beta_n,\ n \geq 2$ is a function only of the time differences. That is,

$$\beta(u; t) = \beta(u), \qquad \beta_2(u_1, u_2; t_1, t_2) = \beta_2(u_1, u_2; t_1 - t_2), \tag{13.25}$$

and so forth.

Stationarity (or the spatial equivalent, homogeneity) can greatly simplify statistical analyses. As we saw in Part II, real situations can often be taken as approximately stationary (or homogeneous), and so in practice, and with justification from the ergodic hypothesis, time or space averages are typically used instead of ensemble averages. Here we will assume that all three give the same value, and continue to use ensemble averages.

13.3 Examples of probability densities

13.3.1 The Gaussian

The probability density of a Gaussian random variable $u(t)$ is

$$\beta(u) = \frac{1}{\sqrt{2\pi}\,\sigma}\, e^{-(u-\overline{u})^2/2\sigma^2}. \tag{13.26}$$

This is completely determined by two parameters, the mean $\overline{u}$ and the variance $\sigma^2 = \overline{(u-\overline{u})^2}$. *Central moments* (moments about the mean) depend only on σ^2:

$$\overline{(u-\overline{u})^n} = \begin{cases} 0 & n \text{ odd}, \\ \dfrac{\sigma^n n!}{2^{n/2}(n/2)!} & n \text{ even}. \end{cases} \tag{13.27}$$

The skewness of a Gaussian random variable is 0 and the flatness factor is 3. Figure 13.2 shows the Gaussian probability density and probability distribution functions. The distribution function involves the *error function* (erf).

The Central Limit Theorem says that the probability distribution of the sum of statistically independent random variables of an arbitrary but identical probability distribution approaches a Gaussian in the limit as the number of elements in the sum becomes large. In particular, the normalized sums of a large number of independent events – for example, the errors made in performing an experiment – have a Gaussian distribution.

If $u(t)$ and $v(t)$ are jointly Gaussian, their joint probability density is

$$\beta(u, v) = \frac{1}{2\pi\sigma_u\sigma_v\sqrt{1-\rho^2}} \exp\left[-\frac{1}{2(1-\rho^2)}\left(\frac{u'^2}{\sigma_u^2} - 2\rho\frac{u'v'}{\sigma_u\sigma_v} + \frac{v'^2}{\sigma_v^2}\right)\right], \tag{13.28}$$

where

$$u' = u - \overline{u}, \quad v' = v - \overline{v}, \quad \rho = \frac{\overline{u'v'}}{\sigma_u\sigma_v}, \quad \sigma_u = (\overline{u'^2})^{1/2}, \quad \sigma_v = (\overline{v'^2})^{1/2}. \tag{13.29}$$

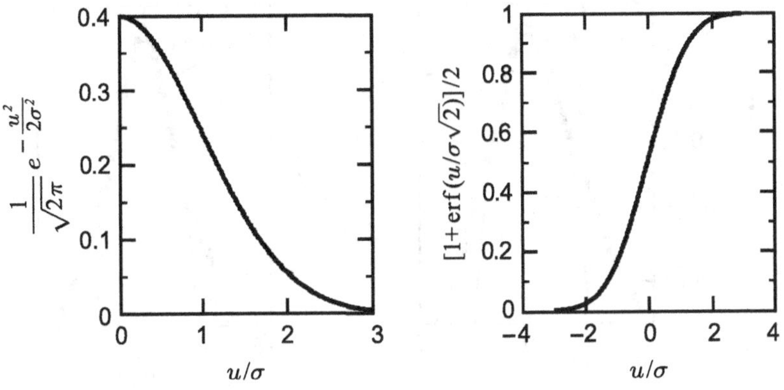

Figure 13.2 A Gaussian probability density made dimensionless with its standard deviation (left) and the corresponding probability distribution (right). Adapted from Lumley and Panofsky (1964).

13.3.2 Observed probability densities

We defined the probability density through ensemble averaging, but in dealing with turbulence data we typically use time averaging of a stationary time series. One can determine the probability density of a measured signal by dividing its amplitude range into bins, and then, for each sample of the signal, adding a count to the proper bin. As the number of samples becomes large, the result, when properly scaled (Problem 13.1), converges to the probability density. With fields from numerical turbulence simulation one can average over statistically homogeneous directions within a "snapshot" – a field at a given time. Averaging over a small ensemble of such snapshots can be sufficient to generate reliable statistics.

13.3.2.1 Velocity derivatives

Figure 13.3 shows probability densities of the first two time derivatives of the streamwise velocity fluctuation $u(t)$ measured at a point in a laboratory flow of moderate turbulence Reynolds number R_t. We typically interpret these as proportional to streamwise spatial derivatives, as suggested by G. I. Taylor (1938):

> If the velocity of the air stream which carries the eddies is very much greater than the turbulent velocity, one may assume that the sequence of changes in u at the fixed point are simply due to the passage of an unchanging pattern of turbulent motion over the point, i.e. one may assume that
>
> $$u = f(t) = f(x/U), \tag{7}$$
>
> where x is measured upstream at time $t = 0$ from the fixed point where u is measured. In the limit when $u/U \to 0$ Eq. (7) is certainly true.

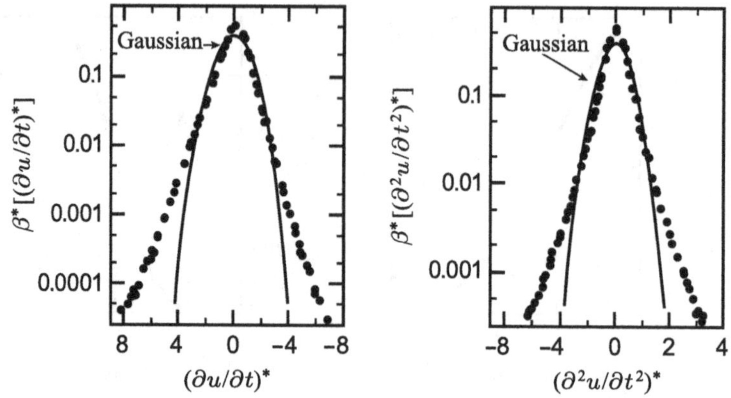

Figure 13.3 Probability densities of the first (left) and second (right) time derivatives of streamwise velocity in a mixing layer at moderate R_t. The tails show strong departures from Gaussianity. The first derivative is slightly skewed but the second is not. The asterisk denotes nondimensionalization by the derivative. From Wyngaard and Tennekes (1970).

For an eddy of velocity and length scales u and ℓ one would expect Taylor's Eq. (7) to be valid if the eddy decay time scale ℓ/u is much greater than the time required to advect the eddy past the probe, ℓ/U. Thus *Taylor's hypothesis* for $u(\mathbf{x}, t)$,

$$\frac{\partial u}{\partial t} \simeq -U\frac{\partial u}{\partial x}, \tag{13.30}$$

should be valid when u/U is sufficiently small. This is one standard way of measuring streamwise derivatives of turbulent velocity components and scalars and, thereby, the rates of dissipation of TKE and scalar variance. Two-point spatial difference measurements are also used (Chapter 16).

As discussed in Chapter 7, Kolmogorov (1941) hypothesized that if the turbulence Reynolds number R_t is large enough, so by Eq. (1.35) the energy-containing and dissipative ranges are sufficiently separated, the dissipative range is in a state of statistical similarity with governing parameters ϵ and ν. Thus under this "universal equilibrium" hypothesis, as Batchelor (1960) termed it, and dimensional analysis the nondimensional moments of velocity derivatives are constants at sufficiently large turbulence Reynolds number.

Batchelor (1960) interpreted early measurements of dissipative-range turbulence statistics as being in accord with the universal equilbrium hypothesis. But beginning in the late 1960s measurements made in very large R_t geophysical turbulence strongly conflicted with it. They also indicated that the instantaneous viscous dissipation is spatially and temporally intermittent, increasingly so as R_t increases. One interpretation (Chapter 7) is that the dissipative regions occupy a decreasing fraction of the flow volume as R_t increases, and so they become more intense with increasing R_t. This is reflected in the broadening of the tails of the pdf of a velocity derivative as R_t increases, which causes the skewness and kurtosis of the velocity derivative to increase with R_t (Chapter 7).

Direct numerical simulation (DNS) of turbulence is currently limited by computer size to about 10^{10} grid points, a few thousand in each of the three spatial directions. Roughly, then, the maximum attainable values of ℓ/η are currently $\sim 10^3$, whereas in the atmospheric boundary layer ℓ/η can be as large as 10^6. Since ℓ/η is a surrogate for the turbulence Reynolds number, Eq. (1.35), fine structure computed from DNS is not a reliable indicator – not even qualitatively – of the fine structure of atmospheric turbulence.

13.3.2.2 *Vertical velocity in the CBL*

In 1975 some intriguing evidence of unusual behavior of neutrally buoyant effluent plumes in the CBL came from experiments done in a laboratory convection tank. As described by Willis and Deardorff (1974), the 114 cm × 122 cm × 76 cm deep tank, using water heated from below, generated turbulent free convection of modest Reynolds number ($R_t \simeq 4200$, $R_\lambda \simeq 140$). In their experiments Deardorff and Willis (1975) used a near-surface, crosswind line source of effluent. The unusual behavior was the ascent of the mean plume as it moved downstream; by a travel time of the order of h/w_* it reached a height of about $0.75h$, with h the CBL depth. As discussed by Lamb (1982), this finding stimulated great interest but also skepticism that a convection tank could model diffusion in the CBL.

Lamb (1978) described a similar phenomenon in a numerical model calculation of plume dispersion from an elevated source in the CBL: the locus of maximum mean concentration downstream of an elevated source descends until it reaches the surface. Motivated by these numerical results, Willis and Deardorff (1978) carried out elevated-source experiments in their convection tank and found plume-descent there. Lamb (1982) describes the two sets of results as being in "remarkably good agreement."†

These findings on effluent-plume dispersion in the CBL stimulated interest in the probability density of vertical velocity fluctuations. Figure 13.4, from Lamb (1982), shows that probability density for w computed by Deardorff through LES in a 5 km × 5 km × 2 km deep domain. It is strongly positively skewed ($\overline{w^3} > 0$). The mode, or most probable velocity, is negative; it is approximately equal to the mean downdraft velocity. About 60% of the area under the density is on the negative half of the w-axis, indicating the higher probability of downdrafts. Field observations

† Lamb (1982) also pointed out that "this is an example in which the predictions of a theoretical model were later confirmed by observations."

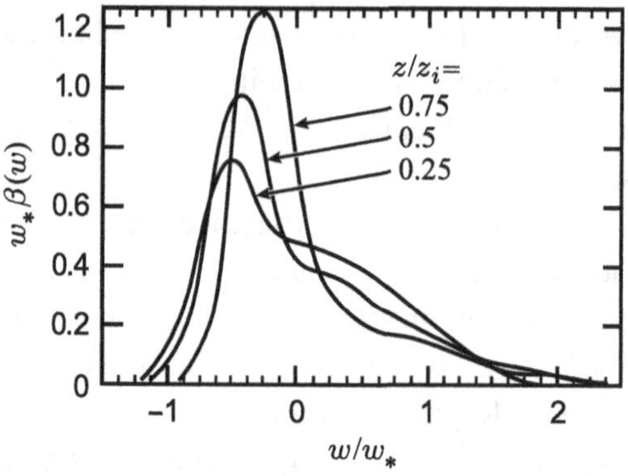

Figure 13.4 The probability density of vertical velocity at three heights in a CBL as determined from LES results. z_i is the boundary-layer depth and w_* is the convective velocity scale. From Lamb (1982).

are generally consistent with this. Lamb (1982) attributes these unusual properties of dispersion in the CBL to the large Lagrangian integral time scale and the highly skewed pdf of its vertical velocity.

13.3.3 Joint probability densities

Wyngaard and Moeng (1992) determined the joint probability density (jpd) of w and mixing ratio c of a passive, conserved scalar through analysis of the fields obtained through LES of a CBL. In general this joint density has the property

$$\int_{-\overline{C}}^{\infty}\int_{-\infty}^{\infty} \beta(w', c')\, dw'\, dc' = 1;$$

$$\int_{-\overline{C}}^{\infty}\int_{-\infty}^{\infty} \beta(w', c') w' c'\, dw'\, dc' = \overline{wc}. \tag{13.31}$$

(The lower limit of the c-integral corresponds to $\tilde{c} = \overline{C} + c = 0, \quad c = -\overline{C}$.)

The jpds for w with the top-down and bottom-up fields c_t and c_b are shown in Figures 13.5 and 13.6. They agree quite well with those measured by Deardorff and Willis (1985) in a laboratory convection tank.[†] The jpds are nondimensionalized so that

$$\beta^* = \sigma_w \sigma_c \beta, \qquad \int\int \beta^* d\frac{w'}{\sigma_w} d\frac{c'}{\sigma_c} = 1. \tag{13.32}$$

[†] Wyngaard and Moeng (1992) commented that the laboratory jpds seem not to integrate to 1.0 over the plane and suggested that an error was made in their scaling. We confirmed that error and corrected it.

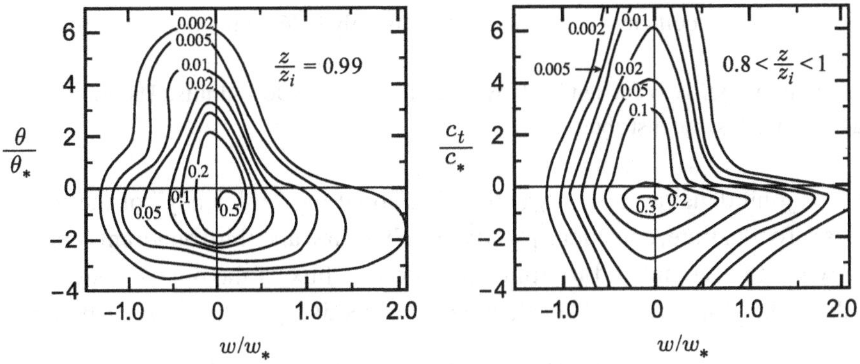

Figure 13.5 Left: The jpd of vertical velocity w and temperature θ measured by Deardorff and Willis (1985) near the top of a convection tank. Right: The jpd of top-down scalar c_t and w calculated through LES by Wyngaard and Moeng (1992) for the same region in a CBL, scaled in the same way for comparison. The jpds are normalized so the integral over this plane is 1.0. From Wyngaard and Moeng (1992).

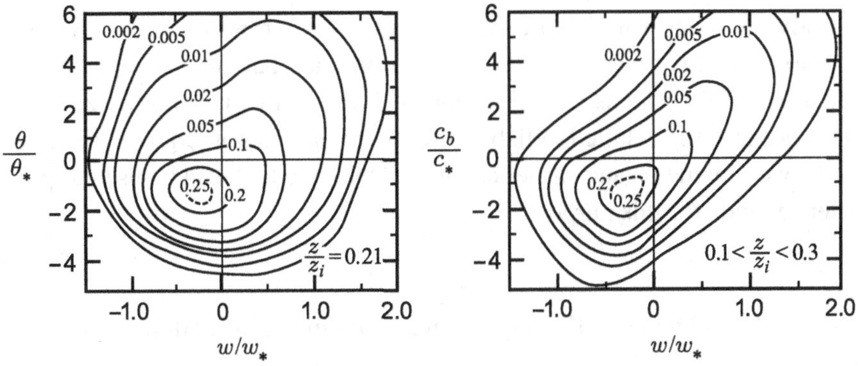

Figure 13.6 As in Figure 13.5 but for measurements near the bottom of the convection tank and for the bottom-up scalar c_b. From Wyngaard and Moeng (1992).

Since β determines all the w and c moments, e.g.,

$$\int\int w'^2 \beta \, dw' \, dc' = \overline{w^2}, \int\int c'^2 \beta \, dw' \, dc' = \overline{c^2}, \tag{13.33}$$

Wyngaard and Moeng were able to extract from these results some of the differences between top-down and bottom-up diffusion in convective turbulence (Chapter 11). For example,

- The correlation coefficient between w and c_t is only 0.26, while that between w and c_b is 0.60.
- The bottom-up flux is carried 70% by updrafts, 30% by downdrafts. The top-down flux has nearly the same distribution.

These joint densities also say something about the "efficiency" of the transport process. To illustrate, consider a positive $\overline{wc}$. The first and third quadrants contribute to it, while the second and fourth contribute to a flux of the opposite sign. Let us call these the "forward" and "back" fluxes, respectively. A "transport efficiency" is

$$e_t = \frac{\text{net flux}}{\text{forward flux}} = \frac{\text{forward flux} + \text{back flux}}{\text{forward flux}}. \tag{13.34}$$

The top-down process, with $\rho = 0.26$, has $e_t = 0.6$; the bottom-up process, with $\rho = 0.6$, has $e_t = 0.9$. This shows again the unusual diffusive properties of convective turbulence. Convective clouds could have similar diffusive properties.

Numerical models of the atmosphere and ocean need concise but reliable submodels of unresolved turbulence. The traditional way to develop such submodels is to study the process in detail experimentally, to interpret its basic physics, and then to express this physics through an approximate but concise equation or set of equations. This process, called *parameterization* in meteorology, can be quite difficult because the detailed, reliable turbulence data that it requires often do not exist. As we saw in this and earlier chapters, DNS and LES can be used to provide turbulence fields for this purpose.

13.4 The evolution equation for the probability density

In Parts I and II we manipulated the fluid equations to produce equations for the evolution of certain statistics of the turbulent fluctuations of velocity and advected scalar constituents. Until fairly recently the probability density of velocity or an advected scalar was not among the statistics for which one could derive an evolution equation. That situation changed with a remarkable paper by Lundgren in 1967. We'll briefly sketch his derivation here.†

If $\beta(v_i; x_i, t)$ is the probability density of the velocity at point x_i and time t, $\beta(v_i; x_i, t)\, dv_i$ is the probability that the velocity at (x_i, t) is within dv_i of v_i. If we define $\beta_\alpha(v_i; x_i, t)$ to be a "single-realization" probability density of velocity – one that produces $v_i(x_i, t) = u_i$ in realization α,‡ then

$$\beta_\alpha(v_i; x_i, t) = \delta(u_i - v_i), \tag{13.35}$$

† We have used tensor notation rather than Lundgren's vector notation.
‡ This is an elaboration of the interpretation offered by Lundgren (1967).

where δ is the Dirac delta function and u_i is the velocity at x_i, t in realization α. Thus the probability that in realization α the velocity at (x_i, t) is within dv_i of v_i is $\beta_\alpha(v_i; x_i, t)\, dv_i = \delta(u_i - v_i)\, dv_i = 1$ if $v_i = u_i$, as required. Averaging this expression over all realizations (and denoting this by angle brackets) yields

$$\beta(v_i; x_i, t) = \langle \beta_\alpha(v_i; x_i, t) \rangle = \langle \delta(u_i - v_i) \rangle. \tag{13.36}$$

In equating the probability density of velocity $u_i(x_i, t)$ to the expected value of a delta function involving velocity, Eq. (13.36) provides a link between the dynamics of the velocity field and those of its probability density. Before showing how Lundgren (1967) used that link, we'll need to generalize Eq. (13.36) to joint probability densities.

The probability density $\beta\left(v_i^{(1)}; x_i^{(1)}, t\right)$ is the "one-point" form. Its product with $dv_i^{(1)}$ is the probability that the velocity at $x_i^{(1)}$ and time t is within dv_i of $v_i^{(1)}$. From Eq. (13.36) it is written as

$$\beta\left(v_i^{(1)}; x_i^{(1)}, t\right) = \left\langle \delta\left(u_i\left(x_i^{(1)}, t\right) - v_i^{(1)}\right) \right\rangle \equiv \left\langle \delta\left(u_i^{(1)} - v_i^{(1)}\right) \right\rangle \equiv \beta(1). \tag{13.37}$$

The "two-point", or joint, probability density $\beta_2\left(v_i^{(1)}, v_i^{(2)}; x_i^{(1)}, x_i^{(2)}; t\right)$ is the probability that at time t the velocity at $x_i^{(1)}$ is within dv_i of $v_i^{(1)}$ and the velocity at $x_i^{(2)}$ is within dv_i of $v_i^{(2)}$. By extension of the arguments leading to Eq. (13.37) it is

$$\beta_2\left(v_i^{(1)}, v_i^{(2)}; x_i^{(1)}, x_i^{(2)}; t\right) = \left\langle \delta\left(u_i^{(1)} - v_i^{(1)}\right) \delta\left(u_i^{(2)} - v_i^{(2)}\right) \right\rangle \equiv \beta(1, 2). \tag{13.38}$$

Lundgren's derivation of the pdf evolution equation proceeds as follows. The time derivative of Eq. (13.37) is

$$\begin{aligned} \frac{\partial}{\partial t}\beta(1) = \left\langle \frac{\partial}{\partial t}\delta\left(u_i^{(1)} - v_i^{(1)}\right) \right\rangle &= \left\langle \frac{\partial \delta\left(u_i^{(1)} - v_i^{(1)}\right)}{\partial\left(u_i^{(1)} - v_i^{(1)}\right)} \frac{\partial\left(u_i^{(1)} - v_i^{(1)}\right)}{\partial t} \right\rangle \\ &= -\left\langle \frac{\partial}{\partial v_i^{(1)}}\delta\left(u_i^{(1)} - v_i^{(1)}\right) \frac{\partial}{\partial t}u_i^{(1)} \right\rangle. \end{aligned} \tag{13.39}$$

Lundgren then evaluated the right side of Eq. (13.39) with the Navier–Stokes and continuity equations through tedious calculations. The resulting evolution equation for the probability density is

$$\frac{\partial \beta(1)}{\partial t} + v_i^{(1)} \frac{\partial \beta(1)}{\partial x_i^{(1)}}$$

$$+ \frac{\partial}{\partial v_i^{(1)}} \left[-\frac{1}{4\pi} \int \left(\frac{\partial}{\partial x_i^{(1)}} \frac{1}{|\mathbf{x}^{(1)} - \mathbf{x}^{(2)}|} \right) \left(v_j^{(2)} \frac{\partial}{\partial x_j^{(2)}} \right)^2 \beta(1,2)\, d\mathbf{x}^{(2)}\, d\mathbf{v}^{(2)} \right]$$

$$+ \frac{\partial}{\partial v_i^{(1)}} \left[\lim_{\mathbf{x}^{(2)} \to \mathbf{x}^{(1)}} \nu \frac{\partial}{\partial x_j^{(2)}} \frac{\partial}{\partial x_j^{(2)}} \int v_i^{(2)} \beta(1,2)\, d\mathbf{v}^{(2)} \right] = 0. \tag{13.40}$$

The conservation equation for $\beta(1)$ involves also the two-point (joint) probability density $\beta(1, 2)$. Lundgren also derived the equation for $\beta(1, 2)$, which involves $\beta(1, 2, 3)$. Thus a closure approximation is needed in order to use the pdf conservation equation. This approach forms the basis of a type of turbulence modeling called *pdf modeling*.

Questions on key concepts

13.1 The terms probability *density* and probability *distribution* are frequently misused. Explain how the two differ.

13.2 Interpret Eq. (13.9) physically.

13.3 Write $\overline{f(u)}$, where u is a random variable, as an integral of the probability density of u. Explain the integral physically.

13.4 What is the Central Limit Theorem? Why is it important in applications?

13.5 Given a joint pdf of two variables, show formally how one obtains a pdf of one of the variables.

13.6 Explain why the broadening of the tails of the pdfs in Figure 13.3 is consistent with the notion of dissipation intermittency discussed in Chapter 7.

13.7 Explain physically, using the concepts of scalar fine structure discussed in Chapter 7, why the kurtosis of velocity derivatives in turbulent flows tends to be large.

13.8 Using the properties of the pdf, explain why a variable with a pdf having a sharp peak at the origin has a larger kurtosis than one with a broad maximum there.

13.9 Interpret the pdf of vertical velocity in a CBL, Figure 13.4, as being consistent with a zero-mean field with narrow, strong updrafts separated by broader regions of weak downdrafts.

13.10 Explain why the behavior of effluent plumes in a CBL with $R_t \sim 10^8$ can evidently be effectively modeled in a convection tank having 4–5 orders of magnitude smaller R_t.

13.11 Use the expression for the convergence of a time average to the ensemble average, Chapter 2, to show why dispersion experiments are much more readily done in the convection tank than in the CBL.

13.12 Defend the argument that laboratory convection-tank studies of dispersion should be phased out in favor of numerical computations through turbulence simulation.

13.13 Refute the argument that laboratory convection-tank studies of dispersion should be phased out in favor of numerical computations through turbulence simulation.

13.14 Discuss the key step, concept, or definition on which Lundgren's derivation of the evolution equation for the pdf of velocity in a turbulent flow is based.

Problems

13.1 Determine the probability density of a sine wave experimentally by sampling it a large number of times and putting each sample into the appropriate amplitude bin. Why must you make sure that your sampling period is not an integral or fractional multiple of the period? Discuss the scaling required to convert the bin counts into the probability density.

13.2 Determine the probability density of the sum of samples taken in Problem 13.1. Let the number of samples in the sum become large. What form does the probability density take?

13.3 Show how the probability density of $u(t)$, say, is related to the joint density for $u(t)$ and $v(t)$ if the processes are stationary. What simplification in the joint density ensues if the processes are statistically independent? Demonstrate the latter with the joint Gaussian probability density.

13.4 Estimate the uncertainty in a probability density determined by binning and counting, as in Problem 13.1. If $\beta^m(u)$ is the measured probability density (determined through binning and counting a total of N samples, say) and $\beta(u)$ is the true density, how does $\overline{[\beta^m(u) - \beta(u)]^2}$ vary with N? (You might want to adapt the averaging-time formula (Chapter 2) for this purpose.) Discuss the benefits of sampling slowly enough that neighboring samples are statistically independent. For what values of u is the probability density least well determined? What are the implications for higher moments of the variable?

13.5 Discuss the proof of the Central Limit Theorem.

13.6 Using Eq. (13.9), show how the value of u at the peak of the integrand defining the n-th moment of a Gaussian variable u depends on n for even values of n.

13.7 Prove Eq. (13.39).

13.8 Discuss, using physical examples, when you would expect turbulence to have nonzero third moments.

13.9 Explain why the characteristic function, Eq. (13.12), can be interpreted as the expected value of $e^{i\kappa u}$.

13.10 Prove Eqs. (13.14) and (13.15).

13.11 Prove Eq. (13.16).

13.12 Explain the interpretation of Eq. (13.19) that is given just below it.

13.13 Interpret the second of Eqs. (13.8) physically.

13.14 Sketch the pdf of the top-down scalar, using its joint pdf with vertical velocity shown in Figure 13.5.

13.15 Show that when $\overline{u} \neq 0$, using $(u - \overline{u})^n$ in the integrand of Eq. (13.9) yields central moments (moments about the mean), $\overline{[u - \overline{u}]^n}$.

References

Batchelor, G. K., 1960: *The Theory of Homogeneous Turbulence.* Cambridge University Press.

Deardorff, J. W., and G. E. Willis, 1975: A parameterization of diffusion into the mixed layer. *J. Appl. Meteor.*, **14**, 1451–1458.

Deardorff, J. W., and G. E. Willis, 1985: Further results from a laboratory model of the convective boundary layer. *Bound.-Layer Meteor.*, **32**, 205–236.

Kolmogorov, A. N., 1941: The local structure of turbulence in incompressible viscous fluid for very large Reynolds numbers. *Doklady ANSSSR*, **30**, 301–305.

Lamb, R., 1978: A numerical simulation of dispersion from an elevated point source in the convective boundary layer. *Atmos. Environ.*, **12**, 1297–1304.

Lamb, R., 1982: Diffusion in the convective boundary layer. In *Atmospheric Turbulence and Air Pollution Modelling*, F. Nieuwstadt and H. van Dop, Eds., Dordrecht: Reidel, pp. 159–229.

Lumley, J. L., and H. A. Panofsky, 1964: *The Structure of Atmospheric Turbulence.* New York: Interscience.

Lundgren, T. S., 1967: Distribution functions in the statistical theory of turbulence. *Phys. Fluids*, **10**, 969–975.

Taylor, G. I., 1938: The spectrum of turbulence. Part I. *Proc. R. Soc. A*, **164**, 476–490.

Willis, G. E., and J. W. Deardorff, 1974: A laboratory model of the unstable planetary boundary layer. *J. Atmos. Sci.*, **31**, 1297–1307.

Willis, G. E., and J. W. Deardorff, 1978: A laboratory study of dispersion from an elevated source within a modeled convective planetary boundary layer. *Atmos. Environ.*, **12**, 1305–1311.

Wyngaard, J. C., and C.-H. Moeng, 1992: Parameterizing turbulent diffusion through the joint probability density. *Bound.-Layer Meteor.*, **60**, 1–13.

Wyngaard, J. C., and H. Tennekes, 1970: Measurements of the small-scale structure of turbulence at moderate Reynolds numbers. *Phys. Fluids*, **13**, 1962–1969.

14

Isotropic tensors

14.1 Introduction

An *isotropic* turbulence field, a concept introduced by G. I. Taylor (1935), has statistical properties that are independent of translation, rotation, and reflection of the coordinate axes.

The covariance equations in Chapter 5 include production terms involving gradients of the mean velocity and mean scalar fields. As we shall see, these mean-field gradients are *anisotropic* and cause the energy-containing range of shear-driven turbulence to be anisotropic as well. Likewise, we'll see that buoyancy-driven turbulence is anisotropic. Thus it appears unlikely that equilibrium isotropic turbulence exists naturally.† It can be produced through DNS, however, by adding a random-force term to the Navier–Stokes equation (Problem 14.14).

As we discussed in Chapter 7, the hypothesis of *local isotropy*, or isotropy of the smallest-scale structure of a large-R_t turbulent flow, is due to Kolmogorov (1941). Because it greatly simplifies the specification of the fine structure of turbulent flows, it is widely used in analytical, observational, and numerical studies. But Kolmogorov pointed out that the arguments for local isotropy, which we'll present in his words in Section 14.5, are largely physical ones. Today the observational evidence against it is compelling, and the physics underlying local *anisotropy* is an active area of research.

14.2 Cartesian tensors

Jeffreys (1961) discusses basic concepts of cartesian tensors. If we have two sets of rectangular axes with the same origin, (Ox, Oy, Oz) and (Ox', Oy', Oz'), then we can write the transformations

† Betchov (1957) did attempt to generate isotropic turbulence with a multiplicity of small, colliding air jets.

$$
\begin{aligned}
x' &= l_1 x + m_1 y + n_1 z,\\
y' &= l_2 x + m_2 y + n_2 z, \qquad (14.1)\\
z' &= l_3 x + m_3 y + n_3 z,
\end{aligned}
$$

where the parameters l, m, and n are *direction cosines* – the cosines of the angles between the old and new axes. (l_1 is the cosine of the angle between Ox and Ox', m_1 that between Oy and Ox', ..., l_2 the cosine of the angle between Ox and Oy', ...) We can also invert Eq. (14.1) to express x, y, and z in terms of the primed quantities.

In cartesian tensor notation we write these transformations more compactly as

$$
x'_j = a_{ij} x_i, \qquad x_i = a_{ij} x'_j; \qquad a_{ik} a_{jk} = \delta_{ij}. \qquad (14.2)
$$

Quantities (such as velocity u_i) that transform in this way are called *tensors of the first order*, or *vectors*. A set of nine quantities w_{ik} referred to a set of axes and transformed to another set by the rule

$$
w'_{jl} = a_{ij} a_{kl} w_{ik} \qquad (14.3)
$$

is called a tensor of the second order.

The covariances we meet in turbulence are tensors. Examples include the scalar-scalar covariance $\overline{ab}$, a zero-order tensor; the scalar flux $\overline{cu_i}$ and the mean scalar gradient $\partial C/\partial x_i$, first-order tensors; and the kinematic Reynolds stress $\overline{u_i u_j}$ and the mean strain rate $\partial U_i/\partial x_j$, second-order tensors. Yet-higher-order tensors appear in the molecular destruction terms in the moment equations of Chapter 5.

These are all *single-point* tensors – tensors that involve only one spatial point. *Multi-point* tensors involve more than one spatial point.

14.3 Determining the form of isotropic tensors

An isotropic tensor is one whose components do not change under rotation, translation, or reflection of the coordinate axes. There are two ways to determine its form: through the process described by Jeffreys, which uses the transformation rules; and by a technique due to Robertson and described by Batchelor (1960). The first was also used by Taylor (1935); it is quite tedious. We will use the latter.

14.3.1 Single-point tensors

Since a vector's components change under coordinate rotation, the only isotropic vector is the zero vector. Put another way, a nonzero isotropic vector would have a direction, and that would be in conflict with isotropy. Thus, the turbulent flux and the mean gradient of a scalar, being vectors, vanish in isotropic turbulence. Likewise any odd-order single-point tensor will change sign under coordinate reflection, so the isotropic form vanishes.

The situation is different for even orders, as we shall illustrate with the Robertson technique. Let T_{ik} be an isotropic, second-order, *single-point* tensor, one that involves properties at one point in space. Contracting it with two arbitrary vectors **A** and **B** yields a scalar that is bilinear in **A** and **B** and depends also on the invariants of T_{ik}:

$$T_{ik}A_i B_k = \alpha(\text{invariants})A_i B_i = \alpha A_i B_k \delta_{ik}. \tag{14.4}$$

Rewriting this gives

$$(T_{ik} - \alpha\delta_{ik})A_i B_k = 0, \quad T_{ik} = \alpha\delta_{ik}. \tag{14.5}$$

Equation (14.5) says that a second-order isotropic single-point tensor has nonzero components only on the diagonal, and those diagonal components are equal. If the diagonal components were not equal, they would change under coordinate rotation (Problem 14.6), so the tensor would be anisotropic.

The pressure-stress tensor is an application of Eq. (14.5). A stress tensor σ_{ik}, say, expresses the force in the i direction per unit area on a surface whose normal is in the k direction. The force per unit area on the surface is the dot product of the stress tensor and n_i, the unit normal to the surface. Since pressure acts normal to a surface, whatever the orientation of that surface, the pressure-stress tensor is isotropic. It must then have the form $\sigma_{ik} = -p\delta_{ik}$, where the scalar p is the magnitude of pressure. The pressure force f_i is

$$f_i = \sigma_{ik}n_k = -p\delta_{ik}n_k = -pn_i, \tag{14.6}$$

which is indeed normal to the surface.

To determine the form of the fourth-order isotropic tensor T_{ijkm}, we contract T_{ijkm} with four arbitrary vectors and write the resulting scalar as

$$\begin{aligned} T_{ijkm}A_i B_j C_k D_m &= \alpha A_i B_i C_j D_j + \beta A_i C_i B_j D_j + \gamma A_i D_i B_j C_j \\ &= \alpha A_i B_j \delta_{ij} C_k D_m \delta_{km} + \beta A_i C_k \delta_{ik} B_j D_m \delta_{jm} \\ &\quad + \gamma A_i D_m \delta_{im} B_j C_k \delta_{jk} \\ &= A_i B_j C_k D_m (\alpha\delta_{ij}\delta_{km} + \beta\delta_{ik}\delta_{jm} + \gamma\delta_{im}\delta_{jk}). \end{aligned} \tag{14.7}$$

Since **A**, **B**, **C**, and **D** are arbitrary we can rewrite this to find

$$T_{ijkm} = \alpha\delta_{ij}\delta_{km} + \beta\delta_{ik}\delta_{jm} + \gamma\delta_{im}\delta_{jk}. \tag{14.8}$$

T_{ijkm} contains the three unknown scalars α, β, and γ.

14.3.2 Two-point tensors

Robertson's technique shows that the following two-point tensors have the isotropic forms (Batchelor, 1960)

$$\begin{aligned}
\overline{a(\mathbf{x})b(\mathbf{x}+\mathbf{r})} &= F(r), \quad r = |\mathbf{r}|; \\
\overline{c(\mathbf{x})u_i(\mathbf{x}+\mathbf{r})} &= G(r)r_i; \\
\overline{u_i(\mathbf{x})u_j(\mathbf{x}+\mathbf{r})} &= H(r)r_ir_j + I(r)\delta_{ij}; \\
\phi_{ij}(\boldsymbol{\kappa}) &= J(\kappa)\kappa_i\kappa_j + K(\kappa)\delta_{ij}, \quad \kappa = |\boldsymbol{\kappa}|.
\end{aligned} \tag{14.9}$$

Here the unknowns F, G, H, I, J, K are functions of the magnitude of the vector on which the tensor depends.

We'll show next that we can determine some of the unknowns in the isotropic tensor forms by using physical constraints such as incompressibility and homogeneity and geometric constraints such as the symmetries of the tensor.

14.4 Implications of isotropy

14.4.1 Rate of shear production of TKE

Equation (14.5) for an isotropic second-order tensor indicates that the kinematic Reynolds stress tensor in isotropic turbulence is

$$\overline{u_iu_k} = \frac{\overline{u_ju_j}}{3}\delta_{ik}, \tag{14.10}$$

so the off-diagonal Reynolds stresses vanish. Thus in isotropic, incompressible turbulence the rate of shear production of TKE, Eq. (5.45), vanishes:

$$-\overline{u_iu_k}\frac{\partial U_i}{\partial x_k} = -\frac{\overline{u_ju_j}}{3}\delta_{ik}\frac{\partial U_i}{\partial x_k} = -\frac{\overline{u_ju_j}}{3}\frac{\partial U_i}{\partial x_i} = 0. \tag{14.11}$$

Under isotropy the rate of buoyant production of turbulence also vanishes (Problem 14.20). Thus, unless it is driven in some other way, such as by stochastic, isotropic forcing in numerical simulation experiments, constant-density isotropic turbulence decays because its rate of TKE production vanishes.

14.4.2 Rates of molecular destruction of covariances

The rates of molecular destruction of $\overline{u_iu_k}$ and TKE, respectively, are

$$\overline{u_iu_k}\colon\ 2\nu\overline{\frac{\partial u_i}{\partial x_j}\frac{\partial u_k}{\partial x_j}} = 2\nu M_{ijkj}, \qquad \epsilon = \nu\overline{\frac{\partial u_i}{\partial x_j}\frac{\partial u_i}{\partial x_j}} = \nu M_{ijij}, \tag{14.12}$$

where from Eq. (14.8) the isotropic form for M is

$$M_{ijkm} = \overline{\frac{\partial u_i}{\partial x_j}\frac{\partial u_k}{\partial x_m}} = \alpha\delta_{ij}\delta_{km} + \beta\delta_{ik}\delta_{jm} + \gamma\delta_{im}\delta_{jk}. \tag{14.13}$$

In order to evaluate these molecular destruction terms through local isotropy we must determine the parameters α, β, and γ in Eq. (14.13). We can identify two constraints on M_{ijkm}:

1. Summing i on j yields zero by incompressibility:

$$M_{iikm} = \overline{\frac{\partial u_i}{\partial x_i}\frac{\partial u_k}{\partial x_m}} = 0 \qquad \text{(constraint 1)}.$$

2. Summing j on k yields zero. This stems from

$$M_{ijjm} = \overline{\frac{\partial u_i}{\partial x_j}\frac{\partial u_j}{\partial x_m}} = \frac{\partial}{\partial x_j}\left(\overline{u_i\frac{\partial u_j}{\partial x_m}}\right),$$

which is the gradient of a mean quantity. An upper limit for its magnitude is (Appendix, Chapter 5)

$$\frac{\partial}{\partial x_j}\left(\overline{u_i\frac{\partial u_j}{\partial x_m}}\right) < \frac{1}{\ell}u\left(\frac{\epsilon}{\nu}\right)^{1/2} = \frac{\epsilon}{\nu}R_t^{-1/2}.$$

Since Eq. (14.12) indicates that M_{ijij} is of order ϵ/ν, M_{ijjm} is negligible at large R_t and we can write

$$M_{ijjm} = \overline{\frac{\partial u_i}{\partial x_j}\frac{\partial u_j}{\partial x_m}} = 0 \qquad \text{(constraint 2)}. \tag{14.14}$$

These two constraints give

$$0 = 3\alpha + \beta + \gamma, \qquad 0 = \alpha + \beta + 3\gamma, \tag{14.15}$$

so that $\alpha = \gamma,\ \beta = -4\gamma$, and

$$M_{ijkm} = \overline{\frac{\partial u_i}{\partial x_j}\frac{\partial u_k}{\partial x_m}} = \gamma\left(\delta_{ij}\delta_{km} - 4\delta_{ik}\delta_{jm} + \delta_{im}\delta_{jk}\right). \tag{14.16}$$

If we write from Eq. (14.16)

$$M_{1111} = \overline{\left(\frac{\partial u_1}{\partial x_1}\right)^2} = -2\gamma, \tag{14.17}$$

then we can express the dissipation-rate tensor M_{ijkm} as

$$M_{ijkm} = \overline{\frac{\partial u_i}{\partial x_j}\frac{\partial u_k}{\partial x_m}} = -\frac{1}{2}\overline{\left(\frac{\partial u_1}{\partial x_1}\right)^2}\left(\delta_{ij}\delta_{km} - 4\delta_{ik}\delta_{jm} + \delta_{im}\delta_{jk}\right). \tag{14.18}$$

From Eqs. (14.12) and (14.18) we can write ϵ, the rate of dissipation of TKE, under the assumption of local isotropy as (Problem 14.1)

$$\epsilon = \nu M_{ijij} = 15\nu\overline{\left(\frac{\partial u_1}{\partial x_1}\right)^2}. \tag{14.19}$$

The variance of the streamwise derivative of u_1 is typically evaluated from the variance of the time derivative through Taylor's hypothesis (Chapter 2),

$$\overline{\left(\frac{\partial u_1}{\partial x_1}\right)^2} \simeq \frac{1}{U^2}\overline{\left(\frac{\partial u_1}{\partial t}\right)^2}. \tag{14.20}$$

The rate of molecular destruction of $\overline{u_i u_k}$, Eq. (14.12), becomes with the local isotropy form (14.18)

$$2\nu M_{ijkj} = 10\nu\overline{\left(\frac{\partial u_1}{\partial x_1}\right)^2}\delta_{ik}. \tag{14.21}$$

This says that under the assumption of local isotropy the velocity variances but not the turbulent shear stress components are dissipated.

The rate of molecular destruction of scalar flux can be written as (Appendix, Chapter 5)

$$\chi_{u_i c} = (\gamma + \nu)\overline{\frac{\partial u_i}{\partial x_j}\frac{\partial c}{\partial x_j}}. \tag{14.22}$$

As a contraction of a third-order, single-point tensor determined by the dissipative-range structure, this vanishes under local isotropy.

The rate of destruction of scalar variance, Eq. (5.7), involves a contraction of a scalar gradient covariance. Under isotropy this covariance becomes

$$\overline{\frac{\partial c}{\partial x_i}\frac{\partial c}{\partial x_j}} = \overline{\left(\frac{\partial c}{\partial x_1}\right)^2}\delta_{ij}. \tag{14.23}$$

14.4.3 Higher order examples

The number of terms in the expressions for isotropic, single-point velocity derivative tensors increases rapidly with the order of the tensor. We saw that at fourth order there are three terms; at sixth order there are 15. Finding the constraints needed to evaluate the constants and doing the algebra can become quite tedious. As kindly noted by Champagne (1978), the author worked out the isotropic forms of two sixth-order tensors in the budget of vorticity variance (Problem 5.6):

$$\overline{\frac{\partial u_i}{\partial x_k \partial x_m} \frac{\partial u_j}{\partial x_l \partial x_n}} = \overline{\left(\frac{\partial^2 u_1}{\partial x_1 \partial x_1}\right)^2} \Big[\delta_{ij}\delta_{kn}\delta_{lm} + \delta_{ij}\delta_{ln}\delta_{km} + \delta_{ij}\delta_{kl}\delta_{mn}$$
$$- \frac{1}{6}(\delta_{il}\delta_{jn}\delta_{km} + \delta_{il}\delta_{kn}\delta_{jm} + \delta_{jl}\delta_{in}\delta_{km} + \delta_{jl}\delta_{kn}\delta_{im}$$
$$+ \delta_{kl}\delta_{in}\delta_{jm} + \delta_{kl}\delta_{jn}\delta_{im} + \delta_{ik}\delta_{jl}\delta_{mn} + \delta_{jk}\delta_{il}\delta_{mn} + \delta_{ik}\delta_{jn}\delta_{lm}$$
$$+ \delta_{ik}\delta_{ln}\delta_{jm} + \delta_{jk}\delta_{in}\delta_{lm} + \delta_{jk}\delta_{ln}\delta_{im})\Big]. \tag{14.24}$$

$$\overline{\frac{\partial u_i}{\partial x_j} \frac{\partial u_k}{\partial x_l} \frac{\partial u_m}{\partial x_n}} = \overline{\left(\frac{\partial u_1}{\partial x_1}\right)^3} \Big[\delta_{ij}\delta_{kl}\delta_{mn} - \frac{4}{3}(\delta_{ij}\delta_{km}\delta_{ln} + \delta_{ik}\delta_{lj}\delta_{mn} + \delta_{im}\delta_{jn}\delta_{kl})$$
$$- \frac{1}{6}(\delta_{ij}\delta_{kn}\delta_{lm} + \delta_{il}\delta_{kj}\delta_{mn} + \delta_{in}\delta_{kl}\delta_{jm})$$
$$- \frac{3}{4}(\delta_{il}\delta_{kn}\delta_{jm} + \delta_{in}\delta_{kj}\delta_{lm}) + \delta_{il}\delta_{km}\delta_{jn} + \delta_{in}\delta_{km}\delta_{lj}$$
$$+ \delta_{ik}\delta_{lm}\delta_{jn} + \delta_{ik}\delta_{ln}\delta_{jm} + \delta_{im}\delta_{jl}\delta_{kn} + \delta_{im}\delta_{jk}\delta_{ln}\Big]. \tag{14.25}$$

14.5 Local isotropy

14.5.1 The concept

The concept of local isotropy is credited to Kolmogorov (1941). His short paper appears in the Friedlander–Topper (1961) collection as a rather rough English translation. After defining *locally isotropic* turbulence in the formal terms of probability statistics, he wrote:

It is natural that in so general and somewhat indefinite a formulation the just advanced proposition cannot be rigorously proved. We may indicate here only certain general considerations speaking for [it]. For very large R [Reynolds number] the turbulent flow may be thought of in the following way: on the [ensemble] averaged flow … are superposed the "pulsations of the first order" consisting in disorderly displacements of separate fluid volumes, one with respect to another, of diameters of the order of magnitude $\ell^{(1)} = \ell$ (where ℓ is Prandtl's mixing [length]); the order of magnitude … of these relative velocities we denote by $v^{(1)}$. The pulsations of the first order are for very large R in their turn unsteady, and on them are superposed the pulsations of second order with mixing [length] $\ell^{(2)} < \ell^{(1)}$ and relative velocities $v^{(2)} < v^{(1)}$; such a process of successive refinement of turbulent pulsations may be carried until for the pulsations of some sufficiently large order n the Reynolds number

$$R^{(n)} = \frac{\ell^{(n)} v^{(n)}}{\nu}$$

becomes so small that the effect of viscosity on the pulsations of the order n finally prevents the formation of pulsations of the order $n + 1$.

From the energetical point of view it is natural to imagine the process of turbulent mixing in the following way: the pulsations of the first order absorb the energy of the motion and pass it over successively to pulsations of higher orders. The energy of the finest pulsations is dispersed in the energy of heat due to viscosity.

In virtue of the chaotical mechanism of the translation of motion from the pulsations of lower orders to the pulsations of higher orders, it is natural to assume that in domains of space, whose dimensions are small in comparison with $\ell^{(1)}$, the fine pulsations of the higher orders are subjected to [an] approximately space-isotropic statistical regime. Within small time-intervals it is natural to consider this regime approximately steady even in the case, when the flow as a whole is not steady.

Since for very large R the differences

$$w_i(P) = u_i(P) - u_i(P^{(0)})$$

of the velocity components in neighboring points P and $P^{(0)}$ of the four-dimensional space are determined nearly exclusively by pulsations of higher orders, the scheme just exposed leads us to the hypothesis of local isotropy....

In summary, Kolmogorov thought it plausible that the "chaotical mechanism of the translation of motion from the pulsations of lower orders to the pulsations of higher orders" that exists in high-R_t turbulence gives the "fine pulsations of the higher orders ... [an] approximately space-isotropic statistical regime," the state of *local isotropy*.

Today we often state his argument in terms of the rms fluctuating strain rate $u(r)/r$ associated with eddies of spatial scale r. As we discussed in Chapter 2, this increases monotonically from of order u/ℓ in the energy-containing eddies to υ/η in the dissipative eddies. Since the mean strain rate is of order u/ℓ, the ratio of the fluctuating and mean strain rates for an eddy of size r, $u(r)\ell/ur$, increases with decreasing eddy size r; it reaches $(\upsilon/\eta)/(u/\ell) \sim R_t^{1/2}$ in the dissipative eddies. At sufficiently large R_t, the argument goes, eddies beyond the energy-containing range are not influenced by the large-scale, anisotropic mean strain rate because it is much smaller than their own turbulent strain rate. Thus local isotropy is seen as an asymptotic state reached at sufficiently large R_t.

A variant of this argument is that the *energy cascade* from the largest eddies to the smallest ones (Chapters 6 and 7) occurs over such a large range of scales in large-R_t turbulence that the anisotropy inherent in the energy-containing range does not reach the smallest eddies. Equivalently, it is argued that the largest and smallest eddies do not directly interact in large-R_t turbulence.

14.5.2 Evidence

Batchelor (1960, p. 110) wrote that Kolgmogorov's notion of local isotropy "has now received considerable support." He cited early measurements in laboratory flows showing that certain predictions of Eq. (14.18), for example

$$\overline{\left(\frac{\partial u_1}{\partial x_1}\right)^2} = \frac{1}{2}\overline{\left(\frac{\partial u_2}{\partial x_1}\right)^2} = \frac{1}{2}\overline{\left(\frac{\partial u_3}{\partial x_1}\right)^2}, \tag{14.26}$$

were approximately verified across the turbulent wake of a cylinder. But since then numerous deviations from the isotropic forms for derivative covariances, Eq. (14.18) for velocity and Eq. (14.23) for a conserved scalar, have been documented in low and moderate Reynolds number shear flows. For example, in a plane jet Antonia *et al.* (1986) found (in comma notation)

$$\frac{\overline{u_{1,2}u_{1,2}}}{2\,\overline{u_{1,1}u_{1,1}}} = 1.8, \qquad \frac{\overline{u_{1,1}u_{1,2}}}{(\overline{u_{1,1}u_{1,1}})^{1/2}(\overline{u_{1,2}u_{1,2}})^{1/2}} = 0.23, \tag{14.27}$$

while under local isotropy these values are 1.0 and 0, respectively. In a nearly homogeneous turbulent shear flow Tavoularis and Corrsin (1981) measured values of 2.2 and −0.44, respectively. Antonia *et al.* (1991) found similar evidence of local anisotropy in the dissipation-rate tensor in direct numerical simulation of turbulent duct flow.

Shen and Warhaft (2000) studied the fine structure of the velocity field in grid turbulence having an unusually large Reynolds number – $R_\lambda \sim 10^3$, $R_t \sim 6 \times 10^4$. Second-order statistics (spectra, structure functions) showed local isotropy, but at fifth order, for $\overline{(\partial u/\partial y)^5}/[\overline{(\partial u/\partial y)^2}]^{5/2}$, local anisotropy had emerged.

Mydlarski and Warhaft (1998) found the temperature field in grid turbulence with a lateral gradient of mean temperature to have strong, Reynolds and Péclet number independent local anisotropy. The anisotropy was evident at second order, in deviations from Eq. (14.23) for derivative variances, and in higher, odd-order structure functions. They found no Reynolds or Péclet number dependence of temperature spectra and no qualitative change in the nature of the temperature fluctuations over the range $30 < R_\lambda < 700$, in strong contrast to the behavior of the turbulent velocity field in the same flow (Mydlarski and Warhaft, 1996). They commented that scalar turbulence shows high-R_t behavior even when the turbulent velocity field has low R_t.

As we discussed in Chapter 7, local anisotropy of temperature derivative fields is prominent at third order (Warhaft, 2000). Figure 14.1 is a summary of laboratory and atmospheric observations of the skewness of $\partial\theta/\partial x$ versus R_λ in flows with both mean shear and a mean temperature gradient. The sign of this skewness is given by (Sreenivasan, 1991)

$$\mathrm{sgn}(S_{\partial\theta/\partial x}) = -\mathrm{sgn}\left(\frac{dU}{dz}\right)\,\mathrm{sgn}\left(\frac{d\Theta}{dz}\right). \tag{14.28}$$

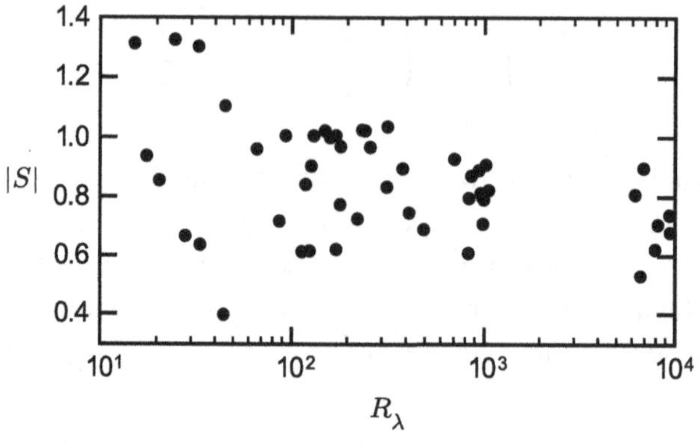

Figure 14.1 Measurements of the skewness of the streamwise derivative of fluctuating temperature as a function of R_λ. These were made in a large number of laboratory and geophysical flows with both mean shear and a mean temperature gradient. From Sreenivasan and Antonia (1997). Reprinted, with permission, from *Annual Review of Fluid Mechanics*, **29**, ©1997 by Annual Reviews, www.annualreviews.org.

In grid turbulence having a lateral (y) gradient of mean temperature, Tong and Warhaft (1994) found even larger skewness ($\sim$1.8) for $\partial\theta/\partial y$.

Sreenivasan (1991) attributes the skewness of the temperature derivative in shear flows with a mean temperature gradient to elongated eddy structures with a preferred orientation. Their signature is the "ramp-cliff" nature of temperature signals (Figure 7.2) observed in heated jets and in the surface layer on a sunny day. In this model the skewness is due largely to the "cliffs" – strong, one-signed temperature gradients. Thus it appears that scalar-derivative skewness is tied to the large-scale structure of the flow, in conflict with the ideas underlying the hypothesis of local isotropy.

14.5.3 The maintenance of local anisotropy

We can gain some insight into local anisotropy of a scalar field† through the conservation equation for scalar derivative skewness. In the process we'll see that the conservation equations for moments of derivative fields, and their scaling rules, differ in important ways from those for the energy-containing-range.

We'll begin at the second-moment level. Taking the gradient of the conserved scalar equation (1.31) and using our tilde notation for full fields yields an equation for $\tilde{g}_i = \partial\tilde{c}/\partial x_i$,

$$\frac{\partial \tilde{g}_i}{\partial t} + \tilde{u}_j \frac{\partial \tilde{g}_i}{\partial x_j} = -\tilde{g}_j \frac{\partial \tilde{u}_j}{\partial x_i} + \gamma \frac{\partial^2 \tilde{g}_i}{\partial x_j \, \partial x_j}. \tag{14.29}$$

† The author is indebted to Chenning Tong for discussions on this topic.

Because of the stretching term on the right side, the scalar gradient, like vorticity, is not a conserved variable. But their dot product $\tilde{g}_i\tilde{\omega}_i$ is conserved (Problem 1.12).

If we decompose the variables in Eq. (14.29) into mean and fluctuating parts,

$$\tilde{g}_i = G_i + g_i, \quad \tilde{u}_i = U_i + u_i, \tag{14.30}$$

the equation for the fluctuating gradient is

$$\frac{\partial g_i}{\partial t} + U_j\frac{\partial g_i}{\partial x_j} + u_j\frac{\partial G_i}{\partial x_j} + u_j\frac{\partial g_i}{\partial x_j} - \overline{u_j\frac{\partial g_i}{\partial x_j}}$$
$$= -G_j\frac{\partial u_j}{\partial x_i} - g_j\frac{\partial U_j}{\partial x_i} - g_j\frac{\partial u_j}{\partial x_i} + \overline{g_j\frac{\partial u_j}{\partial x_i}} + \gamma\frac{\partial^2 g_i}{\partial x_j\,\partial x_j}. \tag{14.31}$$

Multiplying Eq. (14.31) by $2g_i$, averaging, and rewriting the molecular term and dropping its diffusion part yields the evolution equation for the gradient variance:

$$\frac{\partial\overline{g_i g_i}}{\partial t} + U_j\frac{\partial\overline{g_i g_i}}{\partial x_j} + 2\left[\overline{u_j g_i}\frac{\partial G_i}{\partial x_j} + \overline{g_i g_j}\frac{\partial U_j}{\partial x_i} + \overline{g_i\frac{\partial u_j}{\partial x_i}}G_j\right]$$
$$+\frac{\partial\overline{u_j g_i g_i}}{\partial x_j} + 2\overline{\frac{\partial u_j}{\partial x_i}g_i g_j} = -2\gamma\overline{\frac{\partial g_i}{\partial x_j}\frac{\partial g_i}{\partial x_j}}. \tag{14.32}$$

The first four terms (counting the bracketed terms as one) on the left side and the term on the right side of Eq. (14.32) are of types that also appear in the equations for second moments of energy-containing range variables – in order, local time change, mean advection, mean-gradient production, turbulent transport, and (on the right side) molecular destruction. The fifth term on the left side is a new type that we'll call *turbulent production*.

We'll scale the terms in Eq. (14.32) as in Chapter 5, reintroducing the scalar intensity scale s and the Taylor microscale λ:

$$s = (\overline{c^2})^{1/2}, \qquad \epsilon = \nu\left(\overline{\frac{\partial u_i}{\partial x_j}\frac{\partial u_i}{\partial x_j}}\right) \sim \nu\frac{u^2}{\lambda^2} \sim \frac{u^3}{\ell}. \tag{14.33}$$

The fluctuating gradients of the scalar and velocity scale as s/λ and u/λ, respectively, and their mean gradients scale as s/ℓ and u/ℓ. We'll take the time scale of the time-change term as that of the evolution of the turbulent flow structure, the large-eddy time scale ℓ/u. We'll assume that $\gamma \sim \nu$.

Since the far right side of Eq. (14.33) implies that $\ell/\lambda \sim u\lambda/\nu \equiv R_\lambda$, the Taylor-microscale turbulence Reynolds number, the relative importance of the terms on the left side of Eq. (14.32) depends on R_λ. As a result the two dominant terms, which are of order us^2/λ^3, are to first approximation in balance at large R_λ:

$$\overline{\frac{\partial u_j}{\partial x_i} g_i g_j} = -\gamma \overline{\frac{\partial g_i}{\partial x_j}\frac{\partial g_i}{\partial x_j}}. \tag{14.34}$$

This says that to leading order the mean-squared intensity of scalar gradients is in equilibrium, its rate of production through the turbulent strain rate balanced by its rate of removal through molecular diffusion.

To help grasp the meaning of Eq. (14.34), consider the effect of an imbalance I in its two terms. From Eq. (14.32) the time scale of the response in $\overline{g_i g_i}$ would be of order $\overline{g_i g_i}/I \sim \lambda/u \sim \ell/u R_\lambda^{-1}$ – much less than the eddy-turnover time ℓ/u. So Eq. (14.34) says that on the time scale of the large eddies these two leading terms keep themselves in balance.

The other terms in Eq. (14.32) represent finite-R_λ corrections to the balance of the leading pair. Thus we see a fundamental difference in the nature of second-moment budgets for energy-containing-range variables, which we discussed in detail in Part I, and those for dissipative-range variables. In the former, all terms can be of the same order; some involve mean field–turbulence interactions and others represent turbulence–turbulence interactions. By contrast, budgets for dissipative-range quantities are inherently R_λ dependent, so at large R_λ some terms are much larger than others. The leading terms represent turbulence–turbulence interactions.

With that background, we'll now consider the budgets of the third moment of spatial derivatives of a conserved scalar. That for $\overline{(\theta_{,1})^3}$ in a quasi-steady, horizontally homogeneous ABL (Problem 14.23) is

$$\frac{\partial\overline{(\theta_{,1})^3}}{\partial t} = \underset{\frac{u}{\ell}\frac{s^3}{\lambda^3}}{-3\Theta_{,3}\,\overline{u_{3,1}\theta_{,1}\theta_{,1}}} - \underset{\frac{u}{\ell}\frac{s^3}{\lambda^3}}{\overline{(\theta_{,1}\theta_{,1}\theta_{,1}u_3)}_{,3}} - \underset{\frac{u}{\lambda}\frac{s^3}{\lambda^3}}{3\overline{\theta_{,1}\theta_{,1}\theta_{,j}u_{j,1}}} - 6\gamma\overline{\theta_{,1j}\theta_{,1j}\theta_{,1}}. \tag{14.35}$$

The terms on the right are, in order, mean-gradient production, turbulent transport, turbulent production, and molecular destruction; below each of the first three terms is its scaling estimate. The third term is the largest by a factor $\ell/\lambda \sim R_\lambda$. Thus, as in the budget of gradient variance, Eq. (14.34), to lowest order the quasi-steady balance in Eq. (14.35) is between turbulent production and molecular destruction.

However, this lowest-order form of Eq. (14.35) involves only derivatives of velocity and temperature, which qualifies it for the local isotropy assumption. But each of its two leading terms involves an odd-order tensor that vanishes under isotropy.

The third moment in the first term of Eq. (14.35), being odd in x_3 and x_1, also vanishes. Thus we conclude that the $\overline{(\theta_{,1})^3}$ budget is maintained by locally anisotropic mechanisms.

Sreenivasan *et al.* (1977) and Tong and Warhaft (1994) have found that the skewness of the lateral derivative $\theta_{,3}$ can be even larger than that of the streamwise derivative. Here the general third-moment budget is, with scaling of terms indicated (Problem 14.24)

$$\begin{aligned}\frac{\partial\overline{(\theta_{,3})^3}}{\partial t} &= -3\Theta_{,3}\,\overline{u_{3,3}\theta_{,3}\theta_{,3}} - 3U_{,3}\,\overline{\theta_{,3}\theta_{,3}\theta_{,1}} \\ &\qquad\quad \frac{s}{\ell}\frac{us^2}{\lambda^3} \qquad\qquad\quad \frac{u}{\ell}\frac{s^3}{\lambda^3} \\ &\quad -\overline{(\theta_{,3}\theta_{,3}\theta_{,3}u_3)}_{,3} - 3\overline{\theta_{,3}\theta_{,3}\theta_{,j}u_{j,3}} - 6\gamma\overline{\theta_{,3j}\theta_{,3j}\theta_{,3}}. \\ &\qquad\quad \frac{u}{\ell}\frac{s^3}{\lambda^3} \qquad\qquad\quad \frac{u}{\lambda}\frac{s^3}{\lambda^3}\end{aligned} \tag{14.36}$$

Again the lowest-order balance is between turbulent production and molecular destruction, and again both vanish under local isotropy. Of the two mean-gradient production terms, that involving the mean temperature gradient does not vanish under local isotropy, but is smaller than the leading terms by a factor $\ell/\lambda \sim R_\lambda$. It is not clear that it is large enough to maintain the observed values of $\overline{(\theta_{,3})^3}$.

In summary, the observational evidence indicates that shear flows with a mean scalar gradient can have an equilibrium state of local anisotropy of the turbulent scalar field. Scaling arguments suggest that the local anisotropy is produced through the turbulent strain field acting on turbulent scalar gradients and removed at the same mean rate through molecular diffusion. The roots of the anisotropy appear to lie in the mean shear and scalar gradient in the parent flow, but the nature of their statistical link to this local anisotropy is not evident.

On a closely related topic, Durbin and Speziale (1991) derived the evolution equation for the dissipation-rate tensor and found that maintaining the dissipation rate in the presence of a mean strain rate requires locally anisotropic turbulence. This supports the notion that the dissipation-rate tensor can be anisotropic in shear flows, which is consistent with evidence discussed in Section 14.5.2.

14.5.4 Gauging local anisotropy

Relatively few of the myriad implications of local isotropy have been tested. Claims of local anisotropy are often based on an observed ~1 derivative skewness that would vanish under isotropy. The implicit assumption is that $S \sim 1$ is large.

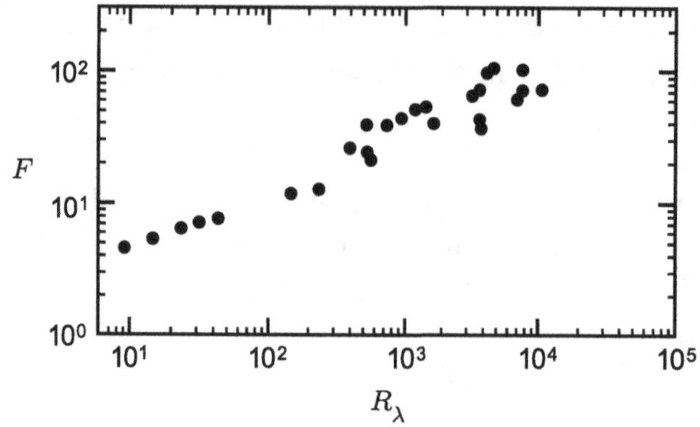

Figure 14.2 The flatness factor F of the streamwise derivative of temperature measured in various laboratory and atmospheric flows. From Sreenivasan and Antonia (1997). Reprinted, with permission, from *Annual Review of Fluid Mechanics*, **29**, ©1997 by Annual Reviews, www.annualreviews.org.

One can show that (Problem 7.16) the skewness S and flatness factor F of a stochastic variable are related by

$$|S| \leq \frac{F+1}{2}. \tag{14.37}$$

This suggests a different criterion for local isotropy – that S be much less than this upper limit:

$$|S| \ll \frac{F+1}{2}. \tag{14.38}$$

Figure 14.2 shows a compilation by Sreenivasan and Antonia (1997) of F for the temperature derivative in laboratory and atmospheric flows; values are as large as 100. Thus at large R_λ one might argue that $S \sim 1$ satisfies the constraint of Eq. (14.38) – i.e., that $S \sim 1$ is small. But the skewness of $\partial u/\partial x$ plays a dynamically important role in turbulence (Problem 14.22), and it is also $\sim$1. From that point of view $S \sim 1$ is not small.

14.5.5 Local isotropy in turbulence models

Given the contrary evidence, what can we say about modeling turbulence as locally isotropic?

In applications to acoustic and electromagnetic signal propagation, for example, the power spectral densities (spectra) of velocity, temperature, and water-vapor fields (Chapter 15) are often modeled as isotropic at large wavenumbers. There are two aspects of such spectra: their behavior in the inertial range, which lies just beyond the energy-containing range; and their behavior in the dissipative range beyond that, which is revealed by derivative moments (Chapter 15). There is now abundant evidence that flux cospectra (Chapter 15) in shear flows with mean scalar gradients do fall faster than κ^{-2} in the inertial subrange. This ensures that the turbulent flux budgets have negligible molecular-destruction terms at large R_t (Chapter 5), as required in local isotropy.

But as we have discussed, measurements show that for moments of velocity and temperature derivatives in shear flows with mean scalar gradients, local isotropy is only a first approximation. The isotropic forms (14.19) and (14.23), while attractive in their simplicity, could introduce non-negligible errors in shear flows.

Questions on key concepts

14.1 Explain the notion of an *isotropic* turbulence field.

14.2 Explain why naturally occurring turbulence is not isotropic in its energy-containing range.

14.3 In Chapter 6 we discussed forcing turbulence with a stochastic body force $\beta_i(t)$, Eq. (6.16). If we wish the turbulence to be isotropic, how should β_i be constrained? Discuss, for example, $\overline{\beta_i}$, $\overline{\beta_i \beta_j}$, and $\overline{\beta_i u_j}$. Explain your reasoning.

14.4 Explain the meaning of local isotropy and the essence of the traditional arguments for it.

14.5 Explain why the notion of local isotropy is useful in dealing with turbulence moment equations.

14.6 What new type of term appears in second-moment budgets for dissipative-range properties?

14.7 Explain the underlying difference in the scaling of second-moment budgets for energy-containing-range properties and for dissipative-range properties.

14.8 Explain physically why the traditional arguments for local isotropy of scalar fields seem not to hold in the presence of mean scalar gradients.

14.9 Explain physically why the skewness of the temperature derivative vanishes under the assumption of local isotropy.

Problems

14.1 Prove Eq. (14.19). Discuss how experimentalists use this result to estimate ϵ from a time series of u_1 measured at a point.

14.2 On the basis of local isotropy, what can you say about the individual components of the molecular destruction rates of scalar-scalar covariance and scalar variance?

14.3 For a coordinate transformation caused by rotation of the x_1 and x_2 axes by an angle α about the x_3 axis, write the transformation equations in terms of α. Then rewrite them in terms of the direction cosines.

14.4 Show that δ_{ij} is unchanged under coordinate rotation and reflection and hence is an isotropic tensor.

14.5 Show that $\epsilon_{ijk} = -\epsilon_{jik}$. Why does this imply that ϵ_{ijk} is unchanged under coordinate rotation but not under reflection?

14.6 Show that a diagonal second-order tensor does not change under a coordinate rotation if and only if its diagonal elements are equal.

14.7 Using the definition of vorticity, relate $\overline{\omega_i \omega_j}$ to M_{ijkm}. Use the isotropic form of M_{ijkm} to develop an expression for $\overline{\omega_i \omega_j}$.

14.8 Define $a_{ik} = \overline{u_i u_k} - \delta_{ik} q^2/3$. Show that the pressure covariance is its only global sink. What is its source term?

14.9 What is the form of the mean velocity gradient $\partial U_i / \partial x_j$ under isotropy? (Use continuity.)

14.10 The *structure function* $\Delta C(\mathbf{r})$ for a conserved scalar c in homogeneous turbulence is defined as

$$\Delta C(\mathbf{r}) = \overline{(c(\mathbf{x}) - c(\mathbf{x} + \mathbf{r}))^2}.$$

What is the dependence of ΔC on separation $\mathbf{r}$ in an isotropic field? How would you expect it to behave in a real turbulence field at separation distances (a) of order ℓ and (b) much less than ℓ?

14.11 Show that the squared and averaged continuity equation is consistent with local isotropy.

14.12 Write the isotropic forms of

$$\overline{\frac{\partial c}{\partial x_i} \frac{\partial c}{\partial x_j} \frac{\partial c}{\partial x_k}}, \quad \overline{\frac{\partial c}{\partial x_i} \frac{\partial c}{\partial x_j} \frac{\partial c}{\partial x_k} \frac{\partial c}{\partial x_l}}.$$

14.13 Use local isotropy to determine the forms of the two leading terms in the equation for conservation of scalar-gradient variance (Problem 5.14).

14.14 What constraints must an added body-force term in the Navier–Stokes equation satisfy in order that numerically calculated turbulence be isotropic?

14.15 Rewrite the molecular terms in the equation for the evolution of mean helicity (Problem 5.7). Under the local-isotropy assumption is mean helicity destroyed molecularly?

14.16 Turbulent *free convection* can be established between parallel horizontal plates, the bottom one heated and the top one cooled. The mean velocity is zero. Why is this flow inhomogeneous in the vertical? When can it approach homogeneity in the horizontal? In the latter case could it also have statistical symmetry about a vertical axis? In this case show that statistics in a horizontal plane follow isotropy but with the separation vector restricted to lie in the plane.

14.17 Use Eqs. (14.24) and (14.25) to express the two leading terms in the vorticity variance equation (Problem 5.6) in terms of streamwise derivatives of u_1 under the assumption of local isotropy.

14.18 Write the isotropic forms of the terms representing the rate of molecular destruction of stress and scalar flux.

14.19 Show that under the hypothesis of local isotropy the rates of molecular destruction of the TKE components are equal.

14.20 Explain why the rate of buoyant production of TKE vanishes under isotropy.

14.21 Under the local isotropy hypothesis the tensor $\overline{u_{i,j}\theta_{,k\ell}}$, as a covariance of derivatives, might be expected to have the isotropic form $\alpha\delta_{ij}\delta_{k\ell} + \beta\delta_{ik}\delta_{j\ell} + \gamma\delta_{i\ell}\delta_{jk}$. Find two constraints on this tensor and use them to solve for two of the coefficients in terms of the third.

14.22 Consider steady isotropic turbulence with zero mean velocity and mean vorticity.
(a) Write the equation for fluctuating vorticity ω_i. Use it to derive the vorticity variance budget.
(b) Identify the turbulent transport term in that budget and explain why it is zero.
(c) Interpret the two remaining terms physically. Write them as velocity derivative moments, and use local isotropy and Eqs. (14.24) and (14.25) to express them as moments of streamwise derivatives of streamwise velocity. Use the physics underlying this budget to explain why $\partial u/\partial x$ has nonzero skewness.

14.23 Derive the conservation equation for $\overline{(\theta_{,1})^3}$ in the ABL, Eq. (14.35).

14.24 Derive the conservation equation for $\overline{(\theta_{,3})^3}$ in the ABL, Eq. (14.36).

14.25 Carry out the scaling of Eq. (14.32) and show that the leading-order form is Eq. (14.34).

References

Antonia, R. A., F. Anselmet, and A. J. Chambers, 1986: Local-isotropy assessment in a turbulent plane jet. *J. Fluid Mech.*, **163**, 365–391.

Antonia, R. A., J. Kim, and L. W. B. Browne, 1991: Some characteristics of small-scale turbulence in a turbulent duct flow. *J. Fluid Mech.*, **233**, 369–388.

Batchelor, G. K., 1960: *The Theory of Homogeneous Turbulence.* Cambridge University Press.

Betchov, R., 1957: On the fine structure of turbulent flows. *J. Fluid Mech.*, **3**, 205–216.

Champagne, F. H., 1978: The fine-scale structure of the turbulent velocity field. *J. Fluid Mech.* **86**, 67–108.

Durbin, P. A., and C. G. Speziale, 1991: Local anisotropy in strained turbulence at high Reynolds numbers. *J. Fluids Eng.*, **113**, 707–709.

Friedlander, S. K., and L. Topper, 1961: *Turbulence: Classic Papers on Statistical Theory.* New York: Interscience Publishers.

Jeffreys, H., 1961: *Cartesian Tensors*. Cambridge University Press.

Kolmogorov, A. N., 1941: The local structure of turbulence in incompressible viscous fluid for very large Reynolds numbers. *Doklady ANSSSR*, **30**, 301–305.

Mydlarski, L., and Z. Warhaft, 1996: On the onset of high Reynolds number grid generated wind tunnel turbulence. *J. Fluid Mech.*, **320**, 331–368.

Mydlarski, L., and Z. Warhaft, 1998: Passive scalar statistics in high-Péclet-number grid turbulence. *J. Fluid Mech.*, **358**, 135–175.

Shen, X., and Z. Warhaft, 2000: The anisotropy of small scale structure in high Reynolds number ($R_\lambda \sim 1000$) turbulent shear flow. *Phys. Fluids*, **12**, 2976–2989.

Sreenivasan, K. R., 1991: On local isotropy of passive scalars in turbulent shear flows. *Proc. Roy. Soc. Lond. A*, **434**, 165–182.

Sreenivasan, K. R., and R. A. Antonia, 1997: The phenomenology of small-scale turbulence. *Ann. Rev. Fluid Mech.*, **29**, 435–472.

Sreenivasan, K. R., R. A. Antonia, and H. Q. Danh, 1977: Temperature dissipation fluctuations in a turbulent boundary layer. *Phys. Fluids*, **26**, 1238–1249.

Tavoularis, S., and S. Corrsin, 1981: Experiments in nearly homogeneous turbulent shear flow with a uniform temperature gradient. Part 2. The fine structure. *J. Fluid Mech.*, **104**, 349–367.

Taylor, G. I., 1935: Statistical theory of turbulence. *Proc. Roy. Soc. A*, **151**, 421–478.

Tong. C., and Z. Warhaft, 1994: On passive scalar derivative statistics in grid turbulence. *Phys. Fluids*, **6**, 2165–2176.

Warhaft, Z., 2000: Passive scalars in turbulent flows. *Ann. Rev. Fluid Mech.*, **32**, 203–240.

15

Covariances, autocorrelations, and spectra

15.1 Introduction

In Chapter 13 we introduced the concept of *moments* of a stochastic, random variable. In this chapter we'll extend that discussion, beginning with scalar functions of a single variable and the autocorrelation function. We'll then introduce its Fourier transform, the *power spectral density* or simply *spectrum*. We'll generalize to three dimensions and vector functions and cover classical material on spectra in isotropic turbulence. Finally we'll introduce spectral formulations tailored for use in the ABL, which is inhomogeneous in the vertical but can be homogeneous in the horizontal plane.

15.2 Scalar functions of a single variable

15.2.1 The autocorrelation function

In Eq. (13.22) we defined cross moments of a stochastic function of time, say $u(t)$. One example is the two-time covariance $\overline{u(t_1)u(t_2)}$, sometimes called the *autocovariance*. For a stationary function of time it depends only on the time separation $t_2 - t_1$. Its *autocorrelation function* is defined by nondimensionalizing with the variance:

$$\frac{\overline{u(t_1)u(t_2)}}{\overline{u^2}} = \rho(t_2 - t_1). \tag{15.1}$$

$\rho(t)$ is an even function (Problem 15.1). Schwartz's inequality (Problem 15.2) implies that $\overline{u(t_1)u(t_2)} \leq \overline{u^2}$, from which it follows that

$$|\rho(t)| \leq \rho(0) = 1. \tag{15.2}$$

For stochastic functions typically $\rho(t) \to 0$ as $t \to \infty$; we interpret this as indicating that $u(t)$ has a "fading memory" at sufficiently large separation in time.

The autocorrelation function often emerges in statistics of derivatives and integrals of $u(t)$. For example (Problem 15.3),

$$\overline{\left[\int_a^b u(t)\,dt\right]^2} = \int_a^b \int_a^b \overline{u(t')u(t'')}\,dt'\,dt'' = \overline{u^2} \int_a^b \int_a^b \rho(t'-t'')\,dt'\,dt''. \quad (15.3)$$

The variance of the derivative is (Problem 15.4)

$$\overline{\left(\frac{du(t)}{dt}\right)^2} = -\overline{u^2}\,\frac{d^2\rho(t)}{dt^2}\bigg|_{t=0}. \quad (15.4)$$

The quantities in (15.3) and (15.4) define two time scales associated with $\rho(t)$:

$$\int_0^\infty \rho(t)\,dt = \tau, \qquad \frac{d^2\rho(t)}{dt^2}\bigg|_{t=0} = -\frac{2}{\lambda^2}. \quad (15.5)$$

τ and λ are known as the integral scale and the microscale, respectively. In turbulence the corresponding scale λ_x of a spatial record $u(x)$ is called the Taylor microscale.

Figure 15.1 shows a typical autocorrelation function and its scales λ and τ. As indicated there, λ is the time of the zero crossing of the parabola fit to ρ at the origin, and τ is defined through the area under the curve. In large-R_t flows λ is small compared to τ (Problem 15.5). Trends in atmospheric data can prevent $\rho(t)$ from going to zero and thus make τ difficult to determine.

15.2.2 Fourier representation of a real, stochastic scalar function

Following Batchelor (1960) and Lumley and Panofsky (1964), we'll use the Fourier–Stieltjes representation for our random, stochastic fields. As the latter

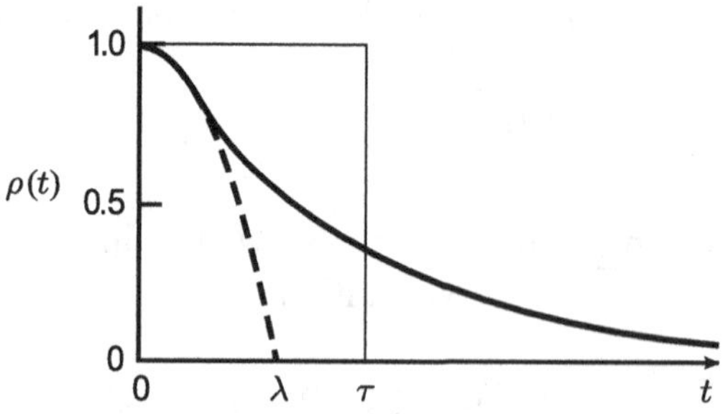

Figure 15.1 A sketch of an autocorrelation function ρ, with its microscale λ and integral scale τ shown. The area of the rectangle is the area under the ρ curve. From Lumley and Panofsky (1964).

explain, because such fields can be neither periodic nor integrable, neither the usual Fourier series nor integrals are formally applicable. However, according to Lumley and Panofsky (1964), under a set of weak assumptions $u(t)$ can be expanded in another random, stochastic process $Z(\omega)$:

$$u(t) = \int_{-\infty}^{+\infty} e^{i\omega t}\, dZ(\omega). \tag{15.6}$$

This is known as a stochastic Fourier–Stieltjes integral. The process $Z(\omega)$ has the property

$$\lim_{T\to\infty} \frac{1}{2\pi} \int_{-T}^{T} \frac{e^{-ibt} - e^{-iat}}{-it} u(t)\, dt = Z(b) - Z(a). \tag{15.7}$$

The integrals are written in this way so that $Z(\omega)$ need not be differentiable.

We can perhaps make Eqs. (15.6) and (15.7) more familiar as follows. If we let $a = \omega,\ b = \omega + \Delta\omega$, then we can write

$$\frac{e^{-ibt} - e^{-iat}}{-it} = \frac{e^{-i\omega t} e^{-i\Delta\omega t} - e^{-i\omega t}}{-it}. \tag{15.8}$$

With the series expansion $e^{-i\Delta\omega t} \simeq 1 - i\Delta\omega t$, this becomes

$$\frac{e^{-ibt} - e^{-iat}}{-it} \simeq \Delta\omega e^{-i\omega t}, \tag{15.9}$$

and Eq. (15.7) can be written, after dividing by $\Delta\omega$ and taking the limit as $\Delta\omega \to 0$,

$$\lim_{\Delta\omega\to 0} \left(\frac{1}{\Delta\omega} \lim_{T\to\infty} \frac{1}{2\pi} \int_{-T}^{T} \frac{e^{-ibt} - e^{-iat}}{-it} u(t)\, dt \right) = \frac{1}{2\pi} \int_{-\infty}^{\infty} e^{-i\omega t} u(t)\, dt = \frac{dZ}{d\omega}, \tag{15.10}$$

assuming the derivative $dZ/d\omega$ exists. Thus in that case Eq. (15.10) is the Fourier transform of $u(t)$. From Eq. (15.6) the other half of this transform pair is

$$u(t) = \int_{-\infty}^{+\infty} e^{i\omega t} \frac{dZ}{d\omega} d\omega, \tag{15.11}$$

the inverse Fourier transform.

We'll see later in the chapter that the Fourier–Stieltjes representation is particularly useful for finding exact statistical solutions to linear stochastic problems.

15.2.3 The power spectral density, or spectrum

The differences or increments $dZ(\omega)$ may be thought of as the complex amplitudes of the Fourier modes of frequency ω. They are *orthogonal*; that is, nonoverlapping members are uncorrelated:

$$\overline{dZ(\omega_1)\,dZ^*(\omega_2)} = 0, \qquad \omega_1 \neq \omega_2, \tag{15.12}$$

with dZ^* the complex conjugate. The variance of the differences is the differential of the power spectral distribution function F,

$$\overline{dZ(\omega)\,dZ^*(\omega)} = dF(\omega) = \phi(\omega)\,d\omega, \qquad F(\omega) = \int_{-\infty}^{\omega} \phi(\omega')\,d\omega', \tag{15.13}$$

with $\phi(\omega)$ the power spectral density or, loosely, the spectrum. The autocorrelation and power spectral distribution functions are related through

$$\overline{u^2}\rho(t) = \int_{-\infty}^{+\infty} e^{i\omega t}\,dF(\omega) = \int_{-\infty}^{+\infty} e^{i\omega t}\phi(\omega)\,d\omega, \tag{15.14}$$

which indicates that the autocorrelation function $\overline{u^2}\rho(t)$ is the Fourier transform of the spectrum ϕ. The inverse transform relation is

$$\phi(\omega) = \frac{1}{2\pi}\int_{-\infty}^{+\infty} e^{-i\omega t}\rho(t)\overline{u^2}\,dt. \tag{15.15}$$

Since ρ is an even function, ϕ is also an even function.

The adjective *power* is used here because when $u(t)$ is a voltage signal, its square is proportional to power. F is a *distribution* function because $F(\omega)$ gives the contribution to the variance $\overline{u^2}$ from frequencies below ω (from (15.13)).

According to Lumley and Panofsky (1964) this set of theorems is called the Wiener–Khintchine theorem, and in effect restores everything that was lost due to the lack of integrability or periodicity of our stochastic function $u(t)$.

15.2.4 Cross correlations and cross spectra

The foregoing analysis can be extended to two different stationary random functions, say $u(t)$ and $v(t)$. Their cross covariance is

$$\overline{u(t_1)v(t_2)} = C_{uv}(t_1 - t_2) = \int\int_{-\infty}^{+\infty} e^{i\omega t_1 - i\omega' t_2}\overline{dZ_u(\omega)\,dZ_v^*(\omega')}. \tag{15.16}$$

The Fourier–Stieltjes covariance here is

$$\overline{dZ_u(\omega)\,dZ_v^*(\omega')} = \begin{cases} 0, & \omega \neq \omega', \\ \phi_{uv}(\omega)d\omega, & \omega = \omega', \end{cases} \tag{15.17}$$

with ϕ_{uv} the cross spectrum. Thus the transform relationships are

$$\begin{aligned} C_{uv}(\tau) &= \int_{-\infty}^{+\infty} e^{i\omega\tau}\phi_{uv}(\omega)\,d\omega, \\ \phi_{uv}(\omega) &= \frac{1}{2\pi}\int_{-\infty}^{+\infty} e^{-i\omega\tau}C_{uv}(\tau)\,d\tau. \end{aligned} \tag{15.18}$$

If we write the cross covariance C_{uv} as the sum of even and odd parts,

$$\begin{aligned} C_{uv}(\tau) &= \frac{1}{2}\Big[C_{uv}(\tau) + C_{uv}(-\tau)\Big] + \frac{1}{2}\Big[C_{uv}(\tau) - C_{uv}(-\tau)\Big] \\ &= E_{uv}(\tau) + O_{uv}(\tau), \end{aligned} \tag{15.19}$$

and then use Eq. (15.18) and write the exponential in trigonometric form, we have

$$\begin{aligned} \phi_{uv}(\omega) &= \frac{1}{2\pi}\int_{-\infty}^{+\infty}(\cos\omega\tau - i\sin\omega\tau)\Big[E_{uv}(\tau) + O_{uv}(\tau)\Big]d\tau \\ &= \frac{1}{2\pi}\int_{-\infty}^{+\infty}\cos\omega\tau\, E_{uv}(\tau)\,d\tau - \frac{i}{2\pi}\int_{-\infty}^{+\infty}\sin\omega\tau\, O_{uv}(\tau)\,d\tau \\ &= Co_{uv}(\omega) - i\,Q_{uv}(\omega). \end{aligned} \tag{15.20}$$

Co_{uv} is the *cospectrum* and Q_{uv} the *quadrature spectrum*. The inverse transform pair is

$$\begin{aligned} E_{uv}(\tau) &= \int_{-\infty}^{+\infty}\cos\omega\tau\, Co_{uv}(\omega)\,d\omega, \\ O_{uv}(\tau) &= \int_{-\infty}^{+\infty}\sin\omega\tau\, Q_{uv}(\omega)\,d\omega. \end{aligned} \tag{15.21}$$

The covariance is given by

$$\overline{uv} = C_{uv}(0) = E_{uv}(0) = \int_{-\infty}^{+\infty} Co_{uv}(\omega)\,d\omega. \tag{15.22}$$

Thus the cospectrum is the density of contributions to the covariance as a function of frequency.

Equation (15.20) says that the quadrature spectrum is zero if C_{uv} is even. If the maximum correlation between u and v occurs at some nonzero time difference, then C_{uv} is not even. For example, u and v might be a conserved scalar at two points separated in the streamwise direction; the maximum correlation would occur at a lag corresponding to the transport time between the points.

If $v(t)$ is simply $u(t)$ delayed by a time interval Δt, we say that $v(t)$ has a "phase shift" of $\theta = \omega \Delta t$ and we can write

$$u(t) = \int_{-\infty}^{+\infty} e^{i\omega t}\, dZ(\omega), \qquad v(t) = \int_{-\infty}^{+\infty} e^{i(\omega t - \theta)}\, dZ(\omega). \tag{15.23}$$

The cross covariance is

$$C_{uv}(\tau) = \int_{-\infty}^{+\infty} e^{i(\omega\tau - \theta)} \phi(\omega)\, d\omega, \tag{15.24}$$

since $\overline{dZ(\omega)\, dZ^*(\omega)} = \phi(\omega)\, d\omega$. Thus we have

$$\begin{aligned} \phi_{uv}(\omega) &= e^{-i\theta}\phi(\omega) = \cos\theta\phi(\omega) - i\sin\theta\phi(\omega), \\ Co_{uv}(\omega) &= \cos\theta\phi(\omega),\ Q_{uv}(\omega) = \sin\theta\phi(\omega), \\ \tan\theta &= \frac{Q_{uv}(\omega)}{Co_{uv}(\omega)}. \end{aligned} \tag{15.25}$$

A time lag $\Delta t(\omega)$ at each frequency is defined by

$$\tan[\omega \Delta t(\omega)] = \frac{Q_{uv}(\omega)}{Co_{uv}(\omega)}. \tag{15.26}$$

Another dimensionless quantity is the coherence, the square of the spectral correlation, or the normalized covariance:

$$\begin{aligned} Coh_{uv}(\omega) &= \frac{|\overline{dZ_u(\omega)\, dZ_v^*(\omega)}|^2}{\overline{dZ_u(\omega)\, dZ_u^*(\omega)}\ \overline{dZ_v(\omega)\, dZ_v^*(\omega)}} \\ &= \frac{|\phi_{uv}|^2}{\phi_u \phi_v} = \frac{Co_{uv}^2(\omega) + Q_{uv}^2(\omega)}{\phi_u(\omega)\, \phi_v(\omega)}. \end{aligned} \tag{15.27}$$

By Schwartz's inequality this cannot exceed 1; that value occurs when the Fourier components of u and v at frequency ω are proportional. The example of (15.23) to (15.25) has a coherence of unity at all frequencies.

15.3 Scalar functions of space and time

15.3.1 Extending the formalism

The concepts we have introduced apply also when our independent variable is space rather than time. Homogeneity in a spatial coordinate then corresponds to stationarity in time. We take as independent variables only those spatial coordinates in which the stochastic function is homogeneous. Spatial coordinates in which it is inhomogeneous, and time in nonstationary conditions, become parameters.

Consider a random scalar function θ (temperature, say) that is homogeneous in three spatial coodinates, $x_i \equiv \mathbf{x}$. Our Fourier representation is now

$$\theta(\mathbf{x}; t) = \int_{-\infty}^{+\infty} e^{i\boldsymbol{\kappa}\cdot\mathbf{x}}\, dZ(\boldsymbol{\kappa}; t), \tag{15.28}$$

where $\boldsymbol{\kappa}$ is a wavenumber vector. We write

$$\overline{dZ(\boldsymbol{\kappa}; t)\, dZ^*(\boldsymbol{\kappa}'; t)} = \begin{cases} 0, & \boldsymbol{\kappa} \neq \boldsymbol{\kappa}' \\ \phi(\boldsymbol{\kappa}, t) d\boldsymbol{\kappa}, & \boldsymbol{\kappa} = \boldsymbol{\kappa}'. \end{cases} \tag{15.29}$$

$\phi(\boldsymbol{\kappa}; t)$ is the power spectral density; it is the Fourier transform of the autocorrelation function:

$$\overline{\theta^2}(t)\rho(\mathbf{r}; t) = \int_{-\infty}^{+\infty} e^{i\boldsymbol{\kappa}\cdot\mathbf{r}} \phi(\boldsymbol{\kappa}; t)\, d\boldsymbol{\kappa},$$

$$\phi(\boldsymbol{\kappa}; t) = \frac{1}{(2\pi)^3} \int_{-\infty}^{+\infty} e^{-i\boldsymbol{\kappa}\cdot\mathbf{r}}\, \overline{\theta^2}(t)\rho(\mathbf{r}; t)\, d\mathbf{r}. \tag{15.30}$$

If θ were homogeneous in only two directions, as in the x_1–x_2 plane in a horizontally homogeneous turbulent boundary layer, for example, we would write

$$\theta(\mathbf{x}, t) = \int_{-\infty}^{+\infty} e^{i(\kappa_1 x_1 + \kappa_2 x_2)}\, dZ(\kappa_1, \kappa_2; x_3, t),$$

with the Fourier–Stieltjes coefficients having the properties

$$\overline{dZ(\kappa_1, \kappa_2; x_3, t) dZ^*(\kappa_1', \kappa_2'; x_3, t)}$$
$$= \begin{cases} 0, & \kappa_1 \neq \kappa_1', \kappa_2 \neq \kappa_2' \\ \phi(\kappa_1, \kappa_2; x_3, t) d\kappa_1 d\kappa_2, & \kappa_1 = \kappa_1', \kappa_2 = \kappa_2' \end{cases} \tag{15.31}$$

The associated transform pair is

$$\overline{\theta^2}(x_3; t)\rho(r_1, r_2; x_3, t) = \int_{-\infty}^{+\infty} e^{i(\kappa_1 r_1 + \kappa_2 r_2)} \phi(\kappa_1, \kappa_2; x_3, t)\, d\kappa_1\, d\kappa_2,$$

$$\phi(\kappa_1, \kappa_2; x_3, t) = \frac{1}{(2\pi)^2} \int_{-\infty}^{+\infty} e^{-i(\kappa_1 r_1 + \kappa_2 r_2)} \overline{\theta^2}(x_3, t) \rho(r_1, r_2; x_3, t)\, dr_1\, dr_2. \tag{15.32}$$

That is, both x_3 and t are parameters, correlations being made in the x_1–x_2 plane at time t. Correlations between two different values of x_3 or t are treated as co- and quadrature spectra and cross covariances were treated.

If a scalar random function depends on three space coordinates and time, and is stationary but inhomogeneous in all three directions, then the analysis of Section 15.1 applies, but with an additional parameter denoting the space point. Joint statistics between two space points are handled as a cross covariance.

If the function were homogeneous in all three directions but nonstationary, then the analysis of Eqs. (15.28) to (15.30) applies, with an additional parameter of time. Statistics at two times would be handled by cross covariances. If the function were both stationary and homogeneous we would have

$$\theta(\mathbf{x}, t) = \int_{-\infty}^{+\infty} e^{i\boldsymbol{\kappa}\cdot\mathbf{x} + i\omega t}\, dZ(\boldsymbol{\kappa}, \omega),$$

$$\overline{dZ(\boldsymbol{\kappa}, \omega)\, dZ^*(\boldsymbol{\kappa}', \omega')} = \begin{cases} 0, & \boldsymbol{\kappa} \neq \boldsymbol{\kappa}', \quad \omega \neq \omega', \\ \phi(\boldsymbol{\kappa}, \omega) d\boldsymbol{\kappa} d\omega, & \boldsymbol{\kappa} = \boldsymbol{\kappa}', \quad \omega = \omega', \end{cases} \tag{15.33}$$

$$\overline{\theta(\mathbf{x}, t)\theta(\mathbf{x} + \boldsymbol{\xi}, t + \tau)} = \overline{\theta^2}\rho(\boldsymbol{\xi}, \tau) = \int_{-\infty}^{+\infty} e^{i\boldsymbol{\kappa}\cdot\boldsymbol{\xi} + i\omega\tau} \phi(\boldsymbol{\kappa}, \omega)\, d\boldsymbol{\kappa}\, d\omega,$$

$$\phi(\boldsymbol{\kappa}, \omega) = \frac{1}{(2\pi)^4} \int_{-\infty}^{+\infty} e^{-i\boldsymbol{\kappa}\cdot\boldsymbol{\xi} - i\omega\tau} \overline{\theta^2} \rho(\boldsymbol{\xi}, \tau)\, d\boldsymbol{\xi} d\tau,$$

where ρ is called a space-time correlation and ϕ is a space-time spectrum.

15.3.2 Application to typical measurements

Say we are measuring a turbulent scalar θ along the x_1 coordinate – e.g., on an aircraft, with x_1 the flight-path direction, or by using Taylor's hypothesis with a stationary sensor at a fixed point. Then in homogeneous turbulence the autocorrelation function, spectrum relation of Eq. (15.30) becomes (hereafter not explicitly indicating dependence on parameters such as t)

$$\begin{aligned} \overline{\theta^2}\rho(\xi_1, 0, 0) &= \int_{-\infty}^{+\infty} e^{i\kappa_1 \xi_1} \phi(\boldsymbol{\kappa})\, d\boldsymbol{\kappa} \\ &= \int_{-\infty}^{+\infty} e^{i\kappa_1 \xi_1} \left(\int \int_{-\infty}^{+\infty} \phi(\boldsymbol{\kappa})\, d\kappa_2\, d\kappa_3 \right) d\kappa_1 \end{aligned}$$

$$= \int_{-\infty}^{+\infty} e^{i\kappa_1 \xi_1} F^1(\kappa_1)\, d\kappa_1, \tag{15.34}$$

$$F^1(\kappa_1) = \iint_{-\infty}^{+\infty} \phi(\boldsymbol{\kappa})\, d\kappa_2\, d\kappa_3.$$

F^1 is called a one-dimensional spectrum. Its superscript indicates the direction of the spatial separation vector $\boldsymbol{\xi}$.† Equation (15.34) indicates that $F^1(\kappa_1)$ and $\overline{\theta^2}\rho(\xi_1, 0, 0)$ are a Fourier transform pair:

$$\int_{-\infty}^{+\infty} e^{i\kappa_1 \xi_1} F^1(\kappa_1)\, d\kappa_1 = \overline{\theta^2}\rho(\xi_1, 0, 0),$$

$$F^1(\kappa_1) = \frac{1}{2\pi} \int_{-\infty}^{+\infty} e^{-i\kappa_1 \xi_1} \overline{\theta^2}\rho(\xi_1, 0, 0)\, d\xi_1. \tag{15.35}$$

Evaluating the second of Eqs. (15.35) at $\kappa_1 = 0$ gives

$$F^1(0) = \frac{\overline{\theta^2}}{2\pi} \int_{-\infty}^{+\infty} \rho(\xi_1, 0, 0)\, d\xi_1 = \frac{\overline{\theta^2}}{\pi} \ell_1, \tag{15.36}$$

where ℓ_1 is the integral scale of θ in the x_1 direction. Equation (15.36) says that the one-dimensional spectrum does not vanish at zero wavenumber. This happens because a Fourier mode of wavenumber magnitude κ, oriented nearly normal to the x_1 direction, appears to have a very small wavenumber in that direction.

The *three-dimensional spectrum* E_c (the subscript c differentiates it from E, the three-dimensional energy spectrum) is the integral of $\phi(\boldsymbol{\kappa})$ over a spherical shell of radius κ:

$$E_c(\kappa) = \iint_{\kappa_i \kappa_i = \kappa^2} \phi(\boldsymbol{\kappa})\, d\sigma. \tag{15.37}$$

As indicated, $E_c(\kappa)$ contains contributions from all Fourier modes of wavenumber magnitude κ regardless of their direction. From (15.30),

$$\overline{\theta^2} = \int_{-\infty}^{+\infty} \phi(\boldsymbol{\kappa})\, d\boldsymbol{\kappa} = \int_0^{\infty} \left(\iint_{\kappa_i \kappa_i = \kappa^2} \phi(\boldsymbol{\kappa})\, d\sigma \right) d\kappa = \int_0^{\infty} E_c(\kappa)\, d\kappa. \tag{15.38}$$

Like E, E_c vanishes at the origin because there is no energy at $\kappa = 0$; we are working with zero-mean variables. Thus, unlike the one-dimensional spectrum, its shape does reflect the relative importance of the contributions of eddies of spatial scale $1/\kappa$ to the variance.

E_c is the scalar spectrum traditionally used in turbulence theory. The Obukhov–Corrsin arguments (Chapter 7) for the behavior of the scalar spectrum in the inertial subrange, for example, are made for E_c.

† Some authors indicate this direction by writing $F(\kappa_1)$, but since a function depends on the *value* of its argument, not the *name*, this is inappropriate.

15.3.3 Isotropy

As we discussed in Chapter 3, isotropy is unlikely to appear in the energy-containing range of "natural" (as opposed to computational) turbulence. But what Batchelor (1960) calls *axisymmetry*, invariance to rotations and reflections about an axis, is possible – as in turbulent free convection, for example, which has axisymmetry about the vertical axis. Axisymmetry is isotropy in the plane normal to the axis of symmetry.

An isotropic scalar field has simple relationships among the functions ϕ, F^1 and E_c discussed in the last section. Under isotropy $\phi(\boldsymbol{\kappa}) = \phi(\kappa)$ and Eq. (15.37) becomes

$$E_c(\kappa) = \int\int_{\kappa_i\kappa_i=\kappa^2} \phi(\kappa)\, d\sigma = 4\pi\kappa^2\phi(\kappa). \tag{15.39}$$

The one-dimensional spectrum $F^1(\kappa_1)$ then is, from (15.34),

$$F^1(\kappa_1) = \int\int_{-\infty}^{\infty} \phi(\kappa)\, d\kappa_2\, d\kappa_3 = \int\int_{-\infty}^{\infty} \frac{E_c(\kappa)}{4\pi\kappa^2} d\kappa_2\, d\kappa_3. \tag{15.40}$$

As indicated by the wavenumber-space diagram in Figure 15.2, we can integrate over circular rings and write (15.40) as

$$F^1(\kappa_1) = \int_0^{\infty} \frac{E_c(\kappa)}{4\pi\kappa^2} 2\pi\kappa\, d\kappa = \int_{\kappa_1}^{\infty} \frac{E_c(\kappa)}{2\kappa} d\kappa, \tag{15.41}$$

from which we find by differentiation with respect to κ_1

$$\frac{\partial F^1(\kappa_1)}{\partial \kappa_1} = -\frac{E_c(\kappa_1)}{2\kappa_1}. \tag{15.42}$$

This equation allows the three-dimensional spectrum E_c to be found from measurements of the one-dimensional spectrum F^1 if the field is isotropic. Furthermore,

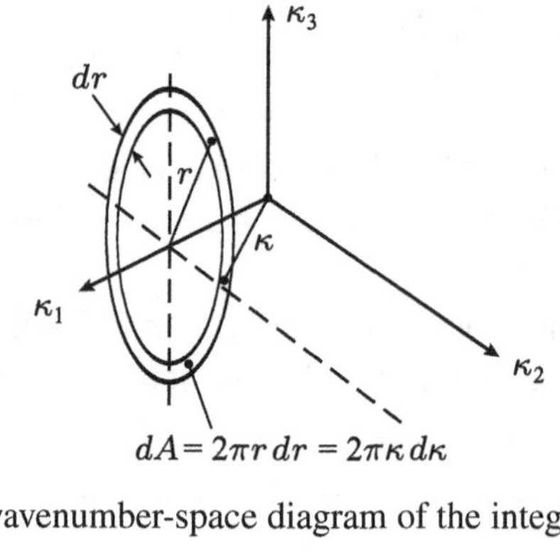

Figure 15.2 A wavenumber-space diagram of the integral in Eq. (15.40).

from (15.34) we see that under isotropy the one-dimensional spectra are all equal: $F^1 = F^2 = F^3$.

15.3.4 The inertial subrange

Conserved scalar fluctuations in the inertial range, i.e., for scales much less than ℓ and much greater than η, are traditionally argued (Chapter 7) to depend only on r, ϵ, and χ_c. If so, then in this range E_c depends only on κ, ϵ, and χ_c. Dimensionally, this implies the inertial-range form (Obukhov, 1949; Corrsin, 1951)

$$E_c(\kappa) = \beta \chi_c \epsilon^{-1/3} \kappa^{-5/3}, \tag{7.9}$$

with β a universal constant. Using this inertial-range form for E_c in Eq. (15.42) gives

$$F^1(\kappa_1) = \frac{3}{10} \beta \chi_c \epsilon^{-1/3} \kappa_1^{-5/3} = \beta_1 \chi_c \epsilon^{-1/3} \kappa_1^{-5/3}, \tag{15.43}$$

with β_1 called the one-dimensional spectral constant.

15.3.5 Conventions in the literature

The one-dimensional spectrum is naturally defined as in Eq. (15.34) – as a function of a variable that ranges from $-\infty$ to ∞. But since it is an even function it can be convenient to use the "one-sided" form that integrates over 0 to ∞ to the variance. This doubles the value of β_1.

There are two conventions for the three-dimensional scalar spectrum E_c. It integrates either to the variance, as in Eq. (15.38), or (as in Tennekes and Lumley, 1972) the half-variance.

15.4 Vector functions of space and time

In order to deal with turbulent velocity fields we shall now extend everything we have done to vectors. The requirements for homogeneity and stationarity given at the beginning of Section 15.2 are the same. We will initially assume that the turbulent velocity field is homogeneous in all three spatial directions, consider only correlations at the same time, and not indicate explicitly the dependence on time or other parameters. Although the results are more complicated notationally, they do have direct and important implications for turbulence measurements.

15.4.1 The covariance, spectral density pair

With the Fourier–Stieltjes formalism we'll write our stationary, homogeneous turbulent velocity field as

$$u_i(\mathbf{x}) = \int_{-\infty}^{+\infty} e^{i\boldsymbol{\kappa}\cdot\mathbf{x}}\, dZ_i(\boldsymbol{\kappa}), \tag{15.44}$$

where the $dZ_i(\boldsymbol{\kappa})$ are random vector functions. As with scalars, they have orthogonal increments:

$$\overline{dZ_i(\boldsymbol{\kappa})\, dZ_j^*(\boldsymbol{\kappa}')} = 0, \quad \boldsymbol{\kappa} \neq \boldsymbol{\kappa}'; \qquad \overline{dZ_i(\boldsymbol{\kappa})\, dZ_j^*(\boldsymbol{\kappa}')} = \phi_{ij}(\boldsymbol{\kappa}) d\boldsymbol{\kappa}, \quad \boldsymbol{\kappa} = \boldsymbol{\kappa}'. \tag{15.45}$$

The transform relationships are

$$\begin{aligned} \overline{u_i(\mathbf{x})u_j(\mathbf{x}+\mathbf{r})} = R_{ij}(\mathbf{r}) &= \int_{-\infty}^{+\infty} e^{i\boldsymbol{\kappa}\cdot\mathbf{r}} \phi_{ij}(\boldsymbol{\kappa})\, d\boldsymbol{\kappa}, \\ \phi_{ij}(\boldsymbol{\kappa}) &= \frac{1}{(2\pi)^3} \int_{-\infty}^{+\infty} e^{-i\boldsymbol{\kappa}\cdot\mathbf{r}} R_{ij}(\mathbf{r})\, d\mathbf{r}. \end{aligned} \tag{15.46}$$

R_{ij} is a *covariance* when $i = j$ and a *cross covariance* when $i \neq j$. It is called the *correlation tensor*. ϕ_{ij} is the *spectral density tensor*.

By definition a homogeneous field is statistically unchanged under a translation of its coordinate axes. From Eq. (15.46) this implies that $R_{\alpha\alpha}$ is an even function; $R_{\alpha\alpha}(\mathbf{r}) = R_{\alpha\alpha}(-\mathbf{r})$. As a result $\phi_{\alpha\alpha}$ is purely real and an even function (Problem 15.12). This is not true for the off-diagonal terms, however; R_{12}, for example, is neither even nor odd, and ϕ_{12} is neither real nor imaginary.

Cospectra and quadrature spectra are defined as for scalars. We first write $R_{ij} = E_{ij} + O_{ij}$, the sum of even and odd parts. The Fourier transform of R_{ij} is, from (15.46),

$$\begin{aligned} \phi_{ij}(\boldsymbol{\kappa}) &= \frac{1}{(2\pi)^3} \int_{-\infty}^{+\infty} e^{-i\boldsymbol{\kappa}\cdot\mathbf{r}} \left[E_{ij}(\mathbf{r}) + O_{ij}(\mathbf{r})\right] d\mathbf{r} \\ &= \frac{1}{(2\pi)^3} \int_{-\infty}^{+\infty} \cos(\boldsymbol{\kappa}\cdot\mathbf{r}) E_{ij}(\mathbf{r})\, d\mathbf{r} - \frac{i}{(2\pi)^3} \int_{-\infty}^{+\infty} \sin(\boldsymbol{\kappa}\cdot\mathbf{r}) O_{ij}(\mathbf{r})\, d\mathbf{r} \\ &= Co_{ij}(\boldsymbol{\kappa}) - i Q_{ij}(\boldsymbol{\kappa}), \end{aligned} \tag{15.47}$$

the sum of real and imaginary parts.

As for scalars, we can obtain the so-called *one-dimensional spectra* that can be determined from measurements. If we measure u_1 along the x_1 axis, for example, then we have from (15.46)

$$R_{11}(r_1, 0, 0) = \int_{-\infty}^{+\infty} e^{i\kappa_1 r_1} \phi_{11}(\boldsymbol{\kappa})\, d\boldsymbol{\kappa}$$
$$= \int_{-\infty}^{+\infty} e^{i\kappa_1 r_1} \left(\iint_{-\infty}^{+\infty} \phi_{11}(\boldsymbol{\kappa})\, d\kappa_2\, d\kappa_3 \right) d\kappa_1. \qquad (15.48)$$

We define the one-dimensional spectrum F_{11}^1 as

$$F_{11}^1(\kappa_1) = \iint_{-\infty}^{+\infty} \phi_{11}(\boldsymbol{\kappa})\, d\kappa_2\, d\kappa_3. \qquad (15.49)$$

From Eq. (15.48) it is the one-dimensional Fourier transform of $R_{11}(r_1, 0, 0)$:

$$R_{11}(r_1, 0, 0) = \int_{-\infty}^{+\infty} e^{i\kappa_1 r_1} F_{11}^1(\kappa_1)\, d\kappa_1. \qquad (15.50)$$

As with the one-dimensional scalar spectrum, the superscript 1 in F_{11}^1 indicates the direction of the separation vector $\mathbf{r}$. Similar one-dimensional spectra can be defined for other combinations of velocity; F_{13}^1, for example, goes with $R_{13}(r_1, 0, 0)$.

If the two lower indices are the same, then the one-dimensional spectrum is a real, even function. If they differ, then the one-dimensional spectrum will not be real, so it can be split into co- and quadrature spectra. The co- and quadrature one-dimensional spectra can be related to the full co- and quadrature spectra in a way similar to (15.49) (Problem 15.13).

The *three-dimensional spectrum* is also defined with the indices contracted,

$$E(\kappa) = \iint_{\kappa_i \kappa_i = \kappa^2} \frac{\phi_{ii}(\boldsymbol{\kappa})}{2} d\sigma, \qquad (15.51)$$

the factor of two being included so it integrates to the turbulence kinetic energy per unit mass (TKE):

$$\frac{\overline{u_i u_i}}{2} = \int_0^\infty E(\kappa) d\kappa. \qquad (2.63)$$

This three-dimensional spectrum represents the contribution of Fourier modes of wavenumber magnitude κ to the TKE, regardless of the orientation of the Fourier modes. Spherical averages of other functions (for example, Co_{ij}, Section 15.4.3) are also used.

15.4.2 Isotropy

We introduced the implications of isotropy for scalar spectra in Section 15.3.3. Here we'll extend the discussion to turbulent velocity fields.

15.4.2.1 The spectral density tensor

As we discussed in Chapter 14, as a second-order tensor function of a vector $\phi_{ij}(\boldsymbol{\kappa})$ has the isotropic form

$$\phi_{ij} = J(\kappa)\kappa_i\kappa_j + K(\kappa)\delta_{ij}, \tag{14.9}$$

with $J(\kappa)$ and $K(\kappa)$ functions to be determined. The requirement of zero velocity divergence implies, from the definition (15.46) of R_{ij}, that $\partial R_{ij}/\partial r_j = 0$. This, in turn, implies from (15.46) that $\kappa_j\phi_{ij} = 0$, which from (14.9) gives the constraint

$$\kappa_i\left[J(\kappa)\kappa^2 + K(\kappa)\right] = 0, \quad \text{so that} \quad J(\kappa) = -\frac{K(\kappa)}{\kappa^2}. \tag{15.52}$$

Thus, ϕ_{ij} becomes

$$\phi_{ij} = K(\kappa)\left[\delta_{ij} - \frac{\kappa_i\kappa_j}{\kappa^2}\right]. \tag{15.53}$$

Using the form (15.53) in (15.51) then gives

$$E(\kappa) = \iint_{\kappa_i\kappa_i=\kappa^2} K(\kappa)\, d\sigma = 4\pi\kappa^2 K(\kappa), \tag{15.54}$$

so $K(\kappa) = E/(4\pi\kappa^2)$ and we have finally

$$\phi_{ij}(\boldsymbol{\kappa}) = \frac{E(\kappa)}{4\pi\kappa^2}\left(\delta_{ij} - \frac{\kappa_i\kappa_j}{\kappa^2}\right) \tag{15.55}$$

as the form of the spectral density tensor in an isotropic turbulent velocity field.

In an isotropic field we have

$$\overline{u_i u_j} = \overline{u_\alpha u_\alpha}\,\delta_{ij},$$

so that, for example, $\overline{u_1u_3} = 0$. From Eq. (15.46) we also have

$$\overline{u_1u_3} = 0 = \int_{-\infty}^{+\infty} \phi_{13}(\boldsymbol{\kappa})\, d\boldsymbol{\kappa}. \tag{15.56}$$

One might expect that Eq. (15.56) is enforced in an isotropic field by making $\phi_{13} = 0$, but Eq. (15.55) shows this is not the case:

$$\phi_{13}(\boldsymbol{\kappa}) = -\frac{E(\kappa)}{4\pi\kappa^2}\left(\frac{\kappa_1\kappa_3}{\kappa^2}\right). \tag{15.57}$$

Instead, the integral (15.56) vanishes because Eq. (15.57) shows that ϕ_{13} is odd in κ_1 and κ_3. Evidently the continuity constraint $u_{i,i} = 0$ prevents ϕ_{13} from vanishing (Problem 15.24).

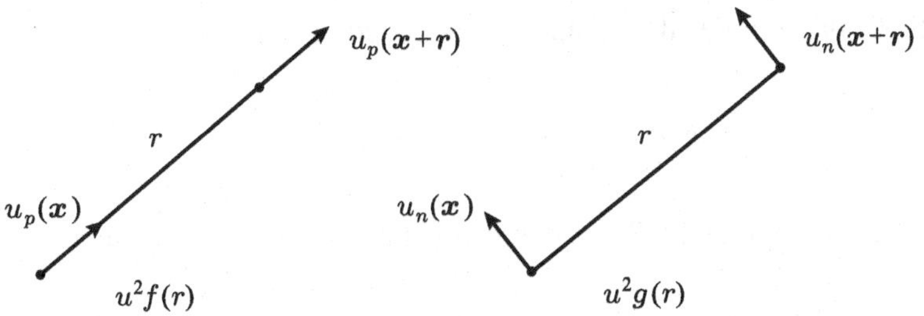

Figure 15.3 A schematic of the longitudinal and transverse correlation functions. From Batchelor, 1960.

15.4.2.2 The correlation tensor

Batchelor (1960) was evidently the first to apply this analysis to the correlation tensor R_{ij} in isotropic turbulence. As a second-order tensor function of a vector, it has the isotropic form

$$R_{ij}(\mathbf{r}) = \overline{u_i(\mathbf{x})u_j(\mathbf{x}+\mathbf{r})} = \alpha(r)r_i r_j + \beta(r)\delta_{ij}, \tag{15.58}$$

with $\alpha(r)$ and $\beta(r)$ to be determined. He expressed them in terms of the longitudinal and transverse correlations f and g, respectively, sketched in Figure 15.3:

$$u^2 f(r) = R_{11}(r,0,0) = \alpha(r)r^2 + \beta(r); \qquad u^2 g(r) = R_{11}(0,r,0) = \beta(r), \tag{15.59}$$

where by isotropy $\overline{u_p^2} = \overline{u_n^2} = \frac{1}{3}\overline{u_i u_i} = u^2$. Solving (15.59) for α and β yields

$$\alpha = u^2\left(\frac{f-g}{r^2}\right), \qquad \beta = u^2 g. \tag{15.60}$$

Thus, in terms of f and g the isotropic form of R_{ij} is

$$R_{ij}(\mathbf{r}) = u^2\left(\frac{f(r)-g(r)}{r^2}r_i r_j + g(r)\delta_{ij}\right). \tag{15.61}$$

The two unknown functions f and g are related through incompressibility. We have

$$\frac{\partial u_i}{\partial x_i} = 0 = \frac{\partial R_{ij}}{\partial r_j}, \tag{15.62}$$

and since

$$r^2 = r_i r_i, \quad \frac{\partial}{\partial r_i} = \frac{\partial r}{\partial r_i}\frac{\partial}{\partial r} = \frac{r_i}{r}\frac{\partial}{\partial r}, \tag{15.63}$$

this yields the constraint

$$\xi\frac{\partial \alpha}{\partial r} + 4\alpha + \frac{1}{r}\frac{\partial \beta}{\partial r} = 0. \tag{15.64}$$

With (15.60) this implies that f and g are related through

$$g(r) = f + \frac{r}{2}\frac{\partial f}{\partial r} = \frac{1}{2r}\frac{\partial}{\partial r}(r^2 f), \tag{15.65}$$

so that in an isotropic field R_{ij} is determined by a single scalar function, either f or g.

G. I. Taylor (1935) defined the length scale λ from the curvature of g at the origin:

$$\lambda = \left(-2 \Big/ \left.\frac{\partial^2 g}{\partial r^2}\right|_0\right)^{1/2}. \tag{15.66}$$

Today we call λ the Taylor microscale. He misinterpreted λ, stating that it "may roughly be regarded as a measure of the diameters of the *smallest* eddies which are responsible for the dissipation of energy." In fact from the definition of g, Figure 15.3, we can express Eq. (15.66) as (Problem 15.20)

$$\lambda^2 = \left(2u^2 \Big/ \overline{\frac{\partial u_n}{\partial x_p}\frac{\partial u_n}{\partial x_p}}\right) \sim \frac{u^2}{\epsilon/\nu}. \tag{15.67}$$

We can rewrite this as

$$\lambda = \ell R_t^{-1/2} = \eta R_t^{1/4}, \tag{15.68}$$

which shows that the length scale λ lies between ℓ and η, much closer to the latter (Problem 15.14). It is considered to be more significant in a time characteristic of the viscous dissipation, $\tau_d \sim u^2/\epsilon \sim \lambda^2/\nu$, than as a length scale.

We can also use isotropy to relate the one-dimensional spectra, which we can now write as

$$u^2 f(r) = \int_{-\infty}^{+\infty} e^{i\kappa r} F_{11}^1(\kappa)\, d\kappa, \qquad u^2 g(r) = \int_{-\infty}^{+\infty} e^{i\kappa r} F_{11}^2(\kappa)\, d\kappa. \tag{15.69}$$

From (15.69) we can write

$$u^2 \frac{\partial f}{\partial r} = \int_{-\infty}^{+\infty} e^{i\kappa r} i\kappa F_{11}^1(\kappa)\, d\kappa. \tag{15.70}$$

Thus, the inverse relation is

$$i\kappa F_{11}^1(\kappa) = \frac{1}{2\pi}\int_{-\infty}^{+\infty} e^{-i\kappa r} u^2 \frac{\partial f}{\partial r} dr. \tag{15.71}$$

Differentiating (15.71) with respect to $i\kappa$ yields

$$\frac{\partial i\kappa F_{11}^1}{\partial i\kappa} = \frac{\partial \kappa F_{11}^1}{\partial \kappa} = \frac{1}{2\pi}\int_{-\infty}^{+\infty} e^{-i\kappa r}\left(-u^2 r\frac{\partial f}{\partial r}\right) dr. \tag{15.72}$$

The inverse of (15.72) is

$$-u^2 r\frac{\partial f}{\partial r} = \int_{-\infty}^{+\infty} e^{i\kappa r}\frac{\partial \kappa F_{11}^1(\kappa)}{\partial \kappa} d\kappa. \tag{15.73}$$

Thus, the physical-space expression (15.65)

$$g(r) = f + \frac{r}{2}\frac{\partial f}{\partial r}$$

transforms to

$$F_{11}^2 = F_{11}^1 - \frac{1}{2}\frac{\partial}{\partial \kappa}\kappa F_{11}^1. \tag{15.74}$$

We can rewrite this as

$$F_{11}^2 = \frac{1}{2}\left(F_{11}^1 - \kappa\frac{\partial F_{11}^1}{\partial \kappa}\right). \tag{15.75}$$

Since under isotropy the statistics are invariant to rotation and reflection of the coordinate system, we also have

$$\begin{gathered} F_{11}^2 = F_{22}^1 = F_{33}^1 = F_{11}^3 = F_{33}^2 = F_{22}^3, \\ F_{11}^1 = F_{22}^2 = F_{33}^3. \end{gathered} \tag{15.76}$$

Finally, under isotropy we can relate these one-dimensional spectra to E. By definition

$$F_{11}^1(\kappa_1) = \iint_{-\infty}^{\infty} \phi_{11}\, d\kappa_2\, d\kappa_3, \tag{15.77}$$

and using the isotropic form (15.55) for ϕ_{ij} gives

$$F_{11}^1(\kappa_1) = \iint_{-\infty}^{\infty} \frac{E(\kappa)}{4\pi\kappa^4}\left(\kappa_2^2 + \kappa_3^2\right) d\kappa_2\, d\kappa_3. \tag{15.78}$$

Figure 15.2 shows the geometry of this integral. Since $d\kappa_2\, d\kappa_3 = 2\pi\kappa\, d\kappa$ and $\kappa_2^2 + \kappa_3^2 = \kappa^2 - \kappa_1^2$, we have

$$F_{11}^1(\kappa_1) = \int_{\kappa_1}^{\infty} \frac{E(\kappa)}{2\kappa^3}\left(\kappa^2 - \kappa_1^2\right) d\kappa. \tag{15.79}$$

Differentiating (15.79) with respect to κ_1 yields

$$\frac{\partial F_{11}^1}{\partial \kappa_1} = -\int_{\kappa_1}^{\infty} \frac{\kappa_1 E(\kappa)}{\kappa^3} d\kappa, \qquad \frac{1}{\kappa_1}\frac{\partial F_{11}^1}{\partial \kappa_1} = -\int_{\kappa_1}^{\infty} \frac{E(\kappa)}{\kappa^3} d\kappa, \tag{15.80}$$

the derivative of the lower limit of the integral not contributing since the integrand vanishes there. Differentiating once more with respect to κ_1 gives

$$\frac{\partial}{\partial \kappa_1}\left(\frac{1}{\kappa_1}\frac{\partial F_{11}^1}{\partial \kappa_1}\right) = \frac{E(\kappa_1)}{\kappa_1^3}, \qquad E(\kappa_1) = \kappa_1^3 \frac{\partial}{\partial \kappa_1}\left(\frac{1}{\kappa_1}\frac{\partial F_{11}^1}{\partial \kappa_1}\right). \tag{15.81}$$

We usually write this result as

$$E(\kappa) = \kappa^3 \frac{\partial}{\partial \kappa}\left(\frac{1}{\kappa}\frac{\partial F_{11}^1}{\partial \kappa}\right). \tag{15.82}$$

15.4.3 The inertial subrange

In the inertial subrange the Kolmogorov (1941) arguments (Chapter 7) imply that

$$E(\kappa) \sim \epsilon^{2/3}\kappa^{-5/3} = \alpha\epsilon^{2/3}\kappa^{-5/3}, \tag{15.83}$$

where $\alpha \simeq 1.5$ is called the Kolmogorov constant. The one-dimensional spectra also have inertial ranges but with different spectral constants. Equation (15.80) (or 15.82) implies that F_{11}^1 behaves in the inertial range as

$$F_{11}^1 = \frac{9}{55}\alpha\epsilon^{2/3}\kappa^{-5/3}. \tag{15.84}$$

Likewise, Eqs. (15.75) and (15.76) show that in the inertial subrange

$$F_{11}^2 = F_{22}^1 = F_{33}^1 = F_{11}^3 = F_{33}^2 = F_{22}^3 = \frac{4}{3}F_{11}^1 = \frac{12}{55}\alpha\epsilon^{2/3}\kappa^{-5/3}. \tag{15.85}$$

The last result – the 4/3 ratio between inertial-subrange levels of "longitudinal" and "transverse" one-dimensional spectra – is often used as a test of the approach to isotropy at small scales. In observational work one typically measures frequency spectra of u and v and/or w, interprets them as the one-dimensional streamwise wavenumber spectra F_{11}^1 and F_{22}^1 and/or F_{33}^1, and examines their ratio at inertial-range κ_1. Figure 15.4 shows results of this procedure applied to the Kansas data.

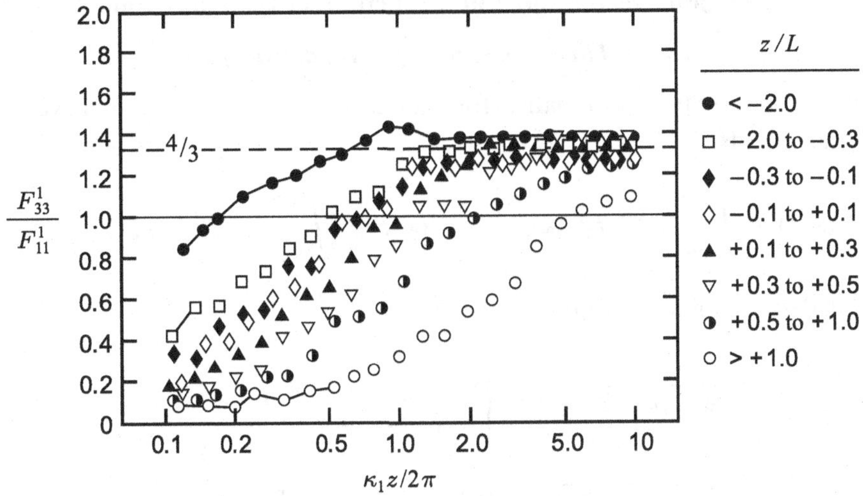

Figure 15.4 The approach to the isotropic (4/3) ratio of F_{33}^1 and F_{11}^1 in the inertial subrange in the 1968 Kansas experiment. z is distance from the surface, L is the Monin–Obukhov length, and κ_1 is streamwise wavenumber. From Wyngaard (1973).

Again, different normalizations of one-dimensional spectra are used. Sometimes a form is used wherein the integral over the half-line is the variance, in which case the factor 9/55 becomes 18/55.

An approach to isotropy in the inertial subrange in a turbulent shear flow should be manifested also in a stress cospectrum that falls faster than the energy spectrum. Lumley (1967) has argued that in the inertial range $Co_{13}(\kappa)$, the spherically averaged cospectrum of kinematic stress $\overline{uw}$, depends on the mean shear U', ϵ, and κ. Since the dependence on U' should be linear, on dimensional grounds we then have

$$Co_{13}(\kappa) \sim \epsilon^{1/3} U' \kappa^{-7/3} f\left[\left(\frac{U'}{(\epsilon\kappa^2)^{1/3}}\right)^2\right]. \qquad (15.86)$$

The argument of the function f here can be interpreted as $(U'/u')^2$, where u' is the strain rate of an eddy of size $1/\kappa$ (Problem 15.23). When this argument is small the turbulent strain rate dominates, $f \to$ constant, and this becomes

$$Co_{13}(\kappa) \sim \epsilon^{1/3} U' \kappa^{-7/3}. \qquad (15.87)$$

This inertial-range behavior has been observed for the one-dimensional cospectrum of u and w (which should be proportional to Co_{13}) in the Kansas experiments (Wyngaard and Coté, 1972) and in high-R_t boundary layers (Saddoughi and Veeravalli, 1994).

15.5 Joint vector and scalar functions of space and time

15.5.1 The covariance, spectral density pair

With the Fourier–Stieltjes formalism for stationary, homogeneous turbulent velocity and scalar fields,

$$u_i(\mathbf{x}; t) = \int_{-\infty}^{+\infty} e^{i\boldsymbol{\kappa}\cdot\mathbf{x}}\, dZ_i(\boldsymbol{\kappa}; t), \qquad \theta(\mathbf{x}; t) = \int_{-\infty}^{+\infty} e^{i\boldsymbol{\kappa}\cdot\mathbf{x}}\, dZ(\boldsymbol{\kappa}; t), \tag{15.88}$$

we can define a joint spectrum:

$$\overline{dZ_i(\boldsymbol{\kappa})\, dZ^*(\boldsymbol{\kappa}')} = \begin{cases} 0, & \boldsymbol{\kappa} \neq \boldsymbol{\kappa}', \\ \phi_i(\boldsymbol{\kappa}) d\boldsymbol{\kappa}, & \boldsymbol{\kappa} = \boldsymbol{\kappa}'; \end{cases}$$

$$\overline{u_i(\mathbf{x}+\mathbf{r})\, \theta(\mathbf{x})} = R_i(\mathbf{r}) = \int_{-\infty}^{+\infty} e^{i\boldsymbol{\kappa}\cdot\mathbf{r}} \phi_i(\boldsymbol{\kappa})\, d\boldsymbol{\kappa},$$

$$\phi_i(\boldsymbol{\kappa}) = \frac{1}{(2\pi)^3} \int_{-\infty}^{+\infty} e^{-i\boldsymbol{\kappa}\cdot\mathbf{r}} R_i(\mathbf{r})\, d\mathbf{r}. \tag{15.89}$$

Again we can write $R_i = E_i + O_i$, the sum of even and odd parts. The Fourier transform of R_i is then, from (15.89),

$$\begin{aligned} \phi_i(\boldsymbol{\kappa}) &= \frac{1}{(2\pi)^3} \int_{-\infty}^{+\infty} e^{-i\boldsymbol{\kappa}\cdot\mathbf{r}} \left[E_i(\mathbf{r}) + O_i(\mathbf{r})\right] d\mathbf{r} \\ &= \frac{1}{(2\pi)^3} \int_{-\infty}^{+\infty} \cos(\boldsymbol{\kappa}\cdot\mathbf{r})\, E_i(\mathbf{r})\, d\mathbf{r} - \frac{i}{(2\pi)^3} \int_{-\infty}^{+\infty} \sin(\boldsymbol{\kappa}\cdot\mathbf{r})\, O_i(\mathbf{r})\, d\mathbf{r} \\ &= Co_i(\boldsymbol{\kappa}) - i Q_i(\boldsymbol{\kappa}), \end{aligned} \tag{15.90}$$

the sum of real and imaginary parts. Since $Q_i(\boldsymbol{\kappa})$ is an odd function, we have from Eq. (15.89)

$$\overline{u_i(\mathbf{x})\, \theta(\mathbf{x})} = \int_{-\infty}^{+\infty} \phi_i(\boldsymbol{\kappa})\, d\boldsymbol{\kappa} = \int_{-\infty}^{+\infty} Co_i(\boldsymbol{\kappa})\, d\boldsymbol{\kappa}. \tag{15.91}$$

Thus $Co_i(\boldsymbol{\kappa})$ represents the spectral density of contributions to the scalar flux $\overline{u_i\theta}$.

15.5.2 Isotropy

In an isotropic field $Co_i(\boldsymbol{\kappa})$ has the form $A\kappa_i$ (Chapter 14), with A a scalar. The zero-divergence constraint on u_i requires that $\kappa_i A \kappa_i = 0$, so that $A = 0$. Hence $Co_i(\boldsymbol{\kappa}) = 0$, and thus $\overline{u_i\theta} = 0$. The last follows more directly in physical space: as an isotropic vector, $\overline{u_i\theta} = 0$.

As we argued for stress, in anisotropic, flux-carrying turbulence a tendency toward isotropy at smaller scales – beginning in the inertial range, say – should be evidenced by a spherically averaged flux cospectrum $Co_i(\kappa)$ that decreases more rapidly with wavenumber than the energy and scalar spectra. As we discussed in Part II, both the streamwise ($i = 1$) and vertical ($i = 3$) components of the one-dimensional forms of this cospectrum are observed to be nonzero in the ABL. Lumley and Panofsky (1964) predicted that Co_3 behaves in the inertial range as

$$Co_3 \sim \frac{\partial \Theta}{\partial z} \epsilon^{1/3} \kappa^{-7/3}. \tag{15.92}$$

This behavior was observed in the one-dimensional cospectrum of u_3 and θ measured in the Kansas experiments (Wyngaard and Coté, 1972). Those authors predicted that the cospectrum of u_1 and θ behaves as

$$Co_1 \sim \frac{\partial \Theta}{\partial z} \frac{\partial U}{\partial z} \kappa^{-3}, \tag{15.93}$$

but the slope of the observed one-dimensional spectrum was $\simeq -2.5$. Bos and Bertoglio (2007) generalized the scaling of Eq. (15.93) to include ϵ, which yields

$$Co_1 \sim \frac{\partial \Theta}{\partial z} \frac{\partial U}{\partial z}^{\alpha} \epsilon^{(1-\alpha)/3} \kappa^{-(7+2\alpha)/3}, \tag{15.94}$$

with α a free parameter. With $\alpha = 1/3$ this yields an inertial-range slope of -2.55, which agrees well with the measurements. Thus the observed behavior of these cospectra is also consistent with the notion of an approach to isotropy at smaller scales.

15.6 Spectra in the plane

15.6.1 The concept

Although they are necessarily inhomogeneous in the vertical, boundary-layer flows can approach homogeneity in the horizontal plane. If so, then from numerically simulated fields, for example, we can calculate spectra as a function of the wavenumber vector $\boldsymbol{\kappa}_{\mathrm{h}} = (\kappa_1, \kappa_2)$ in the horizontal plane; the vertical coordinate is a parameter, as in Eq. (15.32). Unlike one-dimensional wavenumber spectra, these two-dimensional spectra vanish at zero wavenumber and so directly indicate the horizontal spatial scale of eddies contributing to them.

Let's designate the spectrum that integrates over the horizontal plane to the scalar variance as $\phi^{(2)}$:

$$\iint_{-\infty}^{\infty} \phi^{(2)}(\boldsymbol{\kappa}_{\mathrm{h}})\, d\kappa_1\, d\kappa_2 = \overline{c^2}. \tag{15.95}$$

The one-dimensional streamwise wavenumber spectrum F_c is related to $\phi^{(2)}$ by

$$F_c(\kappa_1) = \int_{-\infty}^{\infty} \phi^{(2)}(\boldsymbol{\kappa}_{\mathrm{h}})\, d\kappa_2. \tag{15.96}$$

We'll assume isotropy in the horizontal plane (i.e., axisymmetry in z), so that $\phi^{(2)}(\boldsymbol{\kappa}_{\mathrm{h}}) = \phi^{(2)}(\kappa_{\mathrm{h}})$. We'll define a *two-dimensional spectrum* $E_c^{(2)}$ by integrating over circular rings in the horizontal plane:

$$E_c^{(2)}(\boldsymbol{\kappa}_{\mathrm{h}}) = \int_0^{2\pi} \phi^{(2)}(\kappa_{\mathrm{h}})\kappa_{\mathrm{h}}\, d\theta = 2\pi\kappa_{\mathrm{h}}\phi^{(2)}(\kappa_{\mathrm{h}}). \tag{15.97}$$

It also integrates to the variance,

$$\int_0^{\infty} E_c^{(2)}(\kappa_{\mathrm{h}})\, d\kappa_{\mathrm{h}} = \overline{c^2}. \tag{15.98}$$

From Eqs. (15.96), (15.97), and axisymmetry it follows that

$$F_c(\kappa_1) = \int_{-\infty}^{\infty} \frac{E_c^{(2)}(\kappa_{\mathrm{h}})}{2\pi\kappa_{\mathrm{h}}} d\kappa_2. \tag{15.99}$$

Kelly and Wyngaard (2006) showed that this can be inverted to yield

$$E_c^{(2)}(\kappa_{\mathrm{h}}) = -\frac{d}{d\kappa_{\mathrm{h}}} \int_{\kappa_{\mathrm{h}}}^{\infty} \frac{2\kappa_1 F_c(\kappa_1)}{(\kappa_1^2 - \kappa_h^2)^{1/2}} d\kappa_1. \tag{15.100}$$

Equation (15.100) provides a way to determine the two-dimensional spectrum from measurements of the one-dimensional spectrum, under the assumption of axisymmetry.

15.6.2 The inertial range

At energy-containing wavenumbers spectra depend strongly on the stability state of the ABL, but we've seen that in the inertial subrange and beyond they can approach universality. We'll assume the turbulence at these smaller scales is isotropic in the plane. It is particularly convenient to find variables whose spectra in the plane have axisymmetric forms in an isotropic field, for such spectra depend on a single wavenumber.

Let us assume that at wavenumbers $\kappa \gg 1/h$, where $\kappa = |\boldsymbol{\kappa}| = |(\kappa_1, \kappa_2, \kappa_3)|$ and h is the boundary-layer depth, the vertical inhomogeneity of the boundary layer is unimportant and the spectral density in the horizontal plane is the integral of the full spectral density over all κ_3, as in homogeneous turbulence. Thus, we write the spectral density tensor and the scalar spectral density in the horizontal plane (denoted by a superscript (2)) as

$$\phi_{ij}^{(2)}(\boldsymbol{\kappa}_{\rm h}) = \int_{-\infty}^{\infty} \phi_{ij}(\boldsymbol{\kappa})\, d\kappa_3, \qquad \phi^{(2)}(\boldsymbol{\kappa}_{\rm h}) = \int_{-\infty}^{\infty} \phi(\boldsymbol{\kappa})\, d\kappa_3, \qquad \boldsymbol{\kappa}_h = (\kappa_1, \kappa_2). \tag{15.101}$$

In this large wavenumber region we use the isotropic forms,

$$\phi_{ij}(\boldsymbol{\kappa}) = \frac{E(\kappa)}{4\pi\kappa^2}\left[\delta_{ij} - \frac{\kappa_i\kappa_j}{\kappa^2}\right], \qquad \phi(\boldsymbol{\kappa}) = \phi(\kappa) = \frac{E_c(\kappa)}{4\pi\kappa^2}. \tag{15.102}$$

We'll consider separately the inertial-range spectra of vertical velocity, horizontal velocity, and a conserved scalar in the horizontal plane. Miles *et al.* (2004) have also done the inertial-range spectrum of pressure.

15.6.2.1 Vertical velocity

From Eq. (15.102) we have

$$\phi_{33}(\boldsymbol{\kappa}) = \frac{E(\kappa)}{4\pi\kappa^4}\left(\kappa^2 - \kappa_3^2\right) = \frac{E(\kappa)}{4\pi\kappa^4}\left(\kappa_1^2 + \kappa_2^2\right) = \frac{E(\kappa)}{4\pi\kappa^4}\kappa_{\rm h}^2. \tag{15.103}$$

From Eq. (15.101) the spectrum in the horizontal plane is therefore

$$\phi_{33}^{(2)}(\boldsymbol{\kappa}_{\rm h}) = \int_{-\infty}^{\infty} \frac{E(\kappa)}{4\pi\kappa^4}\kappa_{\rm h}^2 d\kappa_3 = \phi_{33}^{(2)}(\kappa_{\rm h}). \tag{15.104}$$

Since $\phi_{33}^{(2)}$ is axisymmetric, it is natural to define a two-dimensional spectrum for w,

$$E_w^{(2)} = \int_{\boldsymbol{\kappa}_{\rm h}\cdot\boldsymbol{\kappa}_{\rm h}=\kappa_{\rm h}^2} \phi_{33}^{(2)}(\kappa_{\rm h})\, ds = 2\pi\kappa_{\rm h}\phi_{33}^{(2)}(\kappa_{\rm h}) = \int_{-\infty}^{\infty} \frac{E(\kappa)}{2\kappa^4}\kappa_{\rm h}^3\, d\kappa_3. \tag{15.105}$$

The geometry of the integral in (15.105) is shown in Figure 15.5. From $\kappa_3 = \kappa_{\rm h}\tan\theta$ we have $d\kappa_3 = \kappa_{\rm h} d\theta/\cos^2\theta$. We also have $\kappa = \kappa_{\rm h}/\cos\theta$. Using the inertial subrange form of the three-dimensional spectrum

$$E(\kappa) = \alpha\epsilon^{2/3}\kappa^{-5/3} \tag{15.106}$$

in (15.105) then yields

$$E_w^{(2)} = \alpha\epsilon^{2/3}\kappa_{\rm h}^{-5/3}\int_0^{\pi/2} \cos^{11/3}\theta\, d\theta. \tag{15.107}$$

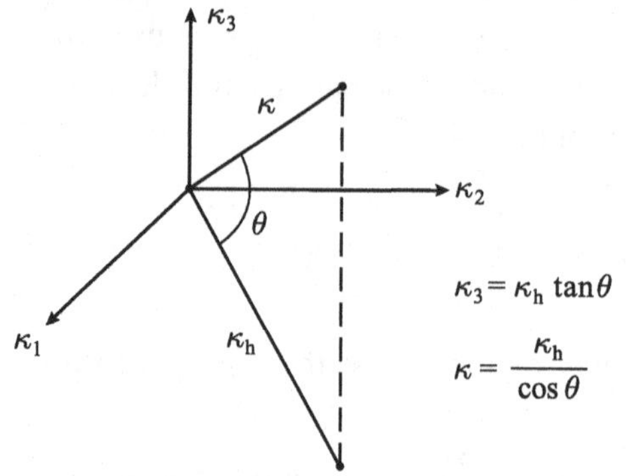

Figure 15.5 The geometry of a common spectral integral.

Integrals of the form of (15.107) appear frequently in these analyses. We can generalize it to†

$$I(\alpha) = \int_0^{\pi/2} \cos^{\alpha}\theta \, d\theta. \tag{15.108}$$

If we define $v = \cos^2\theta$ we can rewrite this as

$$I(\alpha) = \frac{1}{2}\int_0^1 v^{\frac{\alpha-1}{2}}(1-v)^{-\frac{1}{2}}\, dv. \tag{15.109}$$

The Beta function is defined by

$$B(m,n) = \int_0^1 v^{m-1}(1-v)^{n-1}\, dv. \tag{15.110}$$

When m and n are any positive real numbers,

$$B(m,n) = B(n,m) = \frac{\Gamma(m)\Gamma(n)}{\Gamma(m+n)}. \tag{15.111}$$

In our expression (15.109) for $I(\alpha)$ we have $n = 1/2$ and $m = (\alpha+1)/2$, so that

$$I(\alpha) = \frac{\Gamma(\frac{1}{2})\Gamma(\frac{\alpha+1}{2})}{2\Gamma(1+\frac{\alpha}{2})} = \frac{\sqrt{\pi}}{2}\frac{\Gamma\left(\frac{\alpha+1}{2}\right)}{\Gamma\left(1+\frac{\alpha}{2}\right)}, \tag{15.112}$$

from which we find $I(\frac{11}{3}) = 0.61$. Equation (15.107) for the circularly averaged vertical velocity spectrum in the plane then becomes

$$E_w^{(2)} = 0.61\alpha\epsilon^{2/3}\kappa_h^{-5/3}, \tag{15.113}$$

† I am indebted to Ricardo Munoz for this development.

which indicates that the Kolmogorov constant for the two-dimensional spectrum of vertical velocity is 0.61α.

15.6.2.2 Horizontal velocity

From (15.102) the spectral densities of the horizontal velocity components are

$$\phi_{11} = \frac{E(\kappa)}{4\pi\kappa^4}\left(\kappa_2^2 + \kappa_3^2\right), \quad \phi_{22} = \frac{E(\kappa)}{4\pi\kappa^4}\left(\kappa_1^2 + \kappa_3^2\right). \tag{15.114}$$

These are not axisymmetric in the horizontal plane. However, their average

$$\frac{\phi_{11} + \phi_{22}}{2} = \frac{E(\kappa)}{4\pi\kappa^4}\left(\frac{\kappa_h^2 + 2\kappa_3^2}{2}\right) \tag{15.115}$$

is axisymmetric. The spectrum of this average is then

$$\left(\frac{\phi_{11}^{(2)} + \phi_{22}^{(2)}}{2}\right)(\kappa_h) = \int_{-\infty}^{\infty} \frac{E(\kappa)}{4\pi\kappa^4}\left(\frac{\kappa_h^2 + 2\kappa_3^2}{2}\right) d\kappa_3. \tag{15.116}$$

The two-dimensional spectrum is, from the axisymmetry,

$$E_h^{(2)} = 2\pi\kappa_h\left(\frac{\phi_{11}^{(2)} + \phi_{22}^{(2)}}{2}\right) = \int_{-\infty}^{\infty} \frac{\kappa_h E(\kappa)}{2\kappa^4}\left(\frac{\kappa_h^2 + 2\kappa_3^2}{2}\right) d\kappa_3. \tag{15.117}$$

Using the geometry of Figure 15.5, this yields in the inertial subrange

$$E_h^{(2)} = \alpha\epsilon^{2/3}\kappa_h^{-5/3}\left[\int_0^{\pi/2} \cos^{5/3}\theta\, d\theta - \frac{1}{2}\int_0^{\pi/2} \cos^{11/3}\theta\, d\theta\right]. \tag{15.118}$$

Evaluating the integrals through (15.112) yields

$$E_h^{(2)} = 0.54\alpha\epsilon^{2/3}\kappa_h^{-5/3}. \tag{15.119}$$

This says that the Kolmogorov constant for the two-dimensional spectrum of the average of the two horizontal velocity spectra is 0.54α, slightly less than that for vertical velocity.

15.6.2.3 A conserved scalar

From (15.102) the spectral density of a conserved scalar is related to its three-dimensional spectrum through

$$\phi(\kappa) = \frac{E_c(\kappa)}{4\pi\kappa^2}. \tag{15.120}$$

The spectrum in the horizontal plane is

$$\phi^{(2)}(\kappa_{\mathrm{h}}) = \int_{-\infty}^{\infty} \frac{E_c(\kappa)}{4\pi\kappa^2} d\kappa_3. \tag{15.121}$$

Since this is again axisymmetric, the two-dimensional scalar spectrum $E_c^{(2)}$ is

$$E_c^{(2)} = 2\pi\kappa_{\mathrm{h}}\phi^{(2)} = \int_{-\infty}^{\infty} \frac{\kappa_{\mathrm{h}} E_c(\kappa)}{2\kappa^2} d\kappa_3. \tag{15.122}$$

Using the geometry of Figure 15.5 and the inertial subrange form

$$E_c(\kappa) = \beta\chi_c\epsilon^{-1/3}\kappa^{-5/3}, \tag{7.9}$$

this yields

$$E_c^{(2)} = \beta\chi_c\epsilon^{-1/3}\kappa_{\mathrm{h}}^{-5/3} \int_0^{\pi/2} \cos^{5/3}\theta\, d\theta, \tag{15.123}$$

which through (15.112) becomes

$$E_c^{(2)} = 0.84\beta\chi_c\epsilon^{-1/3}\kappa_{\mathrm{h}}^{-5/3}, \tag{15.124}$$

the two-dimensional spectral constant here being 0.84β.

Questions on key concepts

15.1 Outline the notion of an *autocorrelation function.*

15.2 Explain why we say the autocorrelation function is an indicator of the "memory" of a stochastic function.

15.3 How do the integral scale and the microscale of a stochastic process differ?

15.4 Explain in simple terms the meaning of the power spectral density of a function. How does randomness enter, and how is it accommodated?

15.5 Interpret and explain Eq. (15.34) physically.

15.6 Explain physically the notion of a *three-dimensional* spectrum in turbulence and why it is attractive.

15.7 Explain how in relatively low-R_{t} flows the Taylor microscale might be misinterpreted as the scale of the viscous eddies.

15.8 Describe the concept of spectra in the plane.

15.9 Discuss how and why spectra in the plane can be more useful than one-dimensional spectra.

15.10 Discuss some of the spectral implications of the approach to isotropy at small scales.

Problems

15.1 Show that the autocorrelation function is an even function of its argument: $\rho(t) = \rho(-t)$.

15.2 For any two real, random variables u and v one can write

$$\overline{\left(\frac{u}{\sqrt{\overline{u^2}}} - \frac{v}{\sqrt{\overline{v^2}}}\right)^2} \geq 0.$$

Expand this expression and rearrange the result to prove Schwartz's inequality,

$$\overline{uv} \leq \sqrt{\overline{u^2}}\sqrt{\overline{v^2}}.$$

15.3 Show that

$$\overline{\left[\int_a^b u(t)dt\right]^2} = \int_a^b \int_a^b \overline{u(t')u(t'')}dt'dt'' = \overline{u^2} \int_a^b \int_a^b \rho(t' - t'')dt'dt''.$$

15.4 Use the stationarity of $u(t)$ as expressed by

$$\frac{d^2}{dt^2}\overline{u^2(t)} = 0 = 2\overline{u(t)\frac{d^2u(t)}{dt^2}} + 2\overline{\left(\frac{du(t)}{dt}\right)^2}$$

to show that the variance of the derivative is

$$\overline{\left(\frac{du(t)}{dt}\right)^2} = -\overline{u^2}\frac{d^2\rho(t)}{dt^2}\bigg|_{t=0}.$$

15.5 Referring to Figure 15.1, use Taylor's hypothesis to compare the time scales τ and λ of the autocorrelation function of a turbulent velocity signal by relating them to the energy-containing scale ℓ and the Taylor spatial microscale.

15.6 Use the Fourier–Stieltjes representation to relate the spectrum of $\partial u/\partial t$ to the spectrum of $u(t)$.

15.7 Use the Fourier–Stieltjes representation to assess the reliability of the finite-difference approximation to the derivative of $u(t)$,

$$\frac{\partial u}{\partial t}(t) \simeq \frac{u(t + \Delta t) - u(t - \Delta t)}{2\Delta t}.$$

Do this by finding the spectrum of this approximation to the derivative.

15.8 (a) Find the spectral transfer function (the ratio of the spectra of the filtered and unfiltered variables) of the filter defined by

$$u^{\mathrm{f}}(t) + \tau \frac{du^{\mathrm{f}}(t)}{dt} = \tau \frac{du}{dt},$$

where $u(t)$ is a time series and $u^{\mathrm{f}}(t)$ is the filtered series. Sketch the transfer function. What does this filter accomplish?

(b) Find the spectral transfer function of the filter defined by

$$u^{f}(t) + \tau \frac{du^{f}(t)}{dt} = u(t).$$

Sketch its transfer function. What does this filter accomplish?

15.9 In practice when dealing with observational data or the results of numerical simulations of turbulence we estimate spectra through the coefficients of finite Fourier series. Use a stationary, finite time series of a scalar variable to show how this is done.

15.10 Spectra estimated from time series or numerical simulation results are typically "ragged" due to inadequate averaging. Show how averaging the spectral estimates over narrow frequency or wavenumber bands can smooth such spectra. Develop a criterion that gives the maximum width of the averaging band in the inertial subrange for a given distortion of the spectrum there.

15.11 Determine the time constant needed in a fine-wire resistance thermometer in order to measure the finest scales in the turbulent temperature field in a fluid of $Pr \sim 1$.

15.12 Prove that if $R_{\alpha\alpha}$ is an even function of $\boldsymbol{\xi}$, then $\phi_{\alpha\alpha}$ is purely real and an even function of $\boldsymbol{\kappa}$.

15.13 Relate the one-dimensional co- and quadrature spectra to the full ones in a way similar to (15.49).

15.14 Express λ/ℓ and λ/η as functions of the large-eddy Reynolds number R_{t}.

15.15 Prove Eq. (15.121).

15.16 Prove Eq. (15.122).

15.17 Show that in an isotropic field the integral scale determined from f is twice that determined from g. (Hint: use (15.65).)

15.18 Prove Eq. (15.75) by staying completely in wavenumber space.

15.19 Show that Eq. (15.46) implies that $R_{\alpha\alpha}(\boldsymbol{\xi}, t)$ is an even function of $\boldsymbol{\xi}$.

15.20 Derive Eq. (15.67) from Eq. (15.66). Assume homogeneous turbulence.

15.21 If the three-dimensional pressure spectrum has a $\kappa^{-7/3}$ inertial subrange, how do its one- and two-dimensional spectra behave there? Relate the spectral constants.

15.22 Why is the one-dimensional vorticity spectrum unusual? (Hint: sketch it.)

15.23 Interpret the argument of the function f in Eq. (15.86) as a squared ratio of mean and turbulent strain rates.

15.24 Explain why the continuity constraint $u_{i,i} = 0$ prevents the isotropic form of $\phi_{13}(\boldsymbol{\kappa})$, Eq. (15.57), from vanishing.

References

Batchelor, G. K., 1960: *The Theory of Homogeneous Turbulence*. Cambridge University Press.

Bos, W. J. T., and Bertoglio, J.-P., 2007: Inertial range scaling of scalar spectra in uniformly sheared turbulence. *Phys. Fluids*, **19**, 025104-1-025104-8.

Corrsin, S., 1951: On the spectrum of isotropic temperature fluctuations in an isotropic turbulence. *J. Appl. Phys.*, **22**, 469–473.

Kelly, M., and Wyngaard, J. C., 2006: Two-dimensional spectra in the atmospheric boundary layer. *J. Atmos. Sci.*, **63**, 3066–3070.

Kolmogorov, A. N., 1941: The local structure of turbulence in incompressible viscous fluid for very large Reynolds numbers. *Doklady ANSSSR*, **30**, 301–305.

Lumley, J. L., 1967: Similarity and the turbulent energy spectrum. *Phys. Fluids*, **10**, 855–858.

Lumley, J. L., and H. A. Panofsky, 1964: *The Structure of Atmospheric Turbulence*. New York: Interscience.

Miles, N. L., J. C. Wyngaard, and M. Otte, 2004: Turbulent pressure statistics in atmospheric boundary layers from large-eddy simulation. *Bound.-Layer Meteor.*, **113**, 161–185.

Obukhov, A. M., 1949: Structure of the temperature field in turbulent streams. *Izv. Akad. Nauk SSSR, Geogr. Geofiz.*, **13**, 58.

Saddoughi, S. S., and S. V. Veeravalli, 1994: Local isotropy in turbulent boundary layers at high Reynolds number. *J. Fluid Mech.*, **268**, 333–372.

Taylor, G. I., 1935: Statistical theory of turbulence. Parts I–IV. *Proc. R. Soc. London A*, **151**, 421–478.

Tennekes, H., and J. L. Lumley, 1972: *A First Course in Turbulence*. Cambridge, MA: MIT Press.

Wyngaard, J. C., 1973: On surface-layer turbulence. *Workshop on Micrometeorology*, D. A. Haugen, Ed., American Meteorological Society, pp. 101–149.

Wyngaard, J. C., and O. R. Coté, 1972: Cospectral similarity in the atmospheric surface layer. *Quart. J. R. Meteor. Soc.*, **98**, 590–603.

16
Statistics in turbulence analysis

With some reasonable assumptions and the Fourier–Stieltjes representation we can gain analytical insights into a wide range of turbulence problems. We'll discuss several examples in this chapter.

16.1 Evolution equations for spectra

We'll begin with an idealized problem involving a passive, conserved scalar in a field of isotropic turbulence. With the Fourier–Stieltjes representation we'll convert the scalar conservation equation to an evolution equation for its power spectral density, or spectrum. That will give insight into the maintenance of its inertial subrange, and show what motivated Obukhov (1949) and, independently, Corrsin (1951) to propose a Kolmogorov-like similarity hypothesis for it. Then we'll extend the analysis to a conserved scalar in a horizontally homogeneous turbulent boundary layer, where scalar fluctuations are generated by the large-R_t turbulence acting on a mean scalar gradient in the vertical direction.

16.1.1 A scalar in steady, isotropic turbulence

16.1.1.1 The spectral equation

Imagine a volume of steady turbulence of large R_t and Co_t (turbulence Corrsin number, Chapter 7), as in Chapter 6, Section 6.3 but now isotropic. The velocity field, which has only a fluctuating part $u_i(\mathbf{x}, t)$, advects a scalar field $\tilde{c}$:

$$\tilde{c}_{,t} + (\tilde{c}u_j)_{,j} = s(\mathbf{x}, t) + \gamma\tilde{c}_{,jj}. \tag{16.1}$$

$s(\mathbf{x}, t)$ is a stochastic, zero-mean, homogeneous, stationary source term that fluctuates between positive values (a source of $\tilde{c}$) and negative values (a sink). If we think of the scalar as temperature, then the source term represents zero-mean stochastic heating and cooling. We'll assume that the spatial integral scale of the scalar field is of the order of the turbulence scale ℓ.

By this design C, the mean part of the scalar, is independent of $\mathbf{x}$ and t, the turbulent flux $\overline{cu_j}$ vanishes, and the conservation equation for the fluctuating scalar field $c(\mathbf{x}, t)$ is

$$c_{,t} + (cu_j)_{,j} = s(\mathbf{x}, t) + \gamma c_{,jj}. \tag{16.2}$$

Multiplying Eq. (16.2) by $2c$, ensemble averaging, rewriting the molecular term, and dropping its diffusion part yields the scalar variance equation

$$\frac{\partial}{\partial t}\overline{c^2} = 2\overline{sc} - (\overline{c^2 u_j})_{,j} - 2\gamma\overline{c_{,j}c_{,j}} = 0. \tag{16.3}$$

By homogeneity, which is implied by isotropy, the second term on the right, turbulent transport of variance, vanishes so the equation reduces to

$$2\overline{sc} - 2\gamma\overline{c_{,j}c_{,j}} = Pr - \chi_c, \tag{16.4}$$

with Pr the mean rate of production of variance by the fluctuating source term and χ_c its mean rate of destruction through molecular diffusion.

Now we introduce the Fourier–Stieltjes representations for the fluctuating quantities (Chapter 15):

$$\begin{aligned} c(\mathbf{x}, t) &= \int e^{i\boldsymbol{\kappa}\cdot\mathbf{x}}\, dZ(\boldsymbol{\kappa}, t), \qquad & u_i(\mathbf{x}, t) &= \int e^{i\boldsymbol{\kappa}\cdot\mathbf{x}}\, dZ_i(\boldsymbol{\kappa}, t), \\ s(\mathbf{x}, t) &= \int e^{i\boldsymbol{\kappa}\cdot\mathbf{x}}\, dS(\boldsymbol{\kappa}, t), \qquad & cu_j(\mathbf{x}, t) &= \int e^{i\boldsymbol{\kappa}\cdot\mathbf{x}}\, dF_j(\boldsymbol{\kappa}, t). \end{aligned} \tag{16.5}$$

Hereafter we will not explicitly indicate the dependence on t.

Using these representations in Eq. (16.2) gives the evolution equation for the Fourier–Stieltjes components:

$$\frac{\partial}{\partial t} dZ(\boldsymbol{\kappa}) = -i\kappa_j\, dF_j(\boldsymbol{\kappa}) - \gamma\kappa^2\, dZ(\boldsymbol{\kappa}) + dS(\boldsymbol{\kappa}), \tag{16.6}$$

where $\kappa^2 = \kappa_i\kappa_i$.

The power spectral density ϕ of the fluctuating scalar field is defined through

$$\phi(\boldsymbol{\kappa})\, d\boldsymbol{\kappa} = \overline{dZ(\boldsymbol{\kappa})\, dZ^*(\boldsymbol{\kappa})}, \tag{16.7}$$

and its time derivative is

$$\frac{\partial}{\partial t}\phi(\boldsymbol{\kappa})\, d\boldsymbol{\kappa} = \overline{dZ^*(\boldsymbol{\kappa})\frac{\partial}{\partial t} dZ(\boldsymbol{\kappa})} + \overline{dZ(\boldsymbol{\kappa})\frac{\partial}{\partial t} dZ^*(\boldsymbol{\kappa})}. \tag{16.8}$$

Following the steps in Eq. (16.8) yields

$$\frac{\partial \phi(\boldsymbol{\kappa})}{\partial t} d\boldsymbol{\kappa} = -i\kappa_j \overline{dZ^*(\boldsymbol{\kappa})\, dF_j(\boldsymbol{\kappa})} + i\kappa_j \overline{dZ(\boldsymbol{\kappa})\, dF_j^*(\boldsymbol{\kappa})} - \gamma\kappa^2 \left(\overline{dZ^*(\boldsymbol{\kappa})\, dZ(\boldsymbol{\kappa})} + \overline{dZ(\boldsymbol{\kappa})\, dZ^*(\boldsymbol{\kappa})} \right) + \left(\overline{dZ^*(\boldsymbol{\kappa})\, dS(\boldsymbol{\kappa})} + \overline{dZ(\boldsymbol{\kappa})\, dS^*(\boldsymbol{\kappa})} \right). \tag{16.9}$$

We write the first pair of terms on the right side in the notation of Chapter 15,

$$\overline{dZ(\boldsymbol{\kappa})\, dF_j^*(\boldsymbol{\kappa})} = C_{c,cu_j}(\boldsymbol{\kappa})\, d\boldsymbol{\kappa} = Co_{c,cu_j}(\boldsymbol{\kappa})\, d\boldsymbol{\kappa} - i\, Q_{c,cu_j}(\boldsymbol{\kappa})\, d\boldsymbol{\kappa},$$
$$\overline{dZ^*(\boldsymbol{\kappa})\, dF_j(\boldsymbol{\kappa})} = C^*_{c,cu_j}(\boldsymbol{\kappa})\, d\boldsymbol{\kappa} = Co_{c,cu_j}(\boldsymbol{\kappa})\, d\boldsymbol{\kappa} + i\, Q_{c,cu_j}(\boldsymbol{\kappa})\, d\boldsymbol{\kappa}, \tag{16.10}$$

with C_{c,cu_j}, the cross spectrum of c and cu_j, written in terms of its co- and quadrature spectra. The first pair of terms on the rhs of Eq. (16.9) is then

$$-i\kappa_j \overline{dZ^*(\boldsymbol{\kappa})\, dF_j(\boldsymbol{\kappa})} + i\kappa_j \overline{dZ(\boldsymbol{\kappa})\, dF_j^*(\boldsymbol{\kappa})} = 2\kappa_j Q_{c,cu_j}(\boldsymbol{\kappa})\, d\boldsymbol{\kappa}. \tag{16.11}$$

Similarly, the diffusion and source terms in Eq. (16.9) yield

$$-2\gamma\kappa^2 \overline{dZ(\boldsymbol{\kappa})\, dZ^*(\boldsymbol{\kappa})} = -2\gamma\kappa^2 \phi(\boldsymbol{\kappa})\, d\boldsymbol{\kappa},$$
$$\overline{dZ^*(\boldsymbol{\kappa})\, dS(\boldsymbol{\kappa})} + \overline{dZ(\boldsymbol{\kappa})\, dS^*(\boldsymbol{\kappa})} = 2Co_{c,s}(\boldsymbol{\kappa})\, d\boldsymbol{\kappa}, \tag{16.12}$$

with $Co_{c,s}$ the cospectrum of c and s. The resulting spectral evolution equation is

$$\frac{\partial \phi(\boldsymbol{\kappa})}{\partial t} = 2Co_{c,s}(\boldsymbol{\kappa}) + 2\kappa_j Q_{c,cu_j}(\boldsymbol{\kappa}) - 2\gamma\kappa^2 \phi(\boldsymbol{\kappa}). \tag{16.13}$$

The terms in Eq. (16.13) integrate over $\boldsymbol{\kappa}$ to the corresponding terms in Eq. (16.3) for variance. Since the turbulent transport term in Eq. (16.3) vanishes by homogeneity, we see that the corresponding term in Eq. (16.13) integrates to zero:

$$\iiint 2\kappa_j Q_{c,cu_j}(\boldsymbol{\kappa}) d\kappa_1 d\kappa_2 d\kappa_3 = 0. \tag{16.14}$$

This implies that the term represents transfer within wavenumber space.

Under isotropy (Chapter 14) the quadrature spectrum Q_{c,cu_j}, being a vector function of the vector $\boldsymbol{\kappa}$, has the form

$$Q_{c,cu_j}(\boldsymbol{\kappa}) = \kappa_j F(\kappa), \tag{16.15}$$

with $\kappa = |\boldsymbol{\kappa}|$. It follows that $2\kappa_j Q_{c,cu_j}(\boldsymbol{\kappa}) = 2\kappa^2 F(\kappa)$. Under isotropy the other spectra in Eq. (16.13) depend only on wavenumber magnitude:

$$\phi(\boldsymbol{\kappa}) = \phi(\kappa), \qquad Co_{c,s}(\boldsymbol{\kappa}) = Co_{c,s}(\kappa). \tag{16.16}$$

Thus under isotropy the spectral evolution equation (16.13) is

$$\frac{\partial \phi(\kappa)}{\partial t} = 2Co_{c,s}(\kappa) + 2\kappa^2 F(\kappa) - 2\gamma\kappa^2\phi(\kappa). \tag{16.17}$$

In Chapter 15 we integrated ϕ over spherical shells of radius κ, writing

$$E_c(\kappa) = \iint_{\kappa_i\kappa_i=\kappa^2} \phi(\kappa)\, d\sigma = 4\pi\kappa^2\phi(\kappa). \tag{15.39}$$

We'll extend that to the other terms in Eq. (16.17), defining a "production" spectrum $P(\kappa)$ and a "transfer" spectrum $T(\kappa)$:

$$P(\kappa) = \iint_{\kappa_i\kappa_i=\kappa^2} 2Co_{c,s}(\kappa)\, d\sigma = 8\pi\kappa^2 Co_{c,s}(\kappa),$$

$$T(\kappa) = \iint_{\kappa_i\kappa_i=\kappa^2} 2\kappa^2 F(\kappa)\, d\sigma = 8\pi\kappa^4 F(\kappa). \tag{16.18}$$

The spherically integrated version of Eq. (16.17) is then

$$\frac{\partial E_c(\kappa)}{\partial t} = P(\kappa) + T(\kappa) - 2\gamma\kappa^2 E_c(\kappa). \tag{16.19}$$

This says that the equilibrium three-dimensional scalar spectrum $E_c(\kappa)$ results from the balance of three terms: $P(\kappa)$, the rate of gain by production, centered near $\kappa_e \sim 1/\ell$; $T(\kappa)$, the net rate of gain by transfer from other wavenumbers; and the rate of loss by molecular destruction, centered near $\kappa_d \sim 1/\eta_{oc}$, with η_{oc} the Obukhov–Corrsin scale (Chapter 7). These are sketched in Figure 16.1.

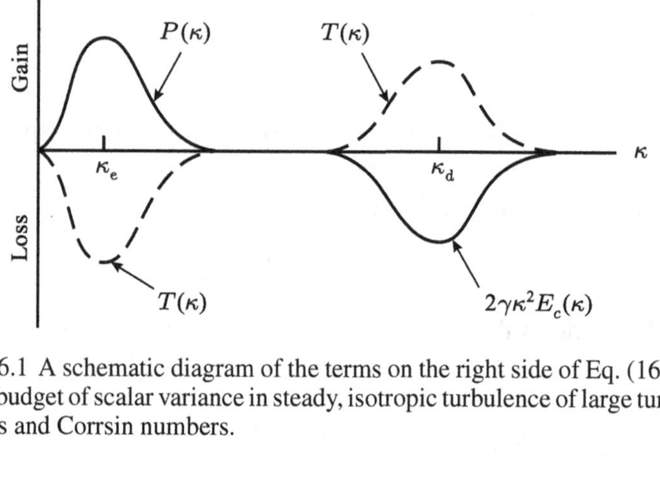

Figure 16.1 A schematic diagram of the terms on the right side of Eq. (16.19), the spectral budget of scalar variance in steady, isotropic turbulence of large turbulence Reynolds and Corrsin numbers.

If we define the mean rate of transfer of variance through wavenumber κ from all smaller wavenumbers as the "cascade rate" $Ca(\kappa)$, then the net rate of gain of variance at wavenumber κ by such transfer is

$$\lim_{\Delta\kappa\to 0} \frac{Ca(\kappa) - Ca(\kappa + \Delta\kappa)}{\Delta\kappa} = -\frac{\partial Ca(\kappa)}{\partial\kappa} = T(\kappa). \tag{16.20}$$

Then the spectral variance budget Eq. (16.19) can be written

$$\frac{\partial E_c(\kappa)}{\partial t} = P(\kappa) - \frac{\partial Ca(\kappa)}{\partial\kappa} - 2\gamma\kappa^2 E_c(\kappa). \tag{16.21}$$

We can briefly explain the behavior of the cascade rate $Ca(\kappa)$, which is sketched in Figure 16.2. Integration of the steady-state form of Eq. (16.21) over the production range, from $\kappa = 0$ to $\kappa = \kappa_{\text{end}}$, say, and using Eq. (16.4) gives

$$Ca(\kappa_{\text{end}}) = \int_0^{\kappa_{\text{end}}} P(\kappa)\, d\kappa = Pr = \chi_c. \tag{16.22}$$

Thus, in the variance-containing range $Ca(\kappa)$ increases from 0 to χ_c.

In the inertial subrange of wavenumbers, $\kappa_e \ll \kappa \ll \kappa_d$, the production and destruction terms in the steady form of the spectral budget (16.19) are negligible so it reduces to

$$\frac{\partial Ca(\kappa)}{\partial\kappa} = 0, \tag{16.23}$$

and $Ca(\kappa) = \chi_c$. This is the counterpart of the inertial subrange in the velocity spectrum; its existence was postulated first by Obukhov (1949) and Corrsin (1951). At yet larger κ, in the dissipative range, molecular destruction of scalar variance is important and the steady spectral variance budget is

$$\frac{\partial Ca(\kappa)}{\partial\kappa} = -2\gamma\kappa^2 E_c(\kappa), \tag{16.24}$$

and $Ca(\kappa)$ decreases to zero at large κ.

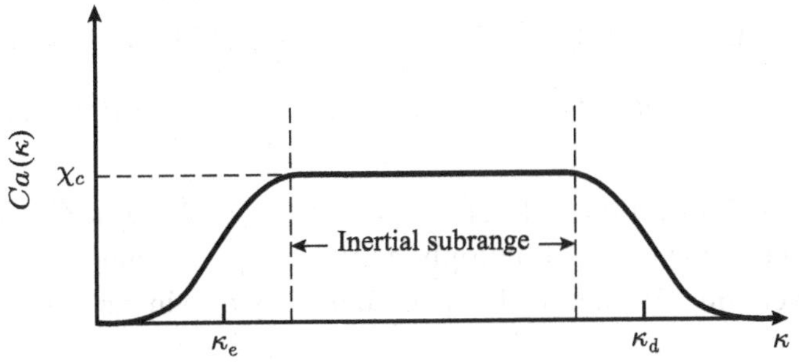

Figure 16.2 The behavior of $Ca(\kappa)$, the cascade rate of scalar variance.

16.1.2 A scalar in a steady, horizontally homogeneous boundary layer

Next we'll consider a physical problem: a conserved scalar with mean vertical gradient $C_{,3}$ in a steady, horizontally homogeneous boundary layer with a mean horizontal velocity U_1. Here the equation for the fluctuating part of the scalar is (Chapter 5)

$$c_{,t} + C_{,3}u_3 + c_{,1}U_1 + (cu_j)_{,j} - (\overline{cu_3})_{,3} = \gamma c_{,jj}. \tag{16.25}$$

Multiplying by $2c$, averaging, using horizontal homogeneity, and rewriting the molecular term (Chapter 5) gives the scalar variance budget:

$$\frac{\partial \overline{c^2}}{\partial t} = -2C_{,3}\,\overline{u_3 c} - (\overline{c^2 u_3})_{,3} - 2\gamma \overline{c_{,j}c_{,j}}. \tag{16.26}$$

The terms on the right side represent, in order, the rates of mean-gradient production, turbulent transport, and molecular destruction.

Because of the vertical inhomogeneity in this problem we must restrict our Fourier–Stieltjes representations to the horizontal plane. As in Chapter 15, we'll use the notation $\boldsymbol{\kappa}_{\mathrm{h}} = (\kappa_1, \kappa_2)$, $\mathbf{x}_{\mathrm{h}} = (x_1, x_2)$:

$$\begin{aligned}
c(\mathbf{x}_{\mathrm{h}}; x_3, t) &= \int e^{i\boldsymbol{\kappa}_{\mathrm{h}}\cdot\mathbf{x}_{\mathrm{h}}} dZ(\boldsymbol{\kappa}_{\mathrm{h}}; x_3, t),\ u_3(\mathbf{x}_{\mathrm{h}}; x_3, t) = \int e^{i\boldsymbol{\kappa}_{\mathrm{h}}\cdot\mathbf{x}_{\mathrm{h}}} dZ_3(\boldsymbol{\kappa}_{\mathrm{h}}; x_3, t);\\
cu_j(\mathbf{x}_{\mathrm{h}}; x_3, t) &= \int e^{i\boldsymbol{\kappa}_{\mathrm{h}}\cdot\mathbf{x}_{\mathrm{h}}} dF_j(\boldsymbol{\kappa}_{\mathrm{h}}; x_3, t),\ \ j = 1, 2; \qquad (16.27)\\
(cu_3 - \overline{cu_3})_{,3} &= \int e^{i\boldsymbol{\kappa}_{\mathrm{h}}\cdot\mathbf{x}_{\mathrm{h}}} dV(\boldsymbol{\kappa}_{\mathrm{h}}; x_3, t).
\end{aligned}$$

We shall not indicate the dependence on x_3 and t hereafter. Substituting these into Eq. (16.25) yields

$$\begin{aligned}
\frac{\partial}{\partial t} dZ(\boldsymbol{\kappa}_{\mathrm{h}}) =& - C_{,3}\, dZ_3(\boldsymbol{\kappa}_{\mathrm{h}}) - i\kappa_1 U_1\, dZ(\boldsymbol{\kappa}_{\mathrm{h}}) - i\kappa_j\, dF_j(\boldsymbol{\kappa}_{\mathrm{h}}) - dV(\boldsymbol{\kappa}_{\mathrm{h}})\\
&- \gamma\kappa_j\kappa_j\, dZ(\boldsymbol{\kappa}_{\mathrm{h}}) + \gamma (dZ)_{,33}(\boldsymbol{\kappa}_{\mathrm{h}}) \qquad (j \text{ summed on } 1, 2). \qquad (16.28)
\end{aligned}$$

We proceed as in Subsection 16.1.1, multiplying Eq. (16.28) by dZ^* and averaging, and adding to that the result of multiplying the complex conjugate of Eq. (16.28) by dZ and averaging. We define the spectral density of c in the horizontal plane,

$$\phi^{(2)}(\boldsymbol{\kappa}_{\mathrm{h}})\, d\boldsymbol{\kappa}_{\mathrm{h}} = \overline{dZ(\boldsymbol{\kappa}_{\mathrm{h}})\, dZ^*(\boldsymbol{\kappa}_{\mathrm{h}})}. \tag{16.29}$$

We also define the quantities

$$\begin{aligned}
&\overline{dZ(\kappa_h)\,dZ_3^*(\kappa_h)} + \overline{dZ^*(\kappa_h)\,dZ_3(\kappa_h)} \\
&\quad = \left(C^{(2)}_{c,u_3}(\kappa_h) + C^{*(2)}_{c,u_3}(\kappa_h)\right) d\kappa_h = 2Co^{(2)}_{c,u_3}\, d\kappa_h, \\
&i\kappa_j\overline{dZ^*(\kappa_h)\,dF_j(\kappa_h)} - i\kappa_j\overline{dZ(\kappa_h)\,dF_j^*(\kappa_h)} = -2\kappa_j Q^{(2)}_{c,cu_j}(\kappa_h)\,d\kappa_h, \\
&\overline{dZ^*(\kappa_h)\,dV(\kappa_h)} + \overline{dZ(\kappa_h)\,dV^*(\kappa_h)} = 2Co^{(2)}_{c,(cu_3),3}(\kappa_h)\,d\kappa_h.
\end{aligned} \tag{16.30}$$

The first of the pair of molecular terms in Eq. (16.28) produces

$$-2\gamma\kappa_j\kappa_j\overline{dZ\,dZ^*} = -2\gamma\kappa_h^2\phi^{(2)}\,d\kappa_h, \tag{16.31}$$

which represents destruction through molecular diffusion in the x_1- and x_2-directions. The second molecular term produces

$$\gamma\left(\overline{dZ_{,33}\,dZ^*} + \overline{dZ^*_{,33}\,dZ}\right) = \gamma(\overline{dZ\,dZ^*})_{,33} - 2\gamma\overline{dZ_{,3}\,dZ^*_{,3}} \simeq -2\gamma\overline{dZ_{,3}\,dZ^*_{,3}}, \tag{16.32}$$

which represents destruction through molecular diffusion in the x_3-direction. If we assume local isotropy the molecular destruction rate is 3/2 of that produced by the horizontal gradients and the spectral evolution equation is (Problem 16.15)

$$\begin{aligned}
\frac{\partial\phi^{(2)}(\kappa_h)}{\partial t} &= -2C_{,3}\,Co^{(2)}_{c,u_3}(\kappa_h) + 2\kappa_j Q^{(2)}_{c,cu_j}(\kappa_h) \\
&\quad -2Co^{(2)}_{c,(cu_3),3}(\kappa_h) - 3\gamma\kappa_h^2\phi^{(2)}(\kappa_h) \qquad (j \text{ summed on } 1, 2).
\end{aligned} \tag{16.33}$$

The terms on the right are ordered as in Eq. (16.13) for three-dimensional, isotropic turbulence. The first term is the rate of production by the interaction of the turbulent flux and the mean gradient. The second term integrates over the horizontal plane to the horizontal part of turbulent transport, which is zero; hence it represents transfer within the horizontal wavenumber plane. The third term integrates to the vertical part of turbulent transport, and the final term is the rate of molecular destruction.

If we assume isotropy in the plane (also called axisymmetry), as in Chapter 15, $\phi^{(2)}$, $Co^{(2)}_{c,u_3}$, and $Co^{(2)}_{c,(cu_3),3}$ depend only on wavenumber magnitude κ_h, not on its direction. Furthermore, the isotropic expression Eq. (16.15) now becomes

$$Q^{(2)}_{c,cu_j}(\kappa_h) = \kappa_j F^{(2)}(\kappa_h), \qquad j = 1, 2, \tag{16.34}$$

so that $2\kappa_j Q^{(2)}_{c,cu_j}(\boldsymbol{\kappa}_\mathrm{h}) = 2\kappa_\mathrm{h}^2 F^{(2)}(\kappa_\mathrm{h})$. Thus under axisymmetry Eq. (16.33) becomes

$$\frac{\partial \phi^{(2)}(\kappa_\mathrm{h})}{\partial t} = -2C_{,3}\, Co^{(2)}_{c,u_3}(\kappa_\mathrm{h}) + 2\kappa_\mathrm{h}^2 F^{(2)}(\kappa_\mathrm{h}) - 2Co^{(2)}_{c,(cu_3),3}(\kappa_\mathrm{h}) - 3\gamma\kappa_\mathrm{h}^2\phi^{(2)}(\kappa_\mathrm{h}). \tag{16.35}$$

We now integrate Eq. (16.35) over circular rings in the horizontal wavenumber plane, reintroducing the two-dimensional scalar spectrum defined in Chapter 15:

$$E_c^{(2)}(\kappa_\mathrm{h}) = \int_0^{2\pi} \phi^{(2)}(\kappa_\mathrm{h})\kappa_\mathrm{h}\, d\theta = 2\pi\kappa_\mathrm{h}\phi^{(2)}(\kappa_\mathrm{h}). \tag{15.97}$$

We define a production spectrum $P^{(2)}(\kappa_\mathrm{h})$,

$$P^{(2)}(\kappa_\mathrm{h}) = -\int_0^{2\pi} 2C_{,3}\, Co^{(2)}_{c,u_3}(\kappa_\mathrm{h})\kappa_\mathrm{h}\, d\theta, \tag{16.36}$$

a horizontal transfer spectrum $T_\mathrm{h}^{(2)}(\kappa_\mathrm{h})$,

$$T_\mathrm{h}^{(2)}(\kappa_\mathrm{h}) = \int_0^{2\pi} 2\kappa_\mathrm{h}^2 F^{(2)}(\kappa_\mathrm{h})\kappa_\mathrm{h}\, d\theta, \tag{16.37}$$

and a spectrum of vertical turbulent transport,

$$T_v^{(2)}(\kappa_\mathrm{h}) = -\int_0^{2\pi} 2Co^{(2)}_{c,(cu_3),3}(\kappa_\mathrm{h})\kappa_\mathrm{h}\, d\theta. \tag{16.38}$$

This gives the spectral scalar variance budget in the horizontal plane,

$$\frac{\partial E_c^{(2)}(\kappa_\mathrm{h})}{\partial t} = P^{(2)}(\kappa_\mathrm{h}) + T_\mathrm{h}^{(2)}(\kappa_\mathrm{h}) + T_v^{(2)}(\kappa_\mathrm{h}) - 3\gamma\kappa_\mathrm{h}^2 E_c^{(2)}(\kappa_\mathrm{h}). \tag{16.39}$$

This is the counterpart of the isotropic turbulence result, Eq. (16.19) (Problem 16.11).

The terms in Eq. (16.39) integrate over κ_h to the terms in the variance budget (16.26):

$$\int_0^\infty P^{(2)}(\kappa_\mathrm{h})\, d\kappa_\mathrm{h} = -2C_{,3}\,\overline{cu_3} = Pr, \qquad \int_0^\infty T_\mathrm{h}^{(2)}(\kappa_\mathrm{h})\, d\kappa_\mathrm{h} = 0;$$
$$\int_0^\infty T_v^{(2)}(\kappa_\mathrm{h})\, d\kappa_\mathrm{h} = -(\overline{c^2u_3})_{,3} = Tr, \qquad -3\gamma\int_0^\infty \kappa_\mathrm{h}^2 E_c^{(2)}(\kappa_\mathrm{h}) = -\chi_c. \tag{16.40}$$

Thus the integrated variance budget is

$$\frac{\partial \overline{c^2}}{\partial t} = Pr + Tr - \chi_c. \tag{16.41}$$

We can define a cascade rate $Ca^{(2)}(\kappa_{\mathrm{h}})$, the mean rate of transfer of scalar variance through wavenumber κ_{h} from all smaller wavenumbers in the horizontal plane. The net rate of gain of variance at wavenumber κ_{h} by such transfer is

$$-\frac{\partial Ca^{(2)}(\kappa_{\mathrm{h}})}{\partial \kappa_{\mathrm{h}}} = T_{\mathrm{h}}^{(2)}(\kappa_{\mathrm{h}}). \tag{16.42}$$

Thus Eq. (16.39) can be written

$$\frac{\partial E_c^{(2)}(\kappa_{\mathrm{h}})}{\partial t} = P^{(2)}(\kappa_{\mathrm{h}}) - \frac{\partial Ca^{(2)}(\kappa_{\mathrm{h}})}{\partial \kappa_{\mathrm{h}}} + T_v^{(2)}(\kappa_{\mathrm{h}}) - 3\gamma\kappa_{\mathrm{h}}^2 E_c^{(2)}(\kappa_{\mathrm{h}}), \tag{16.43}$$

which is the two-dimensional counterpart of Eq. (16.21).

Integration of Eq. (16.43) over the production range gives in steady state

$$Ca^{(2)}(\kappa_{\mathrm{e}}) = \int_0^{\kappa_{\mathrm{e}}} \left(P^{(2)}(\kappa_{\mathrm{h}}) + T_v^{(2)}(\kappa_{\mathrm{h}})\right) d\kappa_{\mathrm{h}} = Pr + Tr. \tag{16.44}$$

Equation (16.41) shows that this is equal to χ_c, the rate of molecular destruction of variance. Thus, in the variance-containing range $Ca^{(2)}(\kappa_{\mathrm{h}})$ increases from 0 to χ_c, as in the three-dimensional case shown in Figure 16.2.

In the range of horizontal wavenumbers κ_{h} such that $\kappa_{\mathrm{e}} \ll \kappa_{\mathrm{h}} \ll \kappa_{\mathrm{d}}$ the production, vertical transfer, and molecular destruction terms in the spectral budget (16.43) are negligible, so it reduces to

$$\frac{\partial Ca^{(2)}(\kappa_{\mathrm{h}})}{\partial \kappa_{\mathrm{h}}} = 0. \tag{16.45}$$

Thus here $Ca^{(2)}(\kappa_{\mathrm{h}}) = \chi_c$, and as in the three-dimensional case we have the possibility of an inertial subrange.† As we discussed in Chapter 15, Section 15.6, the inertial-range spectral constants for these two-dimensional spectra are different from those for three-dimensional spectra.

16.2 The analysis and interpretation of turbulence signals

The Fourier–Stieltjes representation allows us to determine analytically the effects of instrumental or sampling attributes such as spatial and temporal averaging. We shall illustrate with several examples.

† Such inertial subranges can be seen in spectra of fields from high-resolution LES of the ABL.

16.2.1 Spatial averaging of conserved scalars

Flow sensors typically average the measured variable over a small region of space. For example, a resistance-wire temperature sensor averages temperature over its length; infrared-absorption sensors average the mixing ratio of water vapor and carbon dioxide over the length of their transmission path. We can determine analytically how this line averaging impacts some statistics of the signal.

We'll consider first averaging the fluctuating scalar field c over a vector path $\mathbf{L}$ of length L. We'll associate the measurement with the midpoint $\mathbf{x}$ of the path. Denoting c^{m} as the measured value of c, $\mathbf{s}$ as the vector along the path from that midpoint, and s the distance along that path, we have

$$c^{\mathrm{m}}(\mathbf{x}, \mathbf{L}, t) = \frac{1}{L} \int_{-L/2}^{L/2} c(\mathbf{x} + \mathbf{s}, t)\, ds. \tag{16.46}$$

In this path averaging, scalar fluctuations having along-path length scales small compared to L average to nearly zero and, hence, are greatly attenuated. Scalar fluctuations in directions normal to the path are unaffected.

The Fourier–Stieltjes representation of the scalar field allows us to quantify these averaging effects. We assume homogeneous turbulence so we can represent the true and measured fluctuating scalar fields as (we shall hereafter suppress the dependence on time):

$$c(\mathbf{x}) = \int e^{i\boldsymbol{\kappa}\cdot\mathbf{x}}\, dZ(\boldsymbol{\kappa}), \qquad c^{\mathrm{m}}(\mathbf{x}, \mathbf{L}) = \int e^{i\boldsymbol{\kappa}\cdot\mathbf{x}} dZ^{\mathrm{m}}(\boldsymbol{\kappa}, \mathbf{L}). \tag{16.47}$$

Using (16.46) we can write the path-averaged scalar signal as

$$\begin{aligned} c^{\mathrm{m}}(\mathbf{x}, \mathbf{L}) &= \frac{1}{L} \int_{-L/2}^{L/2} \left[\int e^{i\boldsymbol{\kappa}\cdot(\mathbf{x}+\mathbf{s})} dZ(\boldsymbol{\kappa}) \right] ds \\ &= \int \left[\frac{1}{L} \int_{-L/2}^{L/2} e^{i\boldsymbol{\kappa}\cdot(\mathbf{x}+\mathbf{s})} ds \right] dZ(\boldsymbol{\kappa}). \end{aligned} \tag{16.48}$$

The averaging integral is

$$\frac{1}{L} \int_{-L/2}^{L/2} e^{i\boldsymbol{\kappa}\cdot(\mathbf{x}+\mathbf{s})}\, ds = \frac{\sin(\boldsymbol{\kappa}\cdot\mathbf{L}/2)}{\boldsymbol{\kappa}\cdot\mathbf{L}/2} e^{i\boldsymbol{\kappa}\cdot\mathbf{x}}, \tag{16.49}$$

so Eq. (16.48) can be written

$$c^{\mathrm{m}}(\mathbf{x}, \mathbf{L}) = \int e^{i\boldsymbol{\kappa}\cdot\mathbf{x}}\, \frac{\sin(\boldsymbol{\kappa}\cdot\mathbf{L}/2)}{\boldsymbol{\kappa}\cdot\mathbf{L}/2} dZ(\boldsymbol{\kappa}). \tag{16.50}$$

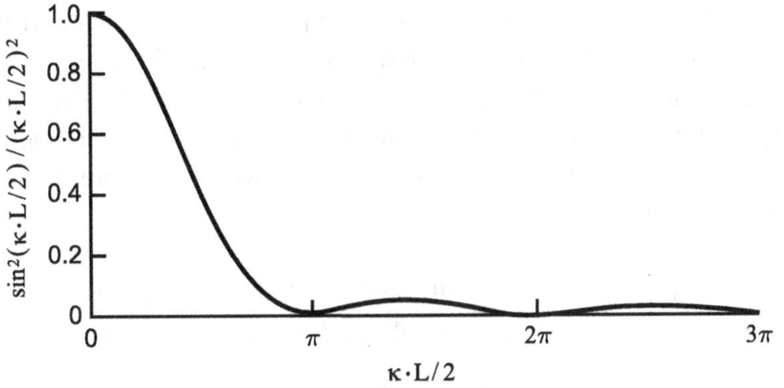

Figure 16.3 The spectral transfer function of line averaging of a scalar field.

Thus we have an expression for the effect of sensor-path averaging on the Fourier–Stieltjes coefficients of the measured scalar:

$$dZ^{\mathrm{m}}(\boldsymbol{\kappa}, \mathbf{L}) = \frac{\sin(\boldsymbol{\kappa} \cdot \mathbf{L}/2)}{\boldsymbol{\kappa} \cdot \mathbf{L}/2} dZ(\boldsymbol{\kappa}). \tag{16.51}$$

This quantifies how this path averaging removes Fourier components of the c field of wavenumber in the path direction much greater than $1/L$, while not affecting those of wavenumber components perpendicular to the path.

Since the scalar spectrum and its measured form are

$$\phi(\boldsymbol{\kappa})\, d\boldsymbol{\kappa} = \overline{dZ(\boldsymbol{\kappa})\, dZ^*(\boldsymbol{\kappa})}, \qquad \phi^m(\boldsymbol{\kappa})\, d\boldsymbol{\kappa} = \overline{dZ^m(\boldsymbol{\kappa})\, dZ^{m*}(\boldsymbol{\kappa})}, \tag{16.52}$$

it follows that the measured spectrum is

$$\phi^{\mathrm{m}}(\boldsymbol{\kappa}) = \frac{\sin^2(\boldsymbol{\kappa} \cdot \mathbf{L}/2)}{(\boldsymbol{\kappa} \cdot \mathbf{L}/2)^2} \phi(\boldsymbol{\kappa}) = T(\boldsymbol{\kappa}, \mathbf{L})\phi(\boldsymbol{\kappa}), \tag{16.53}$$

with $T(\boldsymbol{\kappa}, \mathbf{L})$ the spectral transfer function of line averaging. It is shown in Figure 16.3.

To determine the effect of path averaging on $F^{1\mathrm{m}}(\kappa_1)$, the one-dimensional streamwise wavenumber spectrum that we calculate from a time series measured at a point using Taylor's hypothesis, we integrate over two wavenumbers:

$$\begin{aligned} F^{1\mathrm{m}}(\kappa_1) &= \iint_{-\infty}^{\infty} \phi^{\mathrm{m}}(\boldsymbol{\kappa}) d\kappa_2\, d\kappa_3 \\ &= \iint_{-\infty}^{\infty} \phi(\boldsymbol{\kappa}) \frac{\sin^2(\boldsymbol{\kappa} \cdot \mathbf{L}/2)}{(\boldsymbol{\kappa} \cdot \mathbf{L}/2)^2} d\kappa_2\, d\kappa_3. \end{aligned} \tag{16.54}$$

If the sensor path vector $\mathbf{L}$ is oriented in the x_1 (streamwise) direction the path-averaging transfer function can be taken out of this integral. In order to avoid flow distortion this orientation is not generally used, however, so typically the integral in (16.54) must be carried out over κ_2 and κ_3. This is usually done numerically.

If the sensor averages over a rectangular area ΔA whose sides are Δx_1 and Δx_2, the spectral relation is

$$\phi^{\mathrm{m}}(\boldsymbol{\kappa}, \Delta A) = \phi(\boldsymbol{\kappa})\frac{\sin^2(\kappa_1 \Delta x_1/2)}{(\kappa_1 \Delta x_1/2)^2}\frac{\sin^2(\kappa_2 \Delta x_2/2)}{(\kappa_2 \Delta x_2/2)^2}. \tag{16.55}$$

If the sensor averages over a volume ΔV whose sides are Δx_1, Δx_2, Δx_3, the spectral relation is

$$\phi^{\mathrm{m}}(\boldsymbol{\kappa}, \Delta V) = \phi(\boldsymbol{\kappa})\frac{\sin^2(\kappa_1 \Delta x_1/2)}{(\kappa_1 \Delta x_1/2)^2}\frac{\sin^2(\kappa_2 \Delta x_2/2)}{(\kappa_2 \Delta x_2/2)^2}\frac{\sin^2(\kappa_3 \Delta x_3/2)}{(\kappa_3 \Delta x_3/2)^2}. \tag{16.56}$$

In general the effect of spatial averaging on other statistics is more difficult (or perhaps not possible) to determine analytically.

16.2.2 Response of scalar structure-function sensors

In the same way we can analyze the response of a "structure-function" sensor. Its output is the difference of turbulent quantity at two points separated in space. When the separation distance r falls in the inertial range of scales such two-point difference variances can be used to infer dissipation rates (Chapter 7).

Again suppressing the dependence on time, we write

$$\tilde{c}(\mathbf{x}+\mathbf{r}) = \int e^{i\boldsymbol{\kappa}\cdot(\mathbf{x}+\mathbf{r})}\,dZ(\boldsymbol{\kappa}), \qquad \tilde{c}(\mathbf{x}) = \int e^{i\boldsymbol{\kappa}\cdot\mathbf{x}}\,dZ(\boldsymbol{\kappa}). \tag{16.57}$$

The difference signal is

$$\begin{aligned}\Delta\tilde{c}(\mathbf{x},\mathbf{r}) = \tilde{c}(\mathbf{x}+\mathbf{r}) - \tilde{c}(\mathbf{x}) &= \int e^{i\boldsymbol{\kappa}\cdot\mathbf{x}}\left(e^{i\boldsymbol{\kappa}\cdot\mathbf{r}} - 1\right) dZ(\boldsymbol{\kappa}) \\ &= \int e^{i\boldsymbol{\kappa}\cdot\mathbf{x}}\,dZ_{\mathrm{d}}(\boldsymbol{\kappa},\mathbf{r}).\end{aligned} \tag{16.58}$$

From (16.58) the spectra of the scalar difference and the scalar are related by

$$\begin{aligned}\phi_{\mathrm{d}}(\boldsymbol{\kappa})\,d\boldsymbol{\kappa} = \overline{dZ_{\mathrm{d}}(\boldsymbol{\kappa})\,dZ_{\mathrm{d}}^*(\boldsymbol{\kappa})} &= (e^{i\boldsymbol{\kappa}\cdot\mathbf{r}} - 1)(e^{-i\boldsymbol{\kappa}\cdot\mathbf{r}} - 1)\,\overline{dZ(\boldsymbol{\kappa})\,dZ^*(\boldsymbol{\kappa})} \\ &= 2\left[1 - \cos(\boldsymbol{\kappa}\cdot\mathbf{r})\right]\phi(\boldsymbol{\kappa})\,d\boldsymbol{\kappa}.\end{aligned} \tag{16.59}$$

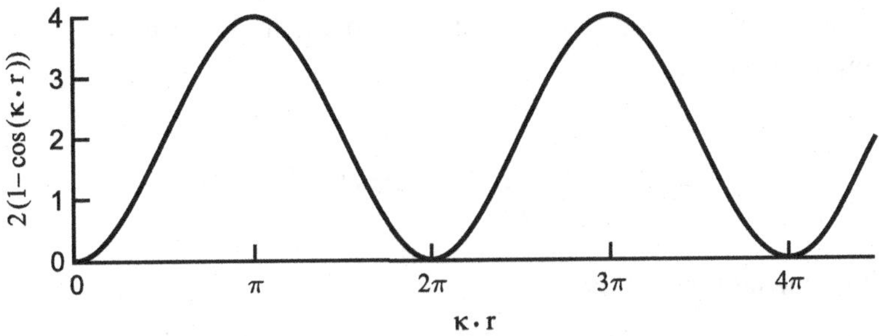

Figure 16.4 The spectral transfer function of the two-point scalar difference operator.

Thus the difference variance is

$$\overline{(\Delta c)^2} = \int \phi_d(\boldsymbol{\kappa})\, d\boldsymbol{\kappa} = \int 2\,[1 - \cos(\boldsymbol{\kappa}\cdot\mathbf{r})]\, \phi(\boldsymbol{\kappa})\, d\boldsymbol{\kappa}. \tag{16.60}$$

The spectral transfer function $2\,[1 - \cos(\boldsymbol{\kappa}\cdot\mathbf{r})]$ of the scalar difference operator is shown in Figure 16.4. Let's interpret it physically. First, Fourier components of wavenumber perpendicular to the separation vector $\mathbf{r}$ have $\boldsymbol{\kappa}\cdot\mathbf{r} = 0$, so that $\cos(\boldsymbol{\kappa}\cdot\mathbf{r}) = 1$ and the transfer function is zero; these components are rejected by the difference filter. More generally, in the small-separation limit $\boldsymbol{\kappa}\cdot\mathbf{r} \to 0$ the two sensors detect essentially the same signal so their difference variance is nearly zero. Put another way, the difference filter rejects eddies that are much larger than the separation distance.

As $\boldsymbol{\kappa}\cdot\mathbf{r}$ increases from zero the magnitude of the transfer function gradually rises to its first maximum of 4 (Figure 16.4), where $\boldsymbol{\kappa}\cdot\mathbf{r} = \pi$. Here Fourier components in the $\mathbf{r}$ direction at the two points are 180 degrees out of phase so they add, not subtract. This happens again at $\boldsymbol{\kappa}\cdot\mathbf{r} = (2n-1)\pi$, $n = 2, 3, \ldots$

The difference array is often used for separations in the inertial range of scales, where the spectrum $\phi(\boldsymbol{\kappa})$ is typically assumed to have its isotropic form $\phi(\kappa)$ (Chapter 15). Let us assume that the separation vector is in the x_1-direction. We can then write (16.60) as

$$\begin{aligned}\overline{(\Delta c)^2} &= \int_{-\infty}^{\infty} 2\,[1 - \cos(\kappa_1 r)] \left(\iint_{-\infty}^{\infty} \phi(\kappa)\, d\kappa_2\, d\kappa_3 \right) d\kappa_1 \\ &= \int_{-\infty}^{\infty} 2\,[1 - \cos(\kappa_1 r)]\, F^1(\kappa_1)\, d\kappa_1.\end{aligned} \tag{16.61}$$

If the separation vector is in the x_2-direction the result is

$$\overline{(\Delta c)^2} = \int_{-\infty}^{\infty} 2\,[1 - \cos(\kappa_2 r)]\, F^2(\kappa_2)\, d\kappa_2, \tag{16.62}$$

but since we saw in Chapter 15 that in an isotropic field $F^1 = F^2 = F^3 = F$, say, we can simply write this as

$$\overline{(\Delta c)^2} = \int_{-\infty}^{\infty} 2\,[1 - \cos(\kappa r)]\, F(\kappa)\, d\kappa. \tag{16.63}$$

In the inertial subrange F is

$$F(\kappa) = \frac{3}{10}\beta \chi_c \epsilon^{-1/3} \kappa^{-5/3}, \tag{15.43}$$

and with this form Eq. (16.63) can be integrated to give (Problem 16.7)

$$\overline{(\Delta c)^2} = 2.4\beta \chi_c \epsilon^{-1/3} r^{2/3} = C_{c^2} r^{2/3}. \tag{16.64}$$

C_{c^2} is known as the structure-function parameter for the scalar c. The $\kappa^{-5/3}$ behavior of the scalar spectrum in the inertial range of wavenumbers corresponds to the $r^{2/3}$ behavior of the difference variance in the inertial range of spatial separations.

Temperature is often used as the scalar in turbulent flows. It typically is measured with a fine-wire sensor of temperature-dependent resistance, operated in a bridge circuit. If the wire has a time constant τ_{w}, the measured temperature fluctuation θ^{m} follows the equation

$$\tau_{\mathrm{w}} \frac{\partial \theta^{\mathrm{m}}}{\partial t} + \theta^{\mathrm{m}}(t) = \theta(t), \tag{16.65}$$

with $\theta(t)$ the fluid temperature at the sensor. A difference filter passes primarily temperature eddies whose scale in the separation direction is smaller than the separation. Since smaller temperature eddies appear in the signal as higher frequency components, the time constant τ_{w} influences the difference filter output. We'd like to know how it affects its variance, for example.

One way to proceed is to use Taylor's hypothesis (Chapter 2) in the form $\partial\theta/\partial t = -U_1 \partial\theta/\partial x_1$, with U_1 the mean velocity in the streamwise direction x_1, to convert Eq. (16.65) into an equation in x_1:

$$-\tau_{\mathrm{w}} U_1 \frac{\partial \theta^{\mathrm{m}}(\mathbf{x})}{\partial x_1} + \theta^{\mathrm{m}}(\mathbf{x}) = \theta(\mathbf{x}). \tag{16.66}$$

The Fourier–Stieltjes components dZ^{m} and dZ of the measured and true signals are then related by

$$(1 - i\kappa_1 \tau_{\mathrm{w}} U_1)\, dZ^{\mathrm{m}}(\boldsymbol{\kappa}) = dZ(\boldsymbol{\kappa}). \tag{16.67}$$

Equation (16.58) for the measured difference signal, generalized to a sensor with a nonzero time constant, becomes

$$\Delta\theta^{\rm m}(\mathbf{x}, \mathbf{r}, \tau_{\rm w}, U_1) = \int e^{i\boldsymbol{\kappa}\cdot\mathbf{x}} \frac{\left(e^{i\boldsymbol{\kappa}\cdot\mathbf{r}} - 1\right)}{(1 - i\kappa_1\tau_{\rm w}U_1)} dZ(\boldsymbol{\kappa})$$
$$= \int e^{i\boldsymbol{\kappa}\cdot\mathbf{x}} dZ_{\rm d}^{\rm m}(\boldsymbol{\kappa}, \mathbf{r}, \tau_{\rm w}, U_1), \tag{16.68}$$

with $dZ_{\rm d}^{\rm m}$ the Fourier–Stieltjes components of the measured difference signal. Likewise, Eq. (16.59) becomes

$$\phi_{\rm d}^{\rm m}(\boldsymbol{\kappa}) = \frac{2\,[1 - \cos(\boldsymbol{\kappa}\cdot\mathbf{r})]}{1 + (\kappa_1\tau_{\rm w}U_1)^2}\phi(\boldsymbol{\kappa}), \tag{16.69}$$

and Eq. (16.60), rewritten for the measured difference variance, becomes

$$\overline{(\Delta\theta^{\rm m})^2} = \int \phi_{\rm d}^{\rm m}(\boldsymbol{\kappa})d\boldsymbol{\kappa} = \int \frac{2\,[1 - \cos(\boldsymbol{\kappa}\cdot\mathbf{r})]}{1 + (\kappa_1\tau_{\rm w}U_1)^2}\phi(\boldsymbol{\kappa})\,d\boldsymbol{\kappa}. \tag{16.70}$$

Experimentalists can be reluctant to separate two sensors only in the streamwise direction because this can put one sensor in the wake of the other. Thus, in general the separation vector $\mathbf{r}$ in Eq. (16.70) has a lateral component, which means that $\overline{(\Delta\theta^{\rm m})^2}$ cannot be reduced to an integral of a transfer function times the one-dimensional spectrum.

One can simulate a sensor separation r in the streamwise direction by using a single sensor sampled at two times separated by r/U_1:

$$c(\mathbf{x} + r, t) - c(\mathbf{x}, t) \simeq c(\mathbf{x}, t - r/U_1) - c(\mathbf{x}, t). \tag{16.71}$$

In this case Eq. (16.70) becomes

$$\overline{(\Delta c^{\rm m})^2} = \int_0^\infty \frac{2\,[1 - \cos(\kappa_1 r)]}{1 + (\kappa_1\tau_{\rm w}U_1)^2}\phi(\boldsymbol{\kappa})\,d\boldsymbol{\kappa} = \int_{-\infty}^\infty \frac{2\,[1 - \cos(\kappa_1 r)]}{1 + (\kappa_1\tau_{\rm w}U_1)^2}F^1(\kappa_1)\,d\kappa_1. \tag{16.72}$$

If the separation r is in the inertial range of scales, the one-dimensional spectrum F^1 falls as $\kappa_1^{-5/3}$ in the region that contributes to the integral (16.72). The transfer function $2\,[1 - \cos(\kappa_1 r)]$ of the difference filter (Figure 16.4) is periodic; the first peak is at $\kappa_1 = \pi/r$, the second at $3\pi/r$. A minimum requirement is that the sensor time constant $\tau_{\rm w}$ be small enough that the transfer function $1 + (\kappa_1\tau_{\rm w}U_1)^2$ is unity in the first pass band of the difference filter. Thus, we require

$$(\kappa_1\tau_{\rm w}U_1)^2 \ll 1 \text{ when } \kappa_1 r = 2\pi, \tag{16.73}$$

so that a requirement for the time constant of the sensor is

$$\tau_w \ll \frac{r}{2\pi U_1}. \tag{16.74}$$

This means that the smaller the separation r and the larger the mean wind speed U_1 the smaller the required time constant. If we interpret "$\ll$" in Eq. (16.74) as meaning a factor of 10 less, then for r = 0.6 m and U_1 = 10 m s^{-1} this criterion says that τ_w should be less than 10^{-3} s.

16.2.3 The two-point difference and the spatial derivative

Since $\cos 2x = 1 - 2\sin^2 x$, we can write Eq. (16.63) for locally isotropic turbulence as

$$\frac{\overline{(\Delta c)^2}}{r^2} = \frac{4}{r^2}\int_{-\infty}^{\infty} \sin^2(\kappa r/2) F(\kappa)\, d\kappa = \int_{-\infty}^{\infty} \frac{\sin^2(\kappa r/2)}{(\kappa r/2)^2}\kappa^2 F(\kappa)\, d\kappa. \tag{16.75}$$

In the limit as r approaches zero the difference variance divided by r^2 is the spatial derivative variance, which in an isotropic field is

$$\lim_{r\to 0} \frac{\overline{(\Delta c)^2}}{r^2} = \overline{\left(\frac{\partial c}{\partial x}\right)^2} = \overline{\left(\frac{\partial c}{\partial y}\right)^2} = \overline{\left(\frac{\partial c}{\partial z}\right)^2}. \tag{16.76}$$

Since the spatial derivative variance is the integral of κ^2 times the one-dimensional spectrum,

$$\overline{\left(\frac{\partial c}{\partial x}\right)^2} = \int_{-\infty}^{\infty} \kappa^2 F(\kappa)\, d\kappa, \tag{16.77}$$

we can write a derivative variance measured as in (16.75) as

$$\frac{\overline{(\Delta c)^2}}{r^2} = \overline{\left(\frac{\partial c}{\partial x}\right)^2}^{\mathrm{m}} = \int_{-\infty}^{\infty} \frac{\sin^2(\kappa r/2)}{(\kappa r/2)^2}\kappa^2 F(\kappa)\, d\kappa. \tag{16.78}$$

Thus $\sin^2(\kappa r/2)/(\kappa r/2)^2$ is the transfer function of the difference approximation to the spatial derivative.

16.2.4 Application to velocity fields

The approach of the last three sections is applicable to turbulent velocity fields as well. Again those applications that concern the smaller-scale structure are traditionally handled with the assumption of local isotropy.

The effects of spatial averaging and sensor separation associated with turbulent velocity measurements tend to be three-dimensional. As a result they tend to appear as spectral transfer functions only in three-dimensional wavenumber space. In the case of three-component sonic anemometers, for example, the spectral transfer functions depend on three acoustic path vectors $\mathbf{L}^i$ and on three path-separation vectors $\mathbf{d}^i$. Thus, to determine the effects of path length and separation on one-dimensional velocity spectra we must integrate the spectral transfer function over two wavenumbers:

$$F_{ij}^{1m} = \int_{-\infty}^{\infty} \int_{-\infty}^{\infty} T(\boldsymbol{\kappa}, \mathbf{L}^1, \mathbf{L}^2, \mathbf{L}^3, \mathbf{d}^1, \mathbf{d}^2, \mathbf{d}^3) \phi_{ij}(\boldsymbol{\kappa}) \, d\kappa_2 \, d\kappa_3. \tag{16.79}$$

This can make the interpretation of velocity measurements from three-dimensional sensor arrays quite complicated at wavenumbers where the path averaging and sensor-separation effects are significant (Kaimal *et al.*, 1968).

Wyngaard (1968, 1969, 1971) has calculated the spectral response of hot-wire velocity and vorticity probes and resistance-wire temperature sensors; Kaimal *et al.* (1968) have done the same for three-component sonic anemometers; and Gal-Chen and Wyngaard (1982) have extended the analysis to multiple-Doppler radars.

16.2.5 Measuring resolved and subfilter-scale variables

Tong *et al.* (1998) developed a technique for measuring resolved and SGS variables in the surface layer. Using high-resolution (256^3) LES data, they first established that filtering in the two horizontal directions is a good surrogate for three-dimensional filtering. They then showed that filtering in time and using Taylor's hypothesis can be an adequate substitute for filtering in the streamwise direction.

These simplifications allow the use of a single linear array of sensors oriented in the cross-stream direction. A variable $\tilde{f}(x, y, t)$, say, at a given height z is filtered in time by using a running average such as

$$\tilde{f}^t(x, y, t) = \frac{1}{2N+1} \sum_{n=-N}^{N} \tilde{f}(x, y, t - n\Delta t). \tag{16.80}$$

These filtered variables from $2N+1$ (five, say) sensors spaced in the lateral direction are then combined with equal weights to simulate lateral filtering:

$$(\tilde{f}^t)^y = \frac{1}{5}\left(\tilde{f}^t(x, y - 2\Delta y, t) + \tilde{f}^t(x, y - \Delta y, t) + \cdots + \tilde{f}^t(x, y + 2\Delta y, t)\right). \tag{16.81}$$

This produces a variable that is in effect spatially averaged over a rectangle of streamwise length $U_1(2N+1)\Delta t$ and width $(2N+1)\Delta y$ in the horizontal plane. We define this as the resolved part of $\tilde{f}$, so we have

$$\tilde{f} = f^{\mathrm{r}} + f^{\mathrm{s}}, \qquad f^{\mathrm{r}} = (\tilde{f}^{t})^{y}, \qquad f^{\mathrm{s}} = \tilde{f} - (\tilde{f}^{t})^{y}. \tag{16.82}$$

Figures 6.4 and 6.5 show subfilter-scale (SFS) momentum and temperature fluxes measured in HATS (Horizontal Array Turbulence Study). The array technique has also been used in several subsequent field studies to give new insights into SFS fluxes and their modeling.

16.3 Probe-induced flow distortion

Figure 16.5 shows a schematic of flow past a circular cylinder. Ahead of such an obstacle the flow streamlines are modified by its blocking effect and the turbulence in the flow is distorted. Since *in-situ* turbulence probes (in general we'll take *probe* to mean the instrument plus its mounting apparatus) necessarily have some bulk, their velocity measurements, in particular, are prone to contamination by this "probe-induced flow distortion."

Hunt (1973) discusses two limits in this problem: $a \gg \ell$, where a is the scale of the body and ℓ is the integral scale of the turbulence; and $a \ll \ell$. The first is quite complicated, but the second is analytically tractable. Fortunately the second applies to the measurement of energy-containing range turbulence structure with flow distortion due to the probe. We'll discuss a simple analysis of it.

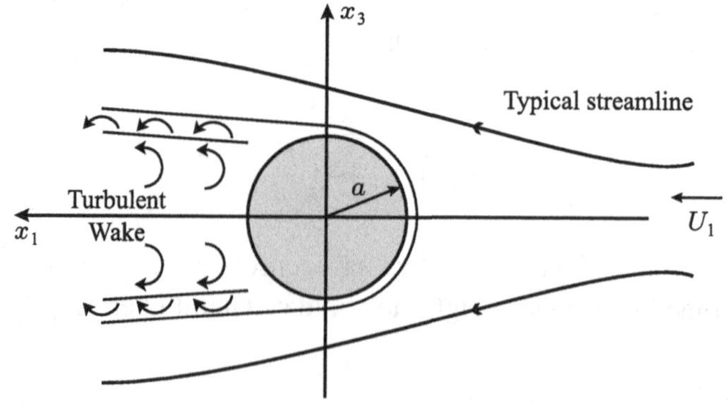

Figure 16.5 Flow past a circular cylinder. From Hunt (1973).

16.3.1 A simple approach for the velocity field

We represent the undistorted flow (that existing in the absence of the body) as $\tilde{u}_i(t) = U_1\delta_{i1} + u_i(t)$, the usual sum of ensemble-mean and fluctuating parts, with the x_1-axis chosen in the mean-flow direction. We denote the distorted flow near the body (but not in its wake) with a superscript d: $\tilde{u}_i^{\rm d}(t) = U_i^{\rm d}(\mathbf{x}) + u_i^{\rm d}(\mathbf{x}, t)$. The mean distorted flow need not be in the x_1-direction.

Now we make a key assumption: that ℓ/U_1, the time scale of the fluctuating velocity signal at a point near the body, is much larger than a/U_1, the time scale of the response of the distorted flow to those velocity fluctuations:

$$\ell/U_1 \gg a/U_1, \tag{16.83}$$

which is equivalent to $\ell \gg a$. This ensures that the distorted flow near the body sees the turbulence in the approach flow as varying so slowly that it "tracks" its time variations perfectly. Thus we can treat the distorted flow near the body as quasi-steady.

We now write the distorted velocity at the measurement position $\mathbf{x}$ near the body in a Taylor series about a base state of steady, nonturbulent approach flow of velocity $(U_1, 0, 0)$:

$$\tilde{u}_i^{\rm d}(\mathbf{x}, t) = \tilde{u}_i^{\rm d}(\mathbf{x})\Big|_0 + \frac{\partial \tilde{u}_i^{\rm d}(\mathbf{x})}{\partial U_j}\Big|_0 u_j(t) + \cdots . \tag{16.84}$$

The subscript 0 means evaluated in the base state. In typical applications we need keep only the linear terms in the expansion, so we truncate Eq. (16.84) and write it as

$$\tilde{u}_i^{\rm d}(\mathbf{x}, t) \simeq \tilde{u}_i^{\rm d}(\mathbf{x})\Big|_0 + a_{ij}(\mathbf{x}) u_j(t). \tag{16.85}$$

The *flow-distortion coefficients* a_{ij} are defined as

$$a_{ij}(\mathbf{x}) = \frac{\partial \tilde{u}_i^{\rm d}(\mathbf{x})}{\partial U_j}\Big|_0 . \tag{16.86}$$

For geometrically simple bodies the a_{ij} can be determined through the analytical solution for potential flow; in this way Wyngaard (1981) found them for the flow ahead of a circular cylinder. For somewhat more complex shapes (e.g., bodies of revolution) one can calculate the a_{ij} numerically (Wyngaard *et al.*, 1985). For the much more complicated geometries of typical *in-situ* probes the a_{ij} can be measured directly (Hogstrom, 1982).

In the absence of flow distortion $\tilde{u}_i^{\rm d}(\mathbf{x})\Big|_0 = U_1\delta_{i1}$, $a_{ij}(\mathbf{x}) = \delta_{ij}$ and Eq. (16.85) reduces to

$$\tilde{u}_i^{\rm d}(\mathbf{x}, t) = U_1\delta_{i1} + \delta_{ij} u_j(t) = U_1\delta_{i1} + u_i(t) = \tilde{u}_i(t), \tag{16.87}$$

as required.

If we ensemble average both sides of the linear expansion (16.85) we find

$$\overline{\tilde{u}_i^{\mathrm{d}}(\mathbf{x},t)} \simeq \tilde{u}_i^{\mathrm{d}}(\mathbf{x})\Big|_0 . \tag{16.88}$$

This says that to first order in the fluctuating velocity, the mean flow at the measurement point is simply that which exists there when the flow is nonturbulent. Differences do appear at second order.

If we subtract Eq. (16.88) from Eq. (16.85) we find an equation for the distorted turbulent velocity:

$$u_i^{\mathrm{d}}(\mathbf{x},t) \simeq a_{ij}(\mathbf{x})u_j(t) = a_{i1}(\mathbf{x})u_1(t) + a_{i2}(\mathbf{x})u_2(t) + a_{i3}(\mathbf{x})u_3(t). \tag{16.89}$$

This says that probe-induced flow distortion can cause attenuation or amplification in a measured turbulence signal as well as crosstalk (signal mixing).

Symmetries of the turbulence probe can simplify the problem, however. For example, symmetry about its x_1, x_2 centerplane (as for the circular cylinder in Figure 16.5) makes $a_{13} = 0$ on that centerplane (Problem 16.12). Similarly, symmetry about a vertical (x_1, x_3) centerplane makes a_{12} and a_{32} vanish there.

The effects of probe-induced flow distortion can be particularly serious for turbulent fluxes. Say, for example, we are measuring $\overline{u_1u_3}$ on a tower in a horizontally homogeneous surface layer. We'll assume a typical situation: lateral symmetry of the probe about a vertical centerplane, but no corresponding vertical symmetry about a horizontal centerplane. Thus when the mean wind vector lies in this vertical centerplane a_{12} and a_{32} are zero. But the asymmetry about the horizontal midplane induces non-zero a_{13} and a_{33} so that

$$\begin{aligned} u_1^{\mathrm{d}}(\mathbf{x},t) &= a_{11}(\mathbf{x})u_1(t) + a_{13}(\mathbf{x})u_3(t), \\ u_3^{\mathrm{d}}(\mathbf{x},t) &= a_{13}(\mathbf{x})u_1(t) + a_{33}(\mathbf{x})u_3(t). \end{aligned} \tag{16.90}$$

Thus the measured stress is

$$\overline{u_1^{\mathrm{d}}u_3^{\mathrm{d}}}(\mathbf{x}) = (a_{11}a_{33} + a_{13}a_{31})\ \overline{u_1u_3} + a_{11}a_{13}\ \overline{u_1^2} + a_{13}a_{33}\ \overline{u_3^2}. \tag{16.91}$$

Figure 16.6 shows the measured stress upstream of a right circular cylinder as calculated with Eq. (16.91). Here the symmetry about the horizontal midplane causes a_{13} to vanish at $x_3 = 0$ and minimizes the errors there. Elsewhere the errors can be quite serious, even several diameters upstream. Hogstrom's (1982) results confirm this.

Wyngaard (1987) showed that the errors in vertical scalar fluxes produced by this distortion of the turbulent velocity field can also be minimized by designing the

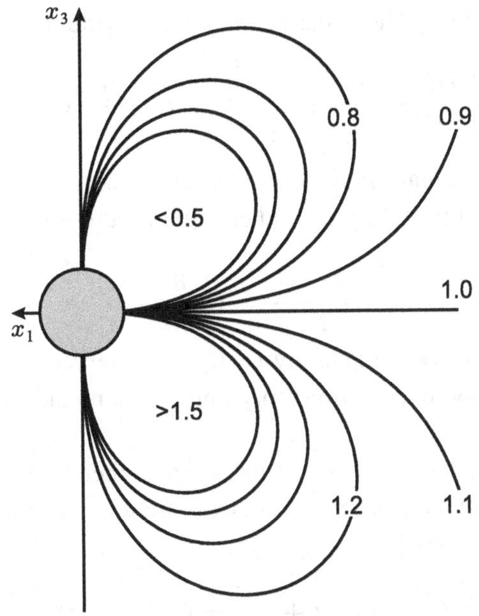

Figure 16.6 Contours of the ratio of distorted kinematic Reynolds stress $\overline{u_1 u_3}$ and the free-stream value for flow upstream of the shaded circular cylinder. $\overline{u_1^2}/u_*^2 = 9$, $\overline{u_3^2}/u_*^2 = 4$, with $-u_*^2$ the free-stream value of $\overline{u_1 u_3}$. From Wyngaard (1981).

probe to be vertically symmetric about its horizontal midplane. He showed also that the flow-blocking effect due to probe bulk causes the scalar flux to be overestimated, but this can be minimized in an array with minimal stagnation loss in streamwise speed at the flux-measurement point.

16.3.2 Conserved scalars

Conserved scalar fields are also affected by flow distortion, but in ways that depend on the nature of the velocity field. We'll consider three limits here.

16.3.2.1 Steady laminar flow

A conserved scalar C is released at a variable rate from a source in a steady, divergence-free, laminar velocity field. We measure the scalar downstream with a sensor that induces flow distortion. How does this influence the measured C signal?

We assume that in the flow-distortion process we can neglect molecular diffusion, so we can write

$$\frac{DC}{Dt} = \frac{\partial C}{\partial t} + U_j C_{,j} = 0, \tag{16.92}$$

with $C_{,j}$ the gradient of C. The scalar gradient follows the equation (Problem 16.13)

$$\frac{DC_{,i}}{Dt} = \frac{\partial C_{,i}}{\partial t} + U_j C_{,ij} = -\frac{\partial U_j}{\partial x_i} C_{,j}, \tag{16.93}$$

which shows that the scalar gradient is not conserved in flow distortion.

If we multiply Eq. (16.93) by U_i and change dummy indices we find (Problem 16.14)

$$\frac{DC_{,i}U_i}{Dt} = C_{,i}\frac{\partial U_i}{\partial t}. \tag{16.94}$$

Thus if the flow is steady the quantity $C_{,i}U_i$ is conserved during flow distortion, any change in velocity along a trajectory being accompanied by a compensating change in the scalar gradient.

If we denote a distorted field with a superscript d, then by integrating Eq. (16.94) in time along a trajectory from $\mathbf{x}_0$ in the free stream to a point $\mathbf{x}_m$ in a region of flow distortion we can write

$$C^{\mathrm{d}}_{,i} U^{\mathrm{d}}_i(\mathbf{x}_m, t + \Delta t) = C_{,i} U_i(\mathbf{x}_0, t), \tag{16.95}$$

where Δt is the travel time between the two points. Using Eq. (16.92) in the free stream and in the flow-distortion region gives

$$\frac{\partial C}{\partial t} + U_i C_{,i} = \frac{\partial C^{\mathrm{d}}}{\partial t} + U^{\mathrm{d}}_i C^{\mathrm{d}}_{,i}, \tag{16.96}$$

which with Eq. (16.95) gives

$$\frac{\partial C^{\mathrm{d}}}{\partial t}(\mathbf{x}_m, t + \Delta t) = \frac{\partial C}{\partial t}(\mathbf{x}_0, t). \tag{16.97}$$

This means that in a steady velocity field the time series of conserved scalar fluctuations in a flow-distortion region has a constant time delay relative to the time series in the free stream. Therefore their frequency spectra are identical.

16.3.2.2 Quasi-steady turbulent flow, low frequencies

In turbulent flow Eq. (16.94) is

$$\frac{D\tilde{c}_{,i}\tilde{u}_i}{Dt} = \tilde{c}_{,i}\frac{\partial \tilde{u}_i}{\partial t} \neq 0, \tag{16.98}$$

so $\tilde{c}_{,i}\tilde{u}_i$ is not conserved under flow distortion.

Wyngaard (1988) discussed the turbulent case when the velocity and scalar signals are low-pass filtered at the "probe frequency" $f_{\mathrm{p}} = U/2\pi a$ associated with the transit time of flow of mean speed U_1 past a body of length a. Using turbulence scaling he argued that if the "error parameter" $p = (u/U_1)(a/\ell)^{1/3} \ll 1$, with u and ℓ the turbulent velocity and length scales of the approach flow, is small Eq. (16.97) holds also for the low-pass-filtered scalar signal:

$$\frac{\partial \tilde{c}^{\mathrm{d}}}{\partial t}(\mathbf{x}_m, t + \Delta t) = \frac{\partial \tilde{c}}{\partial t}(\mathbf{x}_0, t), \qquad p = (u/U_1)(a/\ell)^{1/3} \ll 1. \tag{16.99}$$

If so, then for $p \ll 1$ the low-pass-filtered time series in the free stream and in the distortion region differ only by a fixed time delay, so their frequency spectra are identical below the probe frequency f_{p}.

In an aircraft application with $u \sim 1$ m s^{-1}, $U_1 \sim 100$ m s^{-1}, so that $u/U_1 \sim 10^{-2}$, and $a \sim 1$ m, $\ell \sim 300$ m, so that $(a/\ell)^{1/3} \sim 10^{-1}$, for example, then the error parameter, $p \sim 10^{-3}$, which one expects is small enough for Eq. (16.99) to hold. Thus the frequency spectra of the scalar in the region of flow distortion and in the free stream should be the same. In tower applications the situation is less clear because both u/U_1 and a/ℓ are apt to be larger, making the error parameter p larger. If, for example, we measured at the flight height but from a very tall tower, then in a convective ABL u/U_1 could increase to 3×10^{-1}, say. If the probe were smaller ($a \sim 0.2$ m, say) then $(a/\ell)^{1/3} \sim 0.9 \times 10^{-1}$, making $p \sim 3 \times 10^{-2}$; the error parameter is 30 times the aircraft value. If we measured from a tower but in the surface layer, say, $(a/\ell)^{1/3}$ could be larger yet. Thus p could be still larger for *in-situ* measurements in the unstable surface layer.

16.3.2.3 Quasi-steady turbulent flow, high frequencies

Wyngaard (1988) also extended his analysis to high frequencies (relative to the probe frequency U_1/a). He found that if the parameter $ua/(U_1\lambda)$, with λ the Taylor microscale, is small, the classical rapid-distortion analysis (Batchelor, 1960; Hunt, 1973) can be extended to scalars. As the flow approaches the body along a streamline on its centerline, the decreasing fluid velocity causes fluid particles to decrease their streamwise separation, as in duct flow with a sharp increase in cross-sectional area – a *rapid expansion*. This is found to amplify the streamwise wavenumber spectrum of the scalar in the inertial range. The off-axis flow experiences a *rapid contraction*, which attenuates the inertial-range wavenumber spectrum. However, Taylor's hypothesis in the form $\kappa_1 = 2\pi f/U_1$, plus the constraint that both spectra integrate to the variance, show that the frequency spectrum is unchanged.

Questions on key concepts

16.1 Explain the essence of the process by which a conservation equation for a turbulent field is converted to an evolution equation for its power spectral density. What assumptions are required about the field? What information can be gained by this process?

16.2 Explain Figure 16.1 physically, and in particular discuss why it shows no activity in the wavenumber region between the energy-containing and dissipative ranges.

16.3 Interpret and discuss Figure 16.2 physically.

16.4 Discuss the essence of the arguments in Subsection 16.1.2 that allow us to use the results of DNS or LES to calculate spectra in the homogeneous horizontal plane of the ABL and interpret them in much the same way that we did for isotropic turbulence.

16.5 Discuss the physical interpretation of the response function for line averaging of a scalar field, Figure 16.3.

16.6 Explain physically why Figures 16.3 and 16.4 are so different.

16.7 Explain the concept of the array technique for measuring resolved and subfilter-scale variables.

16.8 Discuss the concept and implications of probe-induced flow distortion, and how and why it can be analyzed quite simply when the scale of the turbulence is much larger than the scale of the probe.

Problems

16.1 Show that the first two terms on the rhs of Eq. (16.33) integrate over the wavenumber plane to turbulent transport.

16.2 Assuming the pressure spectrum behaves as $\kappa^{-7/3}$ in the inertial subrange, contrast the difficulties in measuring the variances of $\partial u/\partial x$ and $\partial p/\partial x$.

16.3 Write an expression for the cross spectrum of a conserved scalar c and vertical velocity w in the homogeneous horizontal plane. In practice we cannot measure w and c at the same point in the plane; if the distance on the plane between their points of measurement is $\mathbf{r}$, write their covariance as an integral of a transfer function times their cross spectrum. Show how both the cospectrum and the quadrature spectrum contribute to the measured covariance.

16.4 Express the two-point velocity difference variance $\overline{[u_1(\mathbf{x}+\mathbf{r}) - u_1(\mathbf{x})]^2}$ as an integral of the spectral density tensor times a transfer function, under the assumption of isotropy. Write the integral for the particular cases $\mathbf{r} = (r, 0, 0)$ and $\mathbf{r} = (0, r, 0)$. Evaluate the integrals. Show how this yields the dissipation rate.

16.5 A linear array of sensors spaced $\mathbf{r}$ apart on the horizontal plane has been used to produce a low-pass filtered variable by adding the sensor outputs with weight w. An example is

$$v^{\mathrm{f}}(\mathbf{x}_{\mathrm{h}}, z, t) = w \times v(\mathbf{x}_{\mathrm{h}} - \mathbf{r}, z, t) + w \times v(\mathbf{x}_{\mathrm{h}}, z, t) + w \times v(\mathbf{x}_{\mathrm{h}} + \mathbf{r}, z, t).$$

For this simple case what is the relation between the spectrum of v^{f} in the plane and that of v, assuming the field is homogeneous in the plane? What criterion should w satisfy to give unity transfer function at small wavenumbers? Choose w that satisfies that criterion and sketch the transfer function.

16.6 To leading order the budget of vorticity variance is (Problem 5.6)

$$\frac{\partial}{\partial t}\frac{\overline{\omega_i \omega_i}}{2} = 0 = \overline{u_{i,j}\omega_i\omega_j} - \nu\overline{\omega_{i,j}\omega_{i,j}}.$$

What is the physical meaning of each of these terms? Use the definition of vorticity and the assumption of local isotropy to evaluate the production term. Write the destruction term as an integral of the velocity spectral density tensor. Simplify and express the skewness S of the streamwise velocity derivative $\partial u_1/\partial x_1$ as an integral of the spectrum of u_1.

16.7 Show that the integral in (16.63) yields the result (16.64).

16.8 Show that $\epsilon = 2\nu \int_0^\infty \kappa^2 E(\kappa)\, d\kappa$.

16.9 Explain why it is (Subsection 16.1.1.1) that isotropy requires that the mean part of the scalar, $C(\mathbf{x}, t)$, not depend on $\mathbf{x}$, and that the turbulent flux $\overline{cu_j}$ vanish.

16.10 Show that if the separation vector $\mathbf{r}$ is in the streamwise direction, Eq. (16.70) can be reduced to a transfer function times the one-dimensional spectrum.

16.11 Sketch the spectral variance budget of Eq. (16.39).

16.12 Explain why a_{13} vanishes on the centerplane ahead of a circular cylinder, Figure 16.6.

16.13 Derive Eq. (16.93).

16.14 Derive Eq. (16.94).

16.15 Derive Eq. (16.33).

References

Batchelor, G. K., 1960: *The Theory of Homogeneous Turbulence*. Cambridge University Press.

Corrsin, S., 1951: On the spectrum of isotropic temperature fluctuations in isotropic turbulence. *J. Appl. Phys.*, **22**, 469–473.

Gal-Chen, T., and J. C. Wyngaard, 1982: Effects of volume averaging on the line spectra of vertical velocity from multiple-Doppler radar observations. *J. Appl. Meteor.*, **21**, 1881–1899.

Hogstrom, U., 1982: A critical evaluation of the aerodynamical error of a turbulence instrument. *J. Appl. Meteor.*, **21**, 1838–1844.

Hunt, J. C. R., 1973: A theory of turbulent flow round two-dimensional bluff bodies. *J. Fluid Mech.*, **61**, 625–706.

Kaimal, J. C., J. C. Wyngaard, and D. A. Haugen, 1968: Deriving power spectra from a three-component sonic anemometer. *J. Appl. Meteor.*, **7**, 727–737.

Obukhov, A. M., 1949: Structure of the temperature field in turbulent streams. *Izvestia ANSSSR, Geogr. Geophys. Ser. 1949*, No. 13, 58–69.

Tong, C., J. C. Wyngaard, S. Khanna, and J. G. Brasseur, 1998: Resolvable- and subgrid-scale measurement in the atmospheric surface layer: technique and issues. *J. Atmos. Sci.*, **55**, 3114–3126.

Wyngaard, J. C., 1968: Measurement of small-scale turbulence structure with hot wires. *J. Sci. Instrum.* (*J. Phys. E.*), **1**, 1105–1108.

Wyngaard, J. C., 1969: Spatial resolution of the vorticity meter and other hot-wire arrays. *J. Sci. Instrum.* (*J. Phys. E.*), **2**, 983–987.

Wyngaard, J. C., 1971: Spatial resolution of a resistance wire temperature sensor. *Phys. Fluids*, **14**, 2052–2054.

Wyngaard, J. C., 1981: The effects of probe-induced flow distortion on atmospheric turbulence measurements. *J. Appl. Meteor.*, **20**, 784–794.

Wyngaard, J. C., 1987: Flow-distortion effects on scalar flux measurements in the surface layer: implications for sensor design. *Bound.-Layer Meteor.*, **42**, 19–26.

Wyngaard, J. C., 1988: The effects of probe-induced flow distortion on atmospheric turbulence measurements: extension to scalars. *J. Atmos. Sci.*, **45**, 3400–3412.

Wyngaard, J. C., L. Rockwell, and C. A. Friehe, 1985: Errors in the measurement of turbulence upstream of an axisymmetric body. *J. Atmos. Ocean. Technol.*, **2**, 605–614.

Index

Printed in the United States
by Bookmasters

Printed in the United States
By Bookmasters